The Biometry of Plant Growth

David R. Causton M.Sc., Ph.D., D.I.C.

Jill C. Venus B.Sc., Ph.D.

Department of Botany and Microbiology,
University College of Wales, Aberystwyth

Edward Arnold

First published 1981
by Edward Arnold (Publishers) Ltd,
41, Bedford Square,
London WC1B 3DQ

British Library Cataloguing in Publication Data

Causton, David R
The biometry of plant growth.
1. Growth (Plants)
I. Title II. Venus, Jill C
581.3′1 QK731

ISBN 0 7131 2812 7

Printed by Butler & Tanner Ltd
Frome and London

Preface

In introducing this work entitled 'the biometry of plant growth', we should start by discussing the meaning of 'biometry'. This word does not appear to be precisely defined at the present time, rather it conveys a general idea. Thus, a dictionary definition is 'the quantitative study of biology' (Chambers's Twentieth Century Dictionary): the Biometric Society is devoted to 'the mathematical and statistical aspects of biology'. However, two well known books (Mather, 1967; Sokal & Rohlf, 1969) imply, by their titles and contents, that 'biometry' is virtually synonymous with 'statistics'; the books are essentially manuals of statistical methods, but using biological data. This is also the view that one obtains on examining the leading biometrical journals – that biometry is statistics in which biology enters only in the form of the original data used to illustrate the methods and/or theory described. On present day evidence, therefore, biometry appears to be a subject much more closely allied with mathematics and statistics than with biology.

We, however, take the view that biometry is a subject in its own right, and the dictionary definition, quoted above, seems to be ideal as a concise description. The aspects of biology requiring quantitative study should form an integral part of biometry, and not merely dismissed once the problem has been put into quantitative form and attention turned to mathematical and statistical theory and methods. This means that a rigorous biological discussion of what has been achieved by quantitative methods should always be incorporated into an account of a biometrical investigation, while mathematical and/or statistical discussion may also be included as required. In our opinion, books such as those of Thornley (1976) and Pielou (1977) are much more truly 'biometrical' in character than those cited above, and the same can be said for many papers which now appear in biological journals as against those found in what purport to be biometrical journals. Of course, this is not to say that in biometry we have only the situation that quantitative methods contribute to the advancement of biology; the opposite is just as prominent and important, namely, the advancement of mathematical science through a consideration of biological phenomena. For example, there was the tremendous development of statistical theory and methods achieved by R. A. Fisher and his colleagues over a relatively short space of time, mainly as a result of assisting biologists in their research work; also, René Thom evolved catastrophe theory in order to quantify and further understand discontinuous events in embryology.

A similar 'bridging' subject is biochemistry and, while the emphasis here is rather more towards chemistry than biology, one gets the impression that there is more integration between the component subjects than there is in biometry. Admittedly, there is a difference: biochemistry deals with the compounds of which organisms are made, and also with metabolic reactions which characterize the 'workings' of living things; whereas an organism is not built out of mathematical symbols, nor does mathematics 'make the organism work'. As in any science, mathematics is a tool, but this does not prevent it being very closely associated with the underlying biology. The subject of biometry should, then, be as near equally concerned as possible with mathematics (particularly statistics) and biology; and a biometrician should ideally be a person who has been trained in the elements of both component subjects.

Growth has long been the subject of quantitative study both by biologists and by statisticians. The question posed by the biologist is whether any useful biological information is provided by a study of growth curves. Assuming an affirmative answer, the extraction of mathematical information must be integrated with biological concepts and investigation in order that such extracted information may have maximum utility to the investigating biologist. A danger inherent in the use of a quantitative approach is to lose sight of one's biological goal and to enter a labyrinth of mathematics without ever emerging into the biological world.

Existing works describing the quantification of plant growth tend to be mainly concerned with the biological applications of various methods of quantification without considering the relative merits and disadvantages of the methods, nor with their statistical validity. Also, the existing literature is usually restricted in terms of level and type of analysis in relation to the plant: thus, Evans (1972) deals at length with classical growth analysis of plants in natural and controlled environments, whereas Hunt (1978b) covers classical and functional growth analysis in both types of environment in a very compact framework.

The present work aims to cover the quantitative aspects of growth at the whole plant, foliage, stem, root, and single leaf levels of organization. Although, inevitably, there is emphasis on theory and methods, the biological basis of the work is highlighted throughout by applications to plant growth data and discussions of the biological relevance of the analytical results. The first chapter sets out the biological background of the work in the form of an essay giving: (i) an historical compendium of the quantitative analysis of plant growth, including many of the past applications of the subject; (ii) some discussion on the nature of growth in plants; and (iii) briefly sets out the links between the methods, their devisers and their users. Whole plant growth analysis (Chapter 2) essentially consists of three broad parts: principles, classical methods, and modern curve-fitting methods. The experimental data to be used throughout the book are also introduced, the classical and curve-fitting methods are applied and contrasted, and there is a discussion on the achievements of the methods of whole plant growth analysis in relation to these data.

Chapters 4 and 5 are the two parts (theory and methods, and applications) of a single theme – the description and analysis of the growth of individual leaves by the Richards function – although the methodology of Chapter 4 is of much wider application. After an introductory section dealing with the relevant mathematical properties of sigmoid functions in general, and of the Richards function in particular, the statistical problems of fitting the Richards function to data are considered, using standard assumptions. Although suitable for many purposes, investigation of the properties of single leaf growth data reveals that the assumptions do not altogether accord with these properties, and so an alternative fitting method is developed and employed for the description of single leaf growth.

In contrast to Chapters 2, 4 and 5, Chapter 6 deals with relationships between plant parts (allometry) rather than the isolated growth of single components. The mathematical consequences of linear allometric growth are first presented, followed by a discussion of the physiological implications. The main statistical estimation problem involved in allometry is that of the 'functional relationship' as opposed to regression, and this is treated in detail. In the applications sections of the chapter, the properties of plant growth data in relation to the assumptions made in the estimation method are considered prior to the presentation and discussion of the analytical results from the experimental data. The final chapter returns to consider whole plant growth, mainly in terms of a synthetic model which draws on the material of previous chapters in its construction. Preceding the methods and applications of the model, which are assessed by comparison with the results given by whole plant growth analysis in Chapter 2, there is a short section on growth mathematics, where, in particular, the consequences of exponential growth are explored.

As we move on from the earlier part of the book to Chapters 6 and 7, the nature of the material changes. In the former chapters, a considerable body of work on the methods now exists, and we are arriving at a position of confidence in the results obtained. Chapters 6 and especially 7, in contrast, are more of an exploratory nature; less work has been done previously on the methods. After making a series of assumptions and going some way to testing these, both statistically and biologically, we then fit a model in Chapter 6, based on our *a priori* assumptions. This is partly because the current incompleteness in the methodology associated with relevant alternative sets of assumptions provides a constraint, and partly owing to the difficulty of defining statistical populations in relation to the biological data. The ideas of Chapter 6 and 7 will provide fuel for a great deal more investigation.

In a multi-disciplinary book of this kind, the level of writing presents a formidable problem. The minimum knowledge required for reading the whole book would be equivalent to G.C.E. 'A' level standard in Biology and Mathematics, plus the elements of statistical theory and methods (Pure Mathematics and Statistics as a single 'A' level subject would be ideal). Better still, of course, would be a knowledge of Botany, Pure Mathematics, and Statistics at first year degree level. We assume that readers of the book, apart from biometricians, would be biologists, mathematicians, or statisticians, and therefore have specialized knowledge of at least one of the three major sciences that the book attempts to unite in the subject of plant growth; but, nevertheless, sufficient explanatory material in each of the three areas has to be given for the benefit of those who are not specialists in all three subject areas. Hence, as far as individual readers are concerned there will be, inevitably, a certain amount of 'talking down to' below the level to which they are accustomed in their own literature, and for which we apologise in advance. This is particularly true of the statistical aspects of our subject, and the reason for devoting a whole chapter (Chapter 3) to the details of standard linear regression theory is because much of Chapters 4 and 6 rely on this as a basis. This is also the reason for reviewing the principles of whole plant growth analysis in Chapter 2 which, to some extent, reiterates aspects of the books by Evans and Hunt, referred to above; but to avoid over-duplication, we have biased the account towards the mathematical and statistical aspects of the principles, while not ignoring the fundamentals. We should stress that, while Chapter 2 forms part of the unfolding biometrical theme of the book, Chapter 3 does not, and could be omitted at a first reading, even by those who are not familiar with its contents if they are willing to take some of the theoretical developments in Chapters 4 and 6 on trust.

Although the book was essentially conceived as a research monograph, since much of the material presented is the result of our own research, it has ultimately been written in a form suitable for use as an advanced text for at least some of the topics covered. Thus, the book may be of interest to research workers, postgraduate students, and final year undergraduate students in the fields of pure and applied botany, biometry, pure and applied mathematics, and statistics. We hope that those who have viewed the subject of growth analysis from one side or the other (i.e. biological or mathematical) will derive benefits from this attempt at synthesis.

We would draw attention to the following points in the matter of notation. Conventional systems of notation in statistics have been adhered to as closely as possible, particularly in respect of populations and samples, expected values and their estimates. With regard to plant growth analysis, certain symbols are deeply entrenched in the literature and these we have retained. Evans (1972) was the first to offer a unified system of notation for the classical methods of whole plant growth analysis, and his system has been adopted and extended by Hunt (1978b). Although the universal adoption of a single notational system is a goal to be aimed at, after much careful consideration we decided that Evans' system was unsuitable in the wider, biometrical, approach to growth analysis of this book. Accordingly, we have evolved a different system of notation, which however, has considerable overlap with Evans' system. The rationale of our system is detailed in the introductory paragraph of the 'Glossary of Symbols' at the back of the book. In a work of this nature, the number of quantities requiring expression in symbolic form is large and, because we have tried as far as possible to have a unique symbol for a single defined quantity, the total number of different symbols appearing throughout the book is rather daunting. The detailed glossary should· assist the reader as a readily available reference, particularly in cases where the same symbol has to do duty for more than one quantity.

Finally, it is a pleasure to acknowledge our indebtedness to various friends and colleagues who have assisted us in diverse ways. Mr. Roger Mead, Dr. Rod Hunt, and Dr. Noel G. Lloyd have read and commented on the typescript, for which we are very grateful; in particular, Dr. Lloyd gave considerable assistance with the proofs of the theorems in Chapters 6 and 7, and Dr. Hunt has allowed us to utilize some of his spline-fitting material prior to publication. Any remaining errors are our own responsibility, and opinions we have expressed are not necessarily always shared by our three reviewers. The postgraduate work of Dr. Chris O. Elias and Dr. Paul Hadley has contributed essential underlying material to some of the topics covered in this book; we also benefitted from the many lively discussions we had while they were working with us. We also thank Miss Hildred M. Bigwood for her excellent draughtsmanship, and Mrs. Joan Crawford and Miss Wendy Wilkins for typing the tables and bibliography. The granting of study leave to one of us (D.R.C.) by the University College of Wales is also gratefully acknowledged, without which the book could not have been written in the short time that it has; and in connection with this, the willing co-operation of our colleagues in the Department of Botany & Microbiology must not be forgotten. Last, but not least, we should like to express our appreciation to the staff of Edward Arnold for their friendly assistance throughout the writting and production stages.

November 1979

D. R. Causton
J. C. Venus

Contents

1

Introduction

Quantification of biology

In comparison with the other sciences, quantitative investigations have only played a minor role in biology until very recently. The physicist is concerned with precise phenomena which lend themselves easily to mathematical analyses; the living organisms with which the biologist works can be very variable and have a more complicated nature than physical material. Also, the quantification of biological processes requires extensive, previously obtained, qualitative information, which may have taken many years to collect, to be biologically useful. These factors, together with the relatively recent widespread availability of high speed computing facilities which have made the handling of large amounts of biological data feasible, are probably responsible for the increased interest in the quantification of various aspects of biology in recent decades.

There is, however, one field of biology in which mathematics has been of greater or lesser importance for many years, and this is in connection with the growth of organisms. Growth, together with differentiation, defines the way in which an organism develops. Differentiation is concerned with the qualitative changes that occur in the formation of the cells, tissues and organs, and the term 'growth' is applied more to the quantitative aspects of development. Thus the concept of growth itself shows the close link that it has with mathematical analyses. Growth can be defined as any irreversible increase in size of an organism or of any of its parts.

In this book we are primarily concerned with the quantification of growth, with methods used to study and analyse the quantitative aspects of growth, and with the biological relevance of quantitative growth studies. The aim of this introductory chapter is to set the biological scene by a review of the historical development, together with the manifold aspects and uses, of the growth analysis methods. We would also recommend the reader to seek amplification of some of the topics discussed in this chapter in Richards (1969).

Levels of growth

The external manifestation of growth cannot be examined in isolation, and the various factors contributing to and having effects upon the growth processes need to be understood. Within the plant, increment in size is brought about by the accruement of new tissues using the products of photosynthesis, together with water and minerals absorbed from the soil; but this seemingly simple process has many facets. To put the growth process into perspective for the higher plant (about which this book is concerned), it is useful to first consider simpler organisms.

Growth considered as the multiplication of individuals

When the growth of a unicellular organism is considered, the ultimate processes consist purely of cell division and enlargement. Differentiation of cells occurs only to a minor extent, if at all, and growth can realistically be described in terms of cell number increases. Every cell in a colony is potentially capable of division, and growth rates are essentially rates of cell division; if division is synchronized until the substrate becomes limiting, then the number of cells increases according to a geometric series. Growth of multicellular organisms is conceptually more involved. If higher plant growth is analysed in a manner analagous to that for unicellular organisms, i.e. as counts of numbers of individuals, it is much less straightforward. The study of the behaviour of populations of organisms is a separate subject and involves the ecological concepts of reproductive strategies, competition between individuals, and problems of niche occupancy in the ecosystem, rather than with growth itself. Growth in the higher plant is usually concerned with the increase in size of individuals or, at the crop or vegetation stand level, with clearly defined groups of plants. It is possible to discuss growth in terms of cell numbers within an individual and, although this can provide useful information, growth rates will always be more difficult to interpret than in the unicellular situation. It is more realistic to break down whole plant growth into a number of lower levels, but not right down to the level of individual cells. However, this topic will be pursued further after a résumé of higher plant organization.

Higher plant organization and growth measurements

Plant organization can be split into seven levels: the whole plant, the main organ systems (root, stem, foliage, inflorescence), the component organs (individual leaves etc.), tissues, cells, organelles, and molecules. The system is detailed in Table 1.1, in which the higher levels are partitioned in greater detail, as it is with these levels that this book is concerned.

Although the plant consists of organs made up of various different tissues, from the viewpoint of growth it is important to remember that the

Table 1.1 Plant organization: a scheme which considers a plant as a system of 'entities' and 'components' at different levels.

Level	Entity	Components
1	Plant	Shoot, root
1a	Plant	Foliage, stem, inflorescence, root
2	Foliage	Main stem leaves, branch leaves
	Stem	Main stem leaves, branch leaves
	Inflorescence	Main stem inflorescence, branch stem inflorescences
	Root	Tap root, lateral roots
2a	As 2	As 2, but individual branches and lateral roots are distinguished
3	Main stem leaves	Individual leaves on main stem
	Branch leaves	Individual leaves on the branches
	Main stem	Individual internodes of main stem
	Branch stems	Individual internodes of branch stems
	Main stem inflorescence	Individual flowers* on main stem
	Branch stem inflorescences	Individual flowers on branches
3a	As 3	As 3, but individual branches distinguished
4	An individual leaf	Upper epidermis, palisade mesophyll, spongy mesophyll, lower epidermis
	An individual internode	Tissues of the stem
	An individual flower	Tissues of the flower
	A root axis	Tissues of the root
5	A tissue	Individual cells
6	An individual cell	Organelles
7	An organelle	Molecules

* Flower includes fruit

actual process of cell division is restricted to the meristematic regions of the plant, and that much of the growth and differentiation of the cells occurs in or near these regions. The meristematic areas of the plant comprise the root and shoot apical meristems, the cambial meristems in stems and roots (absent in monocotyledons), and various meristems within the developing leaves. Thus, if growth in cell numbers is to be followed, certain parts of the plant, or of its main organs, will show very high rates of increase, whereas others will hardly change. Usually, however, one is interested in the growth of a whole organ; in this situation it has to be remembered that only a certain percentage of the cells are dividing, and that this percentage decreases as the organ size increases because the number of dividing cells remains approximately constant (Cutter, 1971).

Growth, up to this point, has been described generally in terms of increase in size, or in terms of cell number for unicellular organisms; but what criterion should be used as a measure of size in the higher plant? Measurements of size can be divided into two types: those that do, and those that do not involve the destructive sampling of the organism. Non-destructive measurements include overall height, numbers of leaves and leaf areas; however, it is not usually possible to assess root growth non-destructively and, in general, a destructive approach is adopted when a complete assessment of plant growth is required. Non-destructive sampling techniques are normally used when examining forest crop growth and sometimes when examining field crop growth, but it must be remembered that measuring leaves involves handling them. This handling will increase respiration rates and thus decrease growth in comparison to the growth that would have occurred if the leaves had not been disturbed. It has been shown that, for many different species, a mechanical stimulus can raise the respiration rate by between 25 and 50%, and that this raised level of respiration could be maintained for more than two days (Godwin, 1935; Audus, 1939).

What, therefore, are the quantities that can be measured by destructive sampling? These include fresh weight, dry weight, cell number counts, and carbon content assays. The carbon content forms the basic unit of plant material, and is found in the plant in three main forms: immobile carbon (incorporated into structures such as cell walls), labile carbon (in storage products such as starch), and metabolic carbon (intermediate compounds and products of physiological processes such as ribulose diphosphate). The proportions of the various categories will change during ontogeny and are difficult to determine, but it should be possible to estimate the total carbon content by combustion of dry plant material or by acid digestion (wet combustion). Originally it was thought that the ratio of carbon to total dry matter was constant throughout the ontogeny of an organ, at about 42 to 45% (Terry & Mortimer, 1972; Turgeon & Webb, 1975); but more recently it has been found that percentage carbon content changes during the ontogeny of an organ, and also between organs (Ho, 1976; Hadley, 1978). Dry weight determinations will not, therefore, be accurately proportional to carbon content, but do give a reasonable estimate of the investment of the plant in new material; unfortunately, fresh weight measurements cannot be used in a similar way. The fresh weight of plant material depends upon the water status of the plant, and there is a complicated relationship between dry weight and fresh weight (Evans, 1972, pages 388–97); thus it is difficult to use fresh weight as a measure of plant productivity.

It was stated earlier that cell number could be used as a measure of size, and this has indeed been done (Sunderland, 1960; Lovell & Moore, 1970; Maksymowych, 1973); however, information concerning cell division and expansion would also be required in order to rigorously describe organ growth at this level. It is felt that, while detailed investigation of growth at the cellular level is both necessary and desirable, it is advisable first to concentrate upon dry weight increases when interest centres upon the overall performance of a plant or organ, rather than the processes by which this is attained.

The dry weight increments of plants over periods of time give an indication of the overall growth of the plant, but a measure that will help to determine the plant's 'productive investment' is also required and this is provided by the measurement of leaf areas. The leaves are the most important photosynthetic producers, and light interception and photosynthetic rate depend, to a large extent, upon the available leaf area.

Aspects of growth at the cellular level

The analysis of growth at the cellular level is rather different from that at higher levels, both in terms of methods and aims, and therefore will be discussed separately. Those aspects of cellular work most relevant to this book are concerned with single leaf development, and discussion will be restricted to this area. The interpretation of leaf growth at the cellular level has been pursued since the later years of the 19th century but, like most subjects, has enjoyed phases of popularity and disinterest. Since the early 1930s, work has been concerned with the patterns of cell division and cell expansion within individual leaves. The organization of the meristematic regions of the leaf are well documented (Maksymowych, 1973; Yeoman, 1976), but it is the products of these regions – in terms of duration, production rates, and the relative quantities of the different cell types – that are of greater immediate relevance to the quantitative analysis of plant growth. Work by Avery (1933) concerning the patterns of cell division and expansion of the leaf surface in *Nicotiana tabacum* (tobacco) showed that cell division was prolonged in certain tissues, and that all parts of the leaf expanded at equal rates. This latter conclusion has, more recently, been frequently questioned; for example by Erikson (1966) working with *Xanthium pensylvanicum* (cocklebur) and using vector analysis, and Sauer & Possingham (1970) working with *Spinacia oleracea* (spinach).

Cell number and size studies by Ashby (1948) and Ashby & Wangermann (1950a,b) in *Ipomoea caerulea* (morning glory) indicated that cell size decreases with leaf number up the main stem, and that the relative roles of cell division and expansion also change with leaf number. The original idea of distinct phases of cell division followed by expansion had to give way to the concept that both processes occur simultaneously to a greater or lesser extent (Sunderland, 1960), and that palisade cell division can continue throughout lamina expansion (Steer, 1971).

The contributions of cell division and expansion can be related to the content and activity of ribulose diphosphate carboxylase (Steer, 1972), and hence to the photosynthetic capacity of mature leaves (Wilson & Cooper, 1970). These differences between leaves at the cellular level are manifested in differences at higher levels of plant organization (Milthorpe & Newton, 1963; Wilson & Cooper, 1969; Steer, 1971), and so organ and plant growth can be interpreted in the light of cellular and physiological evidence.

Quantification of growth

Early work

The earliest work on the quantification of plant growth, as we know it today, was performed early this century, and has been fully described by Evans (1972, pages 193–6, 203–4). Work by Blackman (1919), although now outdated in many respects, provided the basis for much of the growth analysis methodology that followed. He proposed that, provided the rate of assimilation per unit area and the rate of respiration per unit plant mass remained constant and that the size of the leaf system bore a constant relationship to the dry weight of the whole plant, then the rate of production of new material would be proportional to the size of the plant; that is, the increase of plant dry weight would follow the compound interest law. The rate of interest was termed the efficiency index of production, but it is worth noting that an analagous quantity, called the 'Substanzquotient', had been used by German workers very early in the 20th century. Blackman (1919) also thought that:

> '. . . clearly the efficiency of the plant is greatest at first and then falls somewhat, but the fall is only slight until the formation of the inflorescence, when there is a marked diminution in the efficiency index.'

Blackman's analogies to compound interest provoked a lot of criticism, and many of the clauses in his definition are clearly unrealistic; for example, the concept of the rate of assimilation remaining constant for prolonged periods of time. However, the efficiency index, now called the relative growth rate (page 17), is of fundamental importance in the analysis of plant growth. Blackman's main critics in his time were Briggs, Kidd & West (1920a) who thought that separate estimates of growth rate should be made throughout the course of growth, as it would change considerably. Briggs *et al.* coined the phrase 'relative growth rate', instead of 'efficiency index' with its connotations of constancy. Blackman (1920) replied that, in the absence of detailed information, the use of an average efficiency over a long period could be valuable.

Even earlier than this century, Weber (1879, 1882) and Haberlandt (1884) separately used an estimate, called the 'Assimilationsenergie', to compare species; this was the mean rate of dry weight increase per square metre of leaf per 10 hours of daylight. Much later, Briggs, Kidd & West (1920b) used the mean rate of dry weight increase per unit leaf area as a similar measure of productivity, and called this the unit leaf rate. Gregory (1918) had already used a similar quantity, but did not name it specifically; however, later (Gregory, 1926) he coined the phrase 'net assimilation rate', and this has found widespread favour up to the present, although Evans (1972) convincingly argues for the re-adoption of unit leaf rate.

For full information, it has become generally accepted that relative growth and unit leaf rates should be measured at intervals throughout an experiment; exactly how often is a point of difficulty and will be examined and discussed later (page 37). There are, however, still many examples of work where a single relative growth rate has been calculated. In a physiological context of comparison of treatments upon growth, this approach may ignore important ontogenetic trends (Briggs, 1928; Whitehead & Myerscough, 1962), but has apparently been recently used with success by Quebedeaux & Chollet (1977) to compare the growth of *Panicum* species (millets) with differing rates of photorespiration. In an ecological context, Grime & Hunt (1975) found that when screening large numbers of species of the British flora a knowledge of only average and maximum relative growth rates, over a period of four weeks, could provide useful comparative information. Smith & Walton (1975) used a single average relative growth rate, for the entire growing season, to assess the relative performance of different species under tundra conditions.

The advent of curve fitting

Concurrently with developments by Blackman, Gregory, and Briggs *et al.*, referred to above, other workers were attempting to quantify plant growth by a different approach. This was by fitting a particular mathematical function to growth data and then seeking information in the values of the constants, or parameters, of the function. Most of these attempts were based on the logistic (or autocatalytic) function (e.g. Reed & Holland, 1919; Reed, 1920a,b,c), or a generalization of it (Pearl & Reed, 1923). The underlying philosophy behind the utilization of the logistic function was epitomized by Robertson (1923), namely that the growth rate of an organism was controlled by a single 'master' autocatalytic chemical reaction. Of course, we now realize that this notion is absurdly over-simple and, partly because of this gross over-simplification, and partly because of the lack of adequate statistical theory and computing power to enable non-linear regression analysis to be rigorously undertaken, the curve fitting approach soon petered out. One has only to look at the examples in

Robertson's book to see the inadequacies of many of the fittings and, without any estimates of the errors involved in estimating the parameters, the latter would be of little use for comparative purposes. Moreover, the methods developed by Blackman, Gregory, and Briggs *et al.* were designed to be independent of any specific underlying mathematical functions, and so these methods had much greater practical utility.

For nearly half a century, from the early 1920s, very little work on growth analysis methodology was undertaken, only certain refinements were made to what has become known as the classical methods of plant growth analysis (Williams, 1946; Coombe, 1960; Evans & Hughes, 1962; Whitehead & Myerscough, 1962; Watson & Hayashi, 1965). However, many physiological and agronomic investigations, based on the methods of classical growth analysis, were conducted: Heath (1937, 1938), Williams (1937), and Heath & Gregory (1938) were concerned with the change of unit leaf rate with time within a species, and with an initial inter-species comparison of this attribute; species comparisons were extended by Blackman & Rutter (1948) and Blackman & Wilson (1951a,b, 1954); from the Rothamsted Experimental Station there was an extensive series of agronomic investigations, for example Watson & Baptiste (1938), Watson (1947a,b, 1952), Watson & Witts (1959), Thorne (1960, 1961); woody species were investigated by, for example, Vyvyan (1957) and Jarvis & Jarvis (1964); eco-physiological studies were carried out by Warren Wilson (1960, 1966a,b,c, 1967) and by Evans & Hughes (1961), Hughes & Evans (1962, 1963, 1964), and Hughes (1965a,b,c,d); agronomic studies under tropical conditions were made, for example by Crowther (1934, 1937); and extensive physiological studies were carried out by Australian workers, for example Ballard & Petrie (1936), Williams (1936, 1939), Petrie, Watson & Ward (1939), Watson & Petrie (1940), Petrie & Arthur (1943), Tiver (1942), and Tiver & Williams (1943).

In the late 1960s, a combination of advancing statistical theory and the increasing availability of electronic computers enabled advances in growth analysis to recommence, not only in terms of methodology, but also in relation to the level of investigation within the plant. The new methods were based upon the fitting of mathematical functions to growth data and the manipulation of these functions to evaluate, among other things, relative growth and unit leaf rates; the merits of this approach are given by Radford (1967) and Hunt (1979). However, interest in the information contained in the parameters of the fitted functions has been revived only very recently (Causton, Elias & Hadley, 1978) and is a major theme in this book.

Initially, curve fitting was to primary growth data (Vernon & Allison, 1963), but statistical examination of growth data reveal that they are log-normally distributed, and curves have subsequently been fitted to logarithmically transformed data (Hughes & Freeman, 1967). Numerous examples of varied investigations, based on function fitting, may be cited: those based on Hughes & Freeman's scheme, employing a third degree polynomial exponential function*, are Goldsworthy (1970), Hughes & Cockshull (1972), Thornley & Hesketh (1972), Hunt & Burnett (1973), Hurd & Thornley (1974); examples using a second degree polynomial, suggested by Causton (1967) as being adequate for many cases, are Eagles (1969, 1971), and Hurd (1977) who suggests that there should rarely be any need to use polynomials of higher degree than the second; however, Nicholls & Calder (1973) and Hunt & Parsons (1974) advocate the use of a statistical test to objectively determine the choice of polynomial for a particular set of data, and their methods have been extensively used, for example by Hall (1977) and Neales & Nicholls (1978). Comparisons of the 'functional approach' with the classical methods seem to be almost non-existent in the literature; the only ones known to us are Sivakumar & Shaw (1978), Hunt & Parsons (1977) and Parsons & Hunt (1980), and we make some comparisons using our own data in the next chapter (pages 43–47 and 57–62). All the examples cited were concerned with the description of growth at Level 1 (Table 1.1), and they used polynomial functions. The growth of a component at Level 1 is indeterminate (page 86) and therefore, at first sight, the most suitable curve to describe the ontogeny of a Level 1 component is one that is unbounded and hence can continue increasing. The polynomial meets these requirements, and has the advantage of being mathematically and statistically simple. The fitted functions in the above examples were aimed at providing smoothed empirical descriptions of growth in order to facilitate comparisons and clarify ontogenetic changes.

Although not strictly conforming with our theme, we should mention at this point the work of Austin and his collaborators in the early 1960s. In an interesting series of papers (Nelder, Austin, Bleasdale & Salter, 1960; Austin, 1965; Austin, Nelder & Berry, 1964) they used a generalization of the logistic function (Chapter 4) to study variations in crop yields, particularly carrots (*Daucus carota*), as affected by different climatic and nutritional regimes in different years. The method was only partly successful (Austin *et al.*, 1964), and does not appear to have been further developed.

Increasing interest in single leaf growth

A further change to occur has been an increasing interest in the determinate growth of single leaves (Amer & Williams, 1957; Erickson & Michelini, 1957; Maksymowych, 1973; Hackett & Rawson, 1974; Williams, 1975; Auld, Dennett & Elston, 1978; Dennett, Auld & Elston, 1978, 1979;

*A polynomial exponential function implies that the natural logarithm of a growth attribute is a polynomial function of time. In this book, 'polynomial' will henceforward stand for 'polynomial exponential'.

Causton *et al.*, 1978). Very much earlier, Richards (1934) had commented on the fact that successive leaves had different developmental patterns when he criticized the work of Hover & Gustafson (1926) who attempted to examine single leaf respiratory changes by measuring respiration rates of leaves of different developmental ages on the same plant. It was not until many years later that these different growth patterns in leaves were actually investigated quantitatively to any great extent, but this was presumably due to the difficulties involved with the fitting of the more suitable, asymptotic, growth curves to leaf data. Work by Gregory (1921, 1928) showed that this area of investigation had great possibilities, but difficulties, both mathematical and statistical, prevented further progress at the time. Gregory (1921) did, however, appreciate the importance of knowledge of the development of single leaves in characterizing the assimilatory status of the plant.

Advantages of curve fitting

The advantages of fitting functions to growth data are manifold and these, together with the rationale of curve fitting, are given by Hunt (1979). Often of primary concern to the biologist is that a curve can provide a convenient summary of the growth course; the estimated growth characteristics can be calculated at as many times as desired, and these estimates are less disturbed by biological variability than the primary growth measurements because, in effect, each point on the curve contains information from the whole experiment for that growth characteristic. The penalty for this increased precision is that a rigid empirical model underlying the data is specified beforehand when the function type is selected. It is important, therefore, that the chosen function should have as much biological realism as possible but, in relation to plant growth, polynomials are distinctly artificial. A relative growth rate curve can be derived from a fitted function, and this usually provides a more definitive criterion as to whether a function is suitable for the description of growth.

The exponential curve, which is the simplest biologically meaningful function to handle mathematically, gives rise to a constant relative growth rate which, even in an artificially static environment, is highly unlikely over a prolonged period. Relative growth rates are usually found, when calculated directly from primary growth data, to decline more or less rapidly from an initial high value attained when the seedling first becomes photosynthetically autonomous. The relative growth rate form of the Richards function (page 96) has these characteristics, together with sufficient flexibility to follow differences in the way relative growth rate declines. This function is probably the simplest empirical model which is both realistic and sufficiently flexible for growth description in a wide variety of instances: it has been used both as a smoothing device in place of polynomials (Venus & Causton, 1979a) and in a situation where the values of the parameters of the fitted functions may be interpreted in a biologically meaningful way (Causton *et al.*,1978).

Biological applications of the quantitative analysis of growth

Interrelationships between growth and photosynthesis

Growth analysis of a particular species is not, of course, an end in itself and usually forms part of an investigation of a particular aspect of plant development, often in relation to environmental factors. The interpretation of changes in growth characteristics in terms of changes at the cellular level have already been discussed, but it must be remembered that growth is the manifestation of many physiological processes, and much work has been performed in relating growth to these processes. Photosynthesis provides the material required for increase in size, and the rate of photosynthesis shows an ontogenetic drift to which growth rates and dry weight attainment can be related. Workers have recently been concerned with the description of these interrelationships at the single leaf level (e.g. Hardwick, Wood & Woolhouse, 1968; Wilson & Cooper, 1969; Kreidmann, Kliewer & Harris, 1970).

It has been found that the rate at which assimilates are utilized within the plant can affect the net rate of photosynthesis (Maggs, 1964; King, Wardlaw & Evans, 1967). Sweet & Wareing (1966a) proposed that this effect might be auxin mediated. Neales & Incoll (1968) could find little evidence of a biochemical controlling mechanism, although they did demonstrate a correlation between photosynthesis and assimilate utilization; but they did not rule out the possibility of auxin mediation. In the light of evidence for auxin effects upon plant growth, the growth regulator triiodobenzoic acid (TIBA) has been shown to increase yields of *Glycine max* (soybean) (Tanner & Ahmed, 1974), and that it acted by reducing vegetative growth and stimulating reproductive growth. When one starts to be concerned with reproductive growth a great many additional factors come into play, and so we plan to concentrate upon vegetative growth in this book. However, the above example offers a practical illustration of growth regulation.

The effect that the supply of photosynthate may have on the growth of a given leaf may depend to some extent upon vascular connections within the plant. In general, assimilate from lower leaves moves basipetally to the roots, and that from the younger leaves supplies the developing leaves and apex; in *Lolium temulentum* (darnel) up to 90% of the assimilate for the apical meristem is supplied by the latest emerged leaf when growth is vegetative, but 70 to 80% when growth is reproductive (Ryle & Powell, 1972). In rosette plants, patterns of translocation are not as clear cut, although during stem elongation prior to flowering the translocation pattern becomes typical of a non-rosette plant (Thrower, 1977). The distribution pattern of assimilate from a given leaf has been studied by the use of radioactive carbon dioxide in many species; for example for *Triticum aestivum* (wheat) (Doodson, Manners & Myers, 1964) and *Nicotiana tabacum* (Jones, Martin & Porter, 1959), and this type of data can provide valuable information as to how a single leaf may affect the growth of others on the plant. Lovell, Oo & Sagar (1972) interpret assimilate movements in terms of rates controlled by source-sink relationships.

Mechanistic modelling of physiological processes

The processes of growth, photosynthesis, and assimilate translocation are all interrelated, and alteration of any one of these can influence any of the others. Mechanistic mathematical models (Thornley, 1976, page 4) have been proposed that aim to examine these interactions, but most of these are concerned with growth at the crop level (e.g. Acock, Charles-Edwards & Hearn, 1977) and incorporate descriptions of diurnal variation. Thornley (1972a) described a model that demonstrates the applicability of mathematical descriptions of translocation and assimilate utilization, but the model fails because of a total dependence of the model plant upon photosynthetic input. A second, more realistic, model incorporates nitrate uptake and is concerned with the relative shoot and root activities during steady state growth (Thornley, 1972b). Charles-Edwards (1976) takes this model and adapts it to examine the effects of certain environmental variables on growth, and he relates various mathematical constants to environmental parameters. Cooper & Thornley (1976) have also used the model to determine the response of dry matter partitioning and growth in *Lycopersicon esculentum* (tomato) plants to changes in root temperature, and found it gave useful results. However, the model is of limited applicability because it assumes exponential growth, presumably because of the analytical simplicity of this steady state situation.

Various other models of plant growth have been proposed but, again, these are generally at the crop level. Patefield & Austin (1971) produced a useful, intermediate type model (intermediate between mechanistic and empirical) to simulate the growth of a stand of *Beta vulgaris* (red beet), in terms of photosynthesis, respiration, and light interception by the foliage canopy, but it utilized observed distribution patterns of assimilate partitioning within the plant. The model was used to assess the effect on yield of various projected environmental changes, and when compared with real growth data was found to give reasonably realistic results.

Other models of crop growth predict growth increments from photosynthetic systems taking into account various factors, including carbohydrate reserves, water status, and proportions of actively growing tissue (de Wit, Brouwer & Penning de Vries, 1970); also evapotranspiration, dry matter/nitrogen ratios, and assimilate partitioning (McKinion, Jones & Hesketh, 1975; Curry, Baker & Streeter, 1975), and utilize Monteith's (1965) model for photosynthetic input. These models involve hour to hour predictions and, therefore, take into account the diurnal fluctuations of biological processes. In addition to a diurnal rhythm of photosynthesis, growth itself is not a smooth function on an hour to hour basis; structural growth can even occur during darkness using stored photosynthetic products (labile carbon, page 4) (Ball & Dyke, 1954) but, of course, total dry matter content must always decrease when a plant is in darkness owing to dark respiration. When the growth of a whole stand of a crop plant is under investigation, it is important to be able to assess the effects of the environmental changes at different times of the day. However, at a more basic level, where plants are considered as individuals, and often growing under artificially controlled conditions, a day to day knowledge of change is sufficient. Diurnal rhythms must always be remembered, but sampling (harvesting) at similar times each day should be sufficient to eliminate diurnal effects upon growth under cabinet or greenhouse conditions. Modelling methods of the types given above have been reviewed and extended by Thornley (1976).

Empirical models of the growth of plant parts

Bazzaz & Harper (1977) broke away from the traditional aspects of growth description and modelling when they proposed a demographic analysis of the growth of single plants of *Linum usitatissimum* (flax). They stated that traditional growth analysis ignores the fact that the plant can be considered as a population of modules (individual leaves) perfectly suited to studies of age structure, and by this means they investigated the effects of light intensity and planting density upon individual plant growth. Bazzaz & Harper's choice of plant, *Linum usitatissimum*, was probably deliberate, as plants of this species have large numbers of small leaves thus making the situation more amenable to the methods of population dynamics.

Only this last, unique, model has hitherto aimed to investigate the growth of a plant in terms of the growth of its components and their interactions. In a critique of Bazzaz & Harper's proposals, Hunt (1978a) felt that there was a need for a model of this type; he suggested an approach which not only considers the life and death of individual leaves, but also their growth pattern during life. We propose to develop this idea: in the last chapter of this book, a model of whole plant growth is constructed using Richards function descriptions of single leaf growth together with quantitative descriptions of the relationships between plant parts – topics that are dealt with in the intermediate chapters. It is suggested that one use of this model would be in the examination of disturbances to the plant by environmental or artificial means.

Practical applications of growth analysis

Finally in this section on applications, we summarize some of the practical uses of growth analysis which were not specifically mentioned earlier (page 8). The most obvious application is in the determination of the effect of environmental conditions upon a species' performance. For example, Warren Wilson (1966c) investigated the effects of different temperatures on the growth of *Zea mays* (maize), *Brassica napus* (rape), and *Helianthus annuus* (sunflower); Thiagarajah & Hunt (1974) also studied the effect of temperature on the growth of *Zea mays*, while Elias & Causton (1975) did the same with *Impatiens parviflora* (small-flowered balsam). Often, interactions between factors in their effect on growth are examined, especially light intensity and temperature; for example, in *Triticum aestivum* by Friend, Helson & Fisher (1962) in *Zea mays* by Voldeng & Blackman (1973), and in *Gossypium hirsutum* (cotton), *Phaseolus vulgaris* (French bean), *Helianthus annuus*, and *Zea mays* by Rajan & Blackman (1975). For this type of work, classical methods are usually suitable, but there are situations in which curve fitting is more appropriate. Hughes & Freeman (1967) used their curve fitting technique to investigate the effect of differing levels of carbon dioxide on the growth of *Callistephus chinensis* (China aster), and Hunt (1975) applied a similar technique to a study of differing soil nitrogen levels and shading upon root-shoot equilibria in *Lolium perenne* (perennial ryegrass). These two examples involved curve fitting to total leaves, stem and root, but Causton *et al.* (1978) have used Richards functions fitted to single leaves to investigate the effects of different temperatures on leaf growth in *Impatiens parviflora*.

Another area in which growth analysis plays an important role is in the comparison of genotypes of a species, often as part of a breeding programme; for example, Wilson & Cooper (1969) used classical growth analysis to compare contrasting *Lolium perenne* genotypes. Eagles (1971) used fitted curves to compare two populations of *Dactylis glomerata* (cocksfoot grass), and Voldeng & Blackman (1973) used a variety of classical methods to compare selected inbreds and their hybrids of *Zea mays* growing under field conditions. Namkoong & Matzinger (1975) adopted the fitting of the Richards function to plant height of *Nicotiana tabacum* in order to help with the selection of genotypes after different numbers of selection cycles. This type of approach may prove useful in the future to aid with the selection of characteristics at the single leaf level.

The method and the user

Much of the work on the quantification of growth has been concerned with the description of biological processes but less with the methods themselves. Even when new methods have been proposed the statistical applicability of them to growth data has not been fully investigated. Thus, for example, several authors have fitted curves to non-logarithmically transformed data (Vernon & Allison, 1963; Rees & Chapas, 1963; Ledig & Perry, 1969; Moorby, 1970); Voldeng & Blackman (1973) fitted Richards functions by Richards' (1959) original method, which can produce misleading results (page 103); and Stanhill's (1977a,b) use of the allometric relationship is not as rigorous as it should be, although his general approach to experimental design and analysis is good.

The converse has also occurred, namely, the application of rigorous statistical methods to empirical mathematical functions chosen without much regard for the biology of the situation. The classic work here was done by Wishart (1938), and has since been followed and extended by Leech & Healy (1959), Rao (1965), and Sprent (1967). These workers all used orthogonal polynomials in a statistical analysis of growth rate, but their methods may not necessarily be of much use to the biologist wishing to use quantitative methods to gain a deeper insight into biological processes. However, a more recent interesting development in the statistical field has been the introduction of stochastic growth models based on the logistic function (a particular case of the Richards function) (Krause, Siegel & Hurst, 1967; Sandland & McGilchrist, 1979), and it may be that such models could be usefully employed by the biologist in the future.

The overwhelming result of using non-rigorous methods, of the kind exemplified in the penultimate paragraph above, is that totally erroneous conclusions concerning the underlying data could result. Even when attempts at rigour are made, the results are still not always all that they should be owing to some remaining faults in the methodology. Recent work (Gillis & Ratkovsky, 1978), investigating the statistical properties of parameter estimates, highlighted the importance of selecting the most appropriate parameterization of any single model. Also, Davies & Ku (1977) showed that Causton's (1969) method of fitting the Richards function did not always produce a properly converged result, and so an adjustment to the method had to be made (Hadley, 1978). Traps for the unwary abound!

The remainder of this book is, therefore, an attempt to bring together the two aspects of plant growth analysis, namely, the biological and the mathematical (mainly statistical) aspects. Large parts of the book will deal with the relevant underlying statistical models and their application to growth data, while at the same time examining the validity of the models to the biological material that they purport to describe. Where relevant, some basic mathematical theory of growth relationships is given, but throughout the book, applications to real data are described together with biological discussions of the achievements of the analyses.

2

Whole plant growth analysis

The current meaning of the phrase 'plant growth analysis' is not as general as one might think; it has come to imply something rather specific, namely, the scheme based on methods evolved in the years immediately preceding and following 1920 for analysing whole plant growth without any reliance on specific underlying mathematical functional relationships between growth attributes and time. It is true that the last twelve years or so have witnessed the use of curve fitting methods in plant growth analysis (the functional approach, as opposed to the classical methods), but this is a secondary development, and the aims of the analysis have remained essentially the same although the scope has been slightly extended.

The foundations of plant growth analysis were laid by Gregory (1918), Blackman (1919), Briggs, Kidd & West (1920a,b), West, Briggs & Kidd (1920), Fisher (1921), and Gregory (1926), the last of whom coined the term 'net assimilation rate' for what had hitherto been called 'unit leaf rate' by Briggs *et al.*; but Gregory had himself used the same quantity in 1918, and called it merely 'the average rate of assimilation'. The estimation of unit leaf rate has always presented a difficulty in the classical scheme of growth analysis and, although what turned out to be the appropriate formula was first used by Gregory (1926), it was not until 20 years later that Williams (1946) first correctly derived the formula and showed the limitations in its use. Again much later, attempts to resolve these limitations were made by Coombe (1960), Evans & Hughes (1962) and Whitehead & Myerscough (1962); but, as this time was the dawn of the function fitting era, their methods have not been widely used.

A historical perspective of growth analysis has already been given in Chapter 1, and other publications are available which describe the aims and methods of plant growth analysis. These publications include: Hunt (1978b) containing a bird's-eye view of the whole subject in a condensed form, Causton (1977) in which Chapters 10 and 11 deal with the basic mathematics of the subject, and Evans (1972) which is very comprehensive and may be said to be the definitive work on the subject of classical plant growth analysis. Hence, the requirement here is to briefly describe the subject, but include a more detailed discussion of the mathematical relationships and the statistical problems involved.

Principles of whole plant growth analysis

The overall aim of growth analysis is to assist in explaining plant growth from the viewpoint of dry matter production. This is done by analysing total growth into a series of 'components of growth', and it is important to bear this in mind as one proceeds, otherwise the aim of the study may become obscured.

The logical beginning is, therefore, to consider the growth made by, say, an annual plant from seed germination to senescence. It is evident that the total growth made, i.e. the total accruement of dry matter, is the product of the duration of growth and the average absolute rate of growth. Duration of growth is, of course, a very simple concept and is easy to measure, especially under controlled experimental conditions. If, however, growth of plants under natural conditions is being studied, the duration of growth may be somewhat less easy to define but should, nevertheless, be borne in mind. It is the rate of growth component of the above partitioning of total growth made which is almost the sole concern of the growth analyst and forms the subject of growth analysis.

Growth rates

Absolute growth rate

The absolute growth rate of a plant (or, indeed, of any entity or component as defined in Table 1.1, page 3) at any instant is given by $\mathrm{d}W/\mathrm{d}t$, where W is the total dry weight of the plant at time t. As already discussed in Chapter 1, if we neglect ion uptake by plant roots, dry weight is consequent upon photosynthesis, and dry weight loss is continually occurring by respiration of organic compounds. Hence, in darkness a plant's absolute growth rate will be negative; this will also be the case under very low light intensities where the rate of photosynthesis is less than that of respiration. The light intensity at which the photosynthetic and respiratory rates balance, where $\mathrm{d}W/\mathrm{d}t = 0$, is known as the compensation point; light intensities above this point will enable the plant to have a positive absolute growth rate. The dimensions* of absolute growth rate are mt^{-1}, and a typical unit is g day^{-1}.

Relative growth rate

The absolute growth rate does not give much information concerning the physiological performance of the plant in dry matter production, as simple observation shows that very often absolute growth rate is roughly

* The dimensions m, l, and t for mass, length, and time are those defined in basic physics (see e.g. Nelkon and Parker, 1977); the symbol l should not be confused with its main use in this book, defined on p. 96.

proportional to plant size: compare (say) *Helianthus annuus*, having a seed with a large embryo, and *Daucus carota* (carrot), a small seeded plant, over a short period after germination. A better growth rate parameter is the relative growth rate, defined at any instant as

$$\mathrm{R} = \frac{1}{W} \cdot \frac{\mathrm{d}W}{\mathrm{d}t} \tag{2.1}$$

that is, absolute growth rate divided by the existing weight. Notice that R is printed in ordinary Roman type: we make this distinction because relative growth rate is a derived, rather than a measured quantity. Throughout this book, measured quantities will be given italic symbols and derived quantities ordinary Roman symbols. There is a modern tendency, especially among zoologists, to call R the specific growth rate, but as the name 'relative growth rate' is still almost universal in the plant sciences we shall adhere to it. Also, the word 'specific' has a particular association with weight (c.f. specific gravity), and growth analysts often derive the relative growth rates of attributes not involving weight, particularly leaf area.

Since W is always positive, the sign of R is always the same as that of $\mathrm{d}W/\mathrm{d}t$. The relative growth rate may be said to measure the average efficiency of each unit of dry matter in the rate of production of new dry matter, so relative growth rate may indeed be equated with Blackman's (1919) efficiency index. The dimensions of relative growth rate are $mt^{-1}m^{-1}$, i.e. t^{-1}; thus the dimension is independent of mass, and a typical unit is day^{-1}.

Unit leaf rate

Although biochemical reactions are occurring throughout the plant, it is only in certain parts that materials are assimilated from the environment in any quantity. Since the uptake of mineral ions is small compared with the uptake of carbon by photosynthesis – ash content about 6% of total dry matter in *Zea mays* (Bidwell, 1974) – and since photosynthesis in green parts of the shoot other than the foliage is usually small, the leaves may be regarded as the sole assimilatory organs (or the productive part of the plant), and on this definition the inflorescence (if any), stem and root are the unproductive parts. Thus the relative growth rate, which measures the efficiency of the plant *as a whole* in dry matter production, may be resolved into two components, namely, the efficiency of the productive leaves themselves and the ratio of the leaves to the whole plant, i.e.

$$\mathrm{R} = \mathrm{E_W} \cdot \frac{L_\mathrm{W}}{W} \tag{2.2}$$

where L_W is the total leaf dry weight of the plant (foliage dry weight) and L_W/W is known as the leaf weight ratio. The symbol $\mathrm{E_W}$ in equation (2.2) measures the efficiency of the leaf tissue and is called the unit leaf rate.

As defined in equation (2.2), unit leaf rate would be given by

$$\mathrm{E_W} = \frac{1}{L_\mathrm{W}} \cdot \frac{\mathrm{d}W}{\mathrm{d}t} \tag{2.3}$$

that is, the rate of dry matter production per unit dry weight of the leaves. However, leaf dry weight is not the best attribute of the leaves in relation to dry matter increase through photosynthesis. Shorn of all detail, the simplest chemical equation to represent the net reactions of photosynthesis is

$$CO_2 + H_2O \xrightarrow{\text{light}} (CH_2O) + O_2$$

where (CH_2O) represents a carbohydrate. Apart from water, which is present in abundance in an actively photosynthesizing leaf with its stomata open, the two inputs to the system are carbon dioxide and light. For a given ambient light level, the amount of light intercepted is directly proportional to leaf area*. The rate of carbon dioxide uptake by a leaf is partly governed by the amount and level of activity of the carboxylating enzymes which accept the carbon dioxide molecules into the photosynthetic reduction cycle, and also partly by the mesophyll resistance (the resistance to carbon dioxide diffusion through the intercellular spaces of the mesophyll tissue, and through membranes and liquids on the surfaces of membranes). However, the rate of carbon dioxide movement into the leaf also depends upon the number of stomata present, which in turn partly depends on the leaf area. True, the number of stomata also depends on their density, but if they are too close to one another their diffusion shells overlap, which retards the diffusion rate. Hence, it appears that leaf area is a better attribute for the definition of unit leaf rate than is leaf dry weight, and so we define

$$\mathrm{E_A} = \frac{1}{L_\mathrm{A}} \cdot \frac{\mathrm{d}W}{\mathrm{d}t} \tag{2.4}$$

where L_A is total leaf area at time t. The dimensions of unit leaf rate (on a leaf area basis) are $ml^{-2}t^{-1}$, and a typical unit is $\mathrm{g\,m^{-2}\,day^{-1}}$.

Unit leaf rate as a measure of net photosynthetic rate

Gregory (1926) called $\mathrm{E_A}$ the net assimilation rate. This became the more widely used term, and also has something to recommend it since $\mathrm{E_A}$ is

* This is assuming that there is no shading between the leaves of a single plant, and that other plants are sufficiently distant to avoid shading effects.

indeed an approximate measure of net photosynthetic rate if mineral ion uptake is neglected or allowed for. In practice, E_A is a long term measure of photosynthesis as it is derived from measurements made on samples of plants taken from time to time while they are growing (page 26); in effect, E_A estimates the rate of carbohydrate output from the photosynthetic system minus the loss due to respiration. By contrast, the more direct measure of photosynthetic rate is a short term one, using an infra-red gas analyser to measure the rate of carbon dioxide taken up from plants enclosed in an illuminated chamber attached to the apparatus. Unit leaf rate is, therefore, a satisfactory measure of net photosynthetic rate, provided that the long term nature of E_A is understood, and the mineral content of the dried plant material is allowed for.

Despite the popularity of the term 'net assimilation rate' over many years, there is the risk of confusion of this name with what is more truly net photosynthetic rate, and the term 'unit leaf rate' should become standard in the future. The works of Evans (1972) and Hunt (1978b) adopt the latter term; and the former, on pages 205–9, gives a very full discussion of the topics briefly introduced in this subsection.

Other leaf attributes for unit leaf rate

Although leaf area is the most appropriate and widely used attribute on which to express unit leaf rate, for reasons given above (page 19), it is worth mentioning that other attributes have been proposed. Until the availability, comparatively recently, of reliable area meters, leaf area was a time-consuming and often difficult measurement to make. Some workers have, therefore, used leaf weight and evaluated E_W; for example, Sweet & Wareing (1968a,b) based their unit leaf rate measurements on leaf dry weight because they used *Pinus contorta* (lodgepole pine) in their experiments, and it is problematical which of the three narrow surfaces of a pine needle should form the basis of an area measurement. Earlier, other workers (e.g. Crowther, 1934, 1937; Ballard & Petrie, 1936; Williams, 1936; Heath, 1937a,b) used leaf weight purely for convenience.

During the early 1940s, several Australian workers (Williams, 1939; Tiver, 1942; Tiver & Williams, 1943; Petrie & Arthur, 1943) argued that some measure of 'the living machinery' should be used as the basis for unit leaf rate. Leaf protein content was an obvious choice, but since this was relatively difficult to assay, leaf total nitrogen content was suggested as an alternative by Williams (1946) and applied by him to earlier sets of primary data which were available. None of these alternative attributes has withstood the test of time, however, and leaf area is now firmly established. In view of this, we shall use the symbol E in place of E_A in the remainder of the book. Note also that unit leaf rate is another derived quantity, and so is given a Roman rather than an italic symbol.

Leaf area ratio

Because of the revised definition and symbol of unit leaf rate, equation (2.4), (2.2) must be amended to read

$$R = E \cdot \frac{L_A}{W} \tag{2.5}$$

where L_A/W is known as the leaf area ratio. This ratio has dimensions $l^2 m^{-1}$, and the unit, corresponding to the previously mentioned typical units for relative growth and unit leaf rates, is $m^2 \, g^{-1}$. Although simple, the importance of equation (2.5) cannot be overstressed. A plant may have a high relative growth rate either because of a high unit leaf rate or a high leaf area ratio, though not usually both. Thus, a high leaf area in relation to overall plant weight, combined with a moderate unit leaf rate, can make for a highly efficient plant as can the reverse properties. In more colloquial terms, leaf area ratio is an index of the plant's 'leafiness'.

Interdependence of unit leaf rate and leaf area ratio

Although at first sight, unit leaf rate and leaf area ratio seem to be mathematically independent quantities, this is not so, as we shall now demonstrate. In what follows, we assume that the dry matter of a plant is made up wholly of carbon compounds.

At time t let the dry weight of a plant be constituted as follows

$$W = W_0 + W_p + W_r \tag{2.6}$$

where W_0 represents the total plant dry weight at some initial time t_0, W_p represents the dry weight gained by gross photosynthesis between t_0 and t, and W_r represents the dry weight lost by dark respiration over the same duration. The loss due to photorespiration is included in W_p. Differentiation of (2.6) with respect to time gives

$$\frac{dW}{dt} = \frac{dW_p}{dt} - \frac{dW_r}{dt}$$

and so dW_p/dt represents gross photosynthetic rate, and dW_r/dt the dark respiration rate at time t. Dividing throughout by L_A, we have

$$\frac{1}{L_A} \cdot \frac{dW}{dt} = \frac{1}{L_A} \cdot \frac{dW_p}{dt} - \frac{1}{L_A} \cdot \frac{dW_r}{dt}$$

or

$$E = \frac{1}{L_A} \cdot \frac{dW_p}{dt} - \frac{1}{L_A} \cdot \frac{dW_r}{dt} \tag{2.7}$$

The first term on the right-hand side of (2.7) represents photosynthetic rate per unit leaf area: denote this quantity by π. The second term on the right hand side is not very meaningful biologically since respiration occurs in all living parts of the plant. However,

$$\frac{1}{L_A}\cdot\frac{dW_r}{dt}=\frac{W}{L_A}\cdot\left(\frac{1}{W}\cdot\frac{dW_r}{dt}\right)$$

where the term in brackets represents dark respiration rate per unit plant dry weight: denote this quantity by ρ. Substituting into (2.7) we finally have, at time t

$$\mathrm{E}=\pi-\rho\bigg/\left(\frac{L_A}{W}\right) \tag{2.8}$$

Relationship (2.8) can have far-reaching effects in whole plant growth analysis, as will be demonstrated later (page 47), but the existence of the relationship appears to be unknown among growth analysts and its effects only sometimes vaguely glimpsed.

Other morphological ratios

Leaf weight ratio, stem weight ratio, root weight ratio

Leaf weight ratio has already been defined as the ratio of foliage dry weight to total plant dry weight. Similar ratios can be defined for the stem and roots, also for the inflorescence in the reproductive phase; however, only the leaf weight ratio enters into the classical scheme of plant growth analysis *sensu stricto*. For a whole plant, the sum of all these ratios is unity, and being all ratios of weights they are dimensionless.

Specific leaf area

Leaf area ratio is a composite ratio, it can be partitioned thus:

$$\frac{L_A}{W}=\frac{L_W}{W}\cdot\frac{L_A}{L_W} \tag{2.9}$$

The first term on the right-hand side is the leaf weight ratio and shows what proportion of the total assimilate is retained by the foliage. The second term is the ratio of foliage area to foliage dry weight, and is called the specific leaf area; it indicates very broadly what kind of leaf structure is made from the available dry material–a high specific leaf area indicating thin leaves of relatively large area, and *vice versa*. The dimensions and typical unit of specific leaf area are the same as that of leaf area ratio.

Root-shoot ratio

Although not entering into the usual scheme of plant growth analysis, the root-shoot ratio is an important and useful morphological ratio. In a vegetative plant, the ratio is simply $R/C=R/(L+S)$, where R, S, L, and C are, respectively, root, stem, foliage, and shoot dry weights.

The fundamental equation in growth analysis

Substituting equation (2.9) into (2.5), we have

$$R = E \cdot \frac{L_W}{W} \cdot \frac{L_A}{L_W} \tag{2.10}$$

which may be called the fundamental equation of whole plant growth analysis. A common aim of the growth analyst is to explain the differences in plant growth – either differences between species growing under the same environmental conditions, or differences within a species grown in different environments – in terms of the four quantities in equation (2.10), together with leaf area ratio.

In concluding this section on principles, it must be re-emphasized that all the plant attributes, both rates and ratios, have been defined at an instant of time. In the practical application of the principles this is often not the case, and this fact engenders difficulties in the transition from theory to practice.

Classical growth analysis: methods

Ratios

The various growth attributes enumerated in the first section of this chapter can be divided into two types: ratios and rates. The practical evaluation of the ratios presents no problem, even though destructive sampling of the plants is involved; at each harvest a batch of replicate plants is sampled, the leaf areas and the requisite dry weights are determined, and the ratios evaluated for each plant separately. Evaluation of the sample mean and variance of each ratio is straightforward, and the sample mean is assumed to be an unbiased estimate of the ratio in the population from which the sample was drawn. However, this is probably not true, and a better way would be as follows, taking leaf area ratio as an example.

At time t (a particular harvest) let $l_A = \log_e L_A$ and $w = \log_e W$, and we make the usual assumption that the weights and areas are approximately log-normally distributed (page 8).

Then $$\log_e(L_A/W) = l_A - w$$

and so the logarithm of the leaf area ratio is approximately normally distributed, by a standard result that the difference of two normally distributed variates is itself normally distributed. Then, by the theorem of linear combinations of variates,* we have

$$\mathscr{E}\{\log_e(L_A/W)\} = \mathscr{E}(l_A) - \mathscr{E}(w) \tag{2.11}$$

and $$\mathscr{V}\{\log_e(L_A/W)\} = \mathscr{V}(l_A) + \mathscr{V}(w) - 2.\,\mathscr{C}(l_A, w) \tag{2.12}$$

where $\mathscr{E}$ denotes an expected value, $\mathscr{V}$ a variance, and $\mathscr{C}$ a covariance.

* The term 'variate' will be used to denote a random variable throughout the book.

Equation (2.11) shows that the sample mean of the logarithms of the leaf area ratios, obtained by subtracting $\bar{w}$ from $\bar{l}_A$ where $\bar{w}$ and $\bar{l}_A$ are the sample means of w and l_A respectively, is an unbiased estimate of the expected value of the population of the logarithms of the leaf area ratio. Finally, we have

$$\mathscr{E}(L_A/W) = \exp[\mathscr{E}\{\log_e(L_A/W)\} + \tfrac{1}{2}.\mathscr{V}\{\log_e(L_A/W)\}] \quad (2.13)$$

$$\mathscr{V}(L_A/W) = \exp[2.\mathscr{E}\{\log_e(L_A/W)\} + \mathscr{V}\{\log_e(L_A/W)\}] \times (\exp[\mathscr{V}\{\log_e(L_A/W)\}] - 1) \quad (2.14)$$

Hence, if the basic growth data are approximately log-normally distributed, the ratios are as well, and it is better to carry out the analysis of the ratios in logarithmic form.

In the case of ratios of weights, involving the whole plant, there is a slight difference. Taking the leaf weight ratio as an example, we have, at time t

$$\frac{L_W}{W} = \frac{L_W}{L_W + P}$$

where P is the dry weight of the non foliar parts of the plant. Inverting both sides, we have

$$\frac{W}{L_W} = 1 + \frac{P}{L_W} \quad (2.15)$$

Now L_W, P, and W cannot all be log-normally distributed, since if any two are the third is not; the most realistic assumption is that all three quantities are *approximately* log-normally distributed, and so L_W/W is also approximately log-normally distributed.

Rates: mathematical aspects

Two difficulties are encountered in evaluating rates. Firstly, although they are defined at instants of time, instantaneous rates cannot be measured in practice; neither would it be profitable to do so. To appreciate this, recollect that the criterion of size is dry weight, and that the bulk of the dry matter in a plant is formed as a result of photosynthesis. Thus, both the relative growth and unit leaf rates at an instant of time are mostly dependent upon the prevailing environmental conditions, particularly light. This means that the rates are negative during darkness, and during light periods the actual rates at particular times could change dramatically. Figure 2.1 shows a very simple possible course of relative growth rate (or of unit leaf rate) over a three-day period for a plant receiving 12 hours of light on each day. In this hypothetical situation it is further assumed that day and (particularly) night temperatures are the same, that the light intensity (and hence the rates) change in a sinusoidal manner during the light period, that

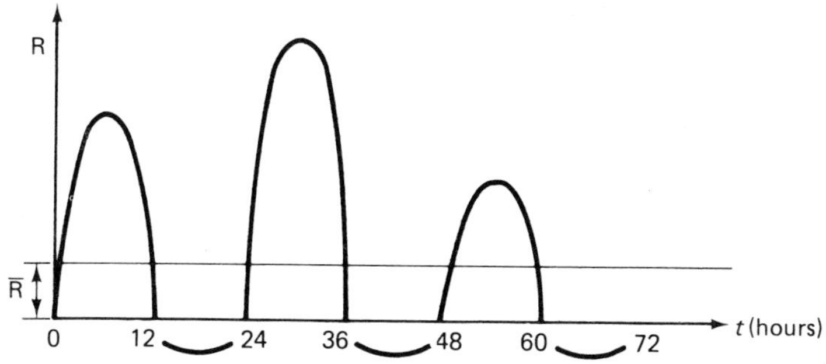

Fig. 2.1 A typical theoretical curve of relative growth rate against time for a plant over a three day period. The daylength is 12 hours and it is assumed that there are no short-term fluctuations in light intensity. The mean relative growth rate $\overline{\mathrm{R}}$, for the 72 hour period, is also shown. (After Causton, 1977.)

the level of daylight is lower on the first and third days than it is on the second day, and that there are no short term fluctuations in light intensity (days are either cloudless or uniformly overcast). Evidently the rates can be continually changing, and so instantaneous rate measurements would be of very little value even if they could be obtained.

However, the concept of a mean rate over a defined interval is valid, and this is shown for the relative growth in the hypothetical example under discussion in Fig. 2.1 as the horizontal line labelled $\overline{\mathrm{R}}$. Between two harvest times at t_1 and t_2, we have

$$\overline{\mathrm{R}} = \frac{1}{t_2 - t_1} \int_{t_1}^{t_2} \mathrm{R}\, \mathrm{d}t \tag{2.16}$$

Substituting for R from equation (2.1),

$$\overline{\mathrm{R}} = \frac{1}{t_2 - t_1} \int_{W_1}^{W_2} \frac{\mathrm{d}W}{W} \tag{2.17}$$

where W_1 and W_2 are the whole plant dry weights at the first and second harvests, respectively. Evaluation of (2.17) yields

$$\overline{\mathrm{R}} = \frac{\log_e W_2 - \log_e W_1}{t_2 - t_1} \tag{2.18}$$

The complete derivation is given in Causton (1977, p. 213). Thus, the mean relative growth rate may be evaluated over an inter-harvest period from a knowledge of the dry weight at each of the two harvests and the time interval involved.

A similar reasoning applies for the evaluation of a mean unit leaf rate, but there is an extra complication. Again, between two harvest times at t_1 and t_2, we have

$$\bar{\mathrm{E}} = \frac{1}{t_2 - t_1} \int_{t_1}^{t_2} \mathrm{E}\,\mathrm{d}t \tag{2.19}$$

and substituting for E from equation (2.4),

$$\bar{\mathrm{E}} = \frac{1}{t_2 - t_1} \int_{W_1}^{W_2} \frac{\mathrm{d}W}{L_\mathrm{A}} \tag{2.20}$$

The integral in (2.20) can be evaluated only if the relationship between W and L_A is known or assumed. If this relationship is linear, i.e. if $W = a + bL_\mathrm{A}$, where a and b are constants, then by the standard result of change of variable, (2.20) becomes

$$\bar{\mathrm{E}} = \frac{b}{t_2 - t_1} \int_{L_{A1}}^{L_{A2}} \frac{\mathrm{d}L_\mathrm{A}}{L_\mathrm{A}}$$

and so

$$\bar{\mathrm{E}} = \frac{b(\log_e L_\mathrm{A2} - \log_e L_\mathrm{A1})}{t_2 - t_1} \tag{2.21}$$

where L_A1 and L_A2 are the foliage areas at the first and second harvests, respectively.

Now, at the second harvest, $W_2 = a + bL_\mathrm{A2}$ and at the first harvest, $W_1 = a + bL_\mathrm{A1}$. Subtracting and re-arranging gives $b = (W_2 - W_1)/(L_\mathrm{A2} - L_\mathrm{A1})$ and substitution into (2.21) finally yields

$$\bar{\mathrm{E}} = \frac{(W_2 - W_1)(\log_e L_\mathrm{A2} - \log_e L_\mathrm{A1})}{(L_\mathrm{A2} - L_\mathrm{A1})(t_2 - t_1)} \tag{2.22}$$

Again, a full derivation is given in Causton (1977, pages 214–16).

Whitehead & Myerscough (1962) generalized the possible relationship between W and L_A, in the form $W = a + bL_A^n$. Substitution into (2.20) gives

$$\bar{\mathrm{E}} = \frac{nb}{t_2 - t_1} \int_{L_\mathrm{A1}}^{L_\mathrm{A2}} L_\mathrm{A}^{(n-2)}\,\mathrm{d}L_\mathrm{A}$$

and so

$$\bar{\mathrm{E}} = \frac{b(L_\mathrm{A2}^{(n-1)} - L_\mathrm{A1}^{(n-1)})n}{(t_2 - t_1)(n-1)} \tag{2.23}$$

Proceeding as before, we have that $b = (W_2 - W_1)/(L^n_{A2} - L^n_{A1})$, and substitution into (2.23) finally yields

$$\overline{E} = \frac{(W_2 - W_1)(L^{(n-1)}_{A2} - L^{(n-1)}_{A1})n}{(L^n_{A2} - L^n_{A1})(t_2 - t_1)(n-1)} \tag{2.24}$$

Although relationship (2.24) appears to be more complicated than (2.22), it can yield some simple results. If $n = 2$, (2.24) becomes

$$\overline{E} = \frac{2(W_2 - W_1)}{(L_{A1} + L_{A2})(t_2 - t_1)} \tag{2.25}$$

Coombe (1960) used this result to show that using the wrong formula for calculating $\overline{E}$ (wrong in the sense that the wrong relationship between W and L_A is assumed) would not result in serious error if harvests were not too far apart. He took a hypothetical situation where $n = 2$ and $L_{A2} = 2L_{A1}$; that is, the foliage had doubled in area between the two harvests. In this situation, Coombe found that the difference between the unit leaf rate calculated from formula (2.25) (the correct formula in this situation) and that calculated from formula (2.22) (the wrong one for this situation) was only 3.8 %. In a normal growth analysis experiment, harvests are taken such that $L_{A2} < 2L_{A1}$, and, coupled with the fact that n in vegetatively growing plants is not usually far from unity (see Chapter 6), formula (2.22) is almost universally used in estimating mean unit leaf rate over an interval of time, and formula (2.24) has presumably been regarded as an unnecessary refinement.

Rates: statistical and biological aspects

Pairing and its problems

Let us suppose for the moment that one could measure the dry weight of a single plant, and its foliage area, on more than one occasion; that is, the measurements are not destructive. Then, it would be possible to calculate the exact value of mean relative growth rate for that plant, using formula (2.18). During the inter-harvest period it would also be possible to determine the relationship between W and L_A, and hence obtain a value for n for the plant in question; then if $n = 1$, the exact value of mean unit leaf rate for the plant between the two harvests would be given by equation (2.22). It is important to realize that all the formulae for calculating both mean relative growth and unit leaf rates are independent of the actual time course of instantaneous values of these rates during the inter-harvest period; this can be seen by following through the derivations.

The same procedure would be carried out for a number, N, of replicate plants for an inter-harvest period. Hence an estimate, $\hat{\overline{R}}$, of the expected value of mean relative growth rate, $\mathscr{E}(\overline{\mathrm{R}})$; and an estimate, $\widehat{\mathscr{V}(\overline{\mathrm{R}})}$, of variance, $\mathscr{V}(\overline{\mathrm{R}})$, would be obtained for the population of plants sampled. Division of $\widehat{\mathscr{V}(\overline{\mathrm{R}})}$ by N would then give the variance of $\hat{\overline{\mathrm{R}}}$, $\mathscr{V}(\hat{\overline{\mathrm{R}}})$. To simplify the printing of notation, $\mathscr{V}(\overline{\mathrm{R}})$ will stand for what is really $\widehat{\mathscr{V}(\overline{\mathrm{R}})}$ in the remainder of this book. A similar system of notation will be used for unit leaf rate.

However, owing to the destructive nature of the measurements, the above procedure is impossible. Until very recently, the universal approach to estimating $\mathscr{E}(\overline{\mathrm{R}})$ and the population variance, using N replicate plants at each harvest, has been to simulate the progress of growth of N hypothetical plants by pairing individual replicates at one harvest with those of the other. Although this has been the standard method for many years, such pairing is an artifact; there is no sound biological basis for pairing, and the whole situation has been very unsatisfactory.

Three criteria may be used as the basis of pairing:

(i) completely random (e.g. Elias & Causton, 1975), statistically sound but no biological validity;
(ii) pairing on the basis of sequential order of size in each of the two samples (e.g. Hunt, 1978b), statistically suspect but with some biological justification (see below);
(iii) a blend of (i) and (ii), to give biological credence and be statistically acceptable (e.g. Sweet & Wareing, 1968a,b).

Criterion (iii) may be used if the experimental design incorporates classified replicates. The simplest appropriate design is the randomized block, and if at the beginning of an experiment the plants are graded for size by eye and then assigned to blocks on the basis of size, we have a criterion of pairing which is statistically sound, because at each harvest a plant is removed from each block.

To assess whether criteria (ii) and (iii) are biologically sound, evidence will be presented from an experiment conducted by Sweet & Wareing (1966b) using seedlings of *Pinus radiata* (Monterey pine) and *Larix leptolepis* (Japanese larch). In the case of *P. radiata* 6 blocks, representing 6 different size classes, were used and 2 harvests only were taken. The logarithms of the whole plant dry weights are shown in Fig. 2.2a, and the gradients of the lines are the mean relative growth rates of the plants in each block between the two harvests. There is no clear evidence of a size effect on relative growth rate. For the *L. leptolepis* (Fig. 2.2b) seedlings the results are very similar, although here there were only 4 size classes, the experimental duration was slightly greater, and intermediate harvests were taken.

These results mean that had the plants been randomly paired, a wider range of mean relative growth rates would have been obtained and, although the average for all the replicates, $\hat{\overline{\mathrm{R}}}$, might be similar, the variance

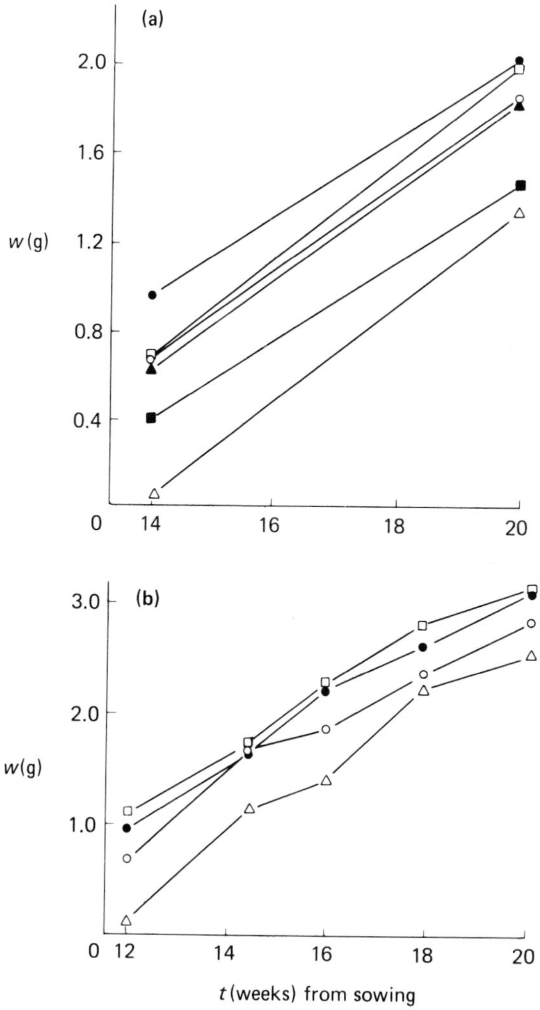

Fig. 2.2 Log$_e$ dry weight plotted against time for first-year seedlings of (**a**) *Pinus radiata* and (**b**) *Larix leptolarix* for six and four (respectively) different sized classes. The slope of the lines gives a measure of relative growth rate, and comparison of the slopes indicates no clear relationship between seedling size and relative growth rate. (After Sweet & Wareing, 1966b.)

of $\overline{\mathrm{R}}$ would be artificially inflated. Thus, at least on the basis of Sweet & Wareing's experiment, pairing on a size criterion is biologically valid.

However the pairing is achieved, the calculations are arranged as follows. Let there be N replicate plants at each harvest, and denote the whole plant dry weight as W_{1j} and the leaf area as $L_{\mathrm{A}ij}(i = 1, 2; j = 1, \ldots, N)$. Then for the jth replicate pair of plants (one plant of the pair from each of the two harvests)

$$\overline{\mathrm{R}}_j = \frac{\log_e W_{2j} - \log_e W_{1j}}{t_2 - t_1} \tag{2.26}$$

and

$$\overline{\mathrm{E}}_j = \frac{(W_{2j} - W_{1j})(\log_e L_{\mathrm{A}2j} - \log_e L_{\mathrm{A}1j})}{(L_{\mathrm{A}2j} - L_{\mathrm{A}1j})(t_2 - t_1)} \tag{2.27}$$

Then the expected values of the mean relative growth and unit leaf rates may be estimated as the arithmetic means of $\overline{R}_j$ and $\overline{\mathrm{E}}_j$ across the N replicates:

$$\hat{\overline{\mathrm{R}}} = (1/N) \,.\, \sum_{j=1}^{N} \overline{\mathrm{R}}_j \tag{2.28}$$

$$\hat{\overline{\mathrm{E}}} = (1/N) \,.\, \sum_{j=1}^{N} \overline{\mathrm{E}}_j \tag{2.29}$$

The sample variances are then given by

$$\mathscr{V}(\overline{\mathrm{R}}) = \{1/(N-1)\} \,.\, \sum_{j=1}^{N} (\overline{\mathrm{R}}_j - \hat{\overline{\mathrm{R}}})^2 \tag{2.30}$$

$$\mathscr{V}(\overline{\mathrm{E}}) = \{1/(N-1)\} \,.\, \sum_{j=1}^{N} (\overline{\mathrm{E}}_j - \hat{\overline{\mathrm{E}}})^2 \tag{2.31}$$

A new method avoiding pairing

Venus & Causton (1979b) have presented a new method which avoids having to pair plants between harvests. Using the above notation, and utilizing standard theorems for a linear combination of variates, we have

$$\hat{\overline{\mathrm{R}}} = \{1/(t_2 - t_1)\}\{\mathscr{E}(\log_e W_2) - \mathscr{E}(\log_e W_1)\} \tag{2.32}$$

and

$$\mathscr{V}(\overline{\mathrm{R}}) = \{1/(t_2 - t_1)^2\}\{\mathscr{V}(\log_e W_2) + \mathscr{V}(\log_e W_1)\} \tag{2.33}$$

where

$$\mathscr{E}(\log_e W_i) = (1/N) \,.\, \sum_{j=1}^{N} \log_e W_{ij}$$

and $\mathscr{V}(\log_e W_i) = \{1/(N-1)\} \,.\, \sum_{j=1}^{N} \{\log_e W_{ij} - \mathscr{E}(\log_e W_i)\}^2$

No covariance term enters into equation (2.33) as the samples taken at each of the two harvests are independent of one another; also, equations (2.32) and (2.33) give an exact estimate of the mean relative growth rate and the variance of that estimate, whereas the pairing process of the previous method causes the variance of the estimate of the mean relative growth rate to be only approximate.

For unit leaf rate, the four variates W_1, W_2, L_{A1}, and L_{A2} are combined in the non-linear form of equation (2.22); hence exact estimates of $\mathscr{E}(\bar{E})$ and $\mathscr{V}(\bar{E})$ are not obtainable, only approximate ones, and the approximations are derived from formulae given by Kendall & Stuart (1977) and in the appendix to this book. The application of these formulae for unit leaf rate involves more extensive mathematics than the derivation of exact formulae for relative growth rate, and the estimators for $\mathscr{E}(\bar{E})$ and $\mathscr{V}(\bar{E})$ are derived below.

The expected value and variance of mean unit leaf rate between two harvests at times t_1 and t_2 are estimated by

$$\hat{\bar{E}} \simeq E + \frac{1}{2}\cdot\frac{\partial^2 E}{\partial W_2^2}\cdot\mathscr{V}(W_2) + \frac{1}{2}\cdot\frac{\partial^2 E}{\partial W_1^2}\cdot\mathscr{V}(W_1)$$

$$+\frac{1}{2}\cdot\frac{\partial^2 E}{\partial L_2^2}\cdot\mathscr{V}(L_2) + \frac{1}{2}\cdot\frac{\partial^2 E}{\partial L_1^2}\cdot\mathscr{V}(L_1)$$

$$+\frac{\partial^2 E}{\partial W_2 \partial L_2}\cdot\mathscr{C}(W_2, L_2) + \frac{\partial^2 E}{\partial W_1 \partial L_1}\cdot\mathscr{C}(W_1, L_1) \qquad (2.34)$$

and $$\mathscr{V}(\bar{E}) \simeq \left(\frac{\partial E}{\partial W_2}\right)^2.\mathscr{V}(W_2) + \left(\frac{\partial E}{\partial W_1}\right)^2.\mathscr{V}(W_1)$$

$$+\left(\frac{\partial E}{\partial L_1}\right)^2.\mathscr{V}(L_1) + \left(\frac{\partial E}{\partial L_2}\right)^2.\mathscr{V}(L_2)$$

$$+2.\frac{\partial E}{\partial W_2}\cdot\frac{\partial E}{\partial L_2}.\mathscr{C}(W_2, L_2)$$

$$+2.\frac{\partial E}{\partial W_1}\cdot\frac{\partial E}{\partial L_1}.\mathscr{C}(W_1, L_1), \qquad (2.35)$$

where $\mathscr{V}(W_i)$ and $\mathscr{V}(L_i)$ are the variances of plant dry weight and leaf area, respectively, at the ith harvest; $\mathscr{C}(W_i, L_i)$ is the covariance between plant weight and leaf area at the ith harvest, and E is given by

$$E = \frac{\{\mathscr{E}(W_2) - \mathscr{E}(W_1)\}\{\log_e \mathscr{E}(L_2) - \log_e \mathscr{E}(L_1)\}}{\{\mathscr{E}(L_2) - \mathscr{E}(L_1)\}\{t_2 - t_1\}}$$

There are also terms involving covariances of W and L between the two harvests, i.e. $\mathscr{C}(W_1, W_2)$, $\mathscr{C}(L_1, L_2)$, $\mathscr{C}(W_1, L_2)$, $\mathscr{C}(L_1, W_2)$ in the equations for $\bar{E}$ and $\mathscr{V}(\bar{E})$; but since the samples at each harvest are independent of one another, these terms are all zero and are consequently omitted.

The various derivatives are as follows.

$$\frac{\partial \mathrm{E}}{\partial W_1} = \frac{-(\log_e L_2 - \log_e L_1)}{(L_2 - L_1)(t_2 - t_1)} \qquad \frac{\partial \mathrm{E}}{\partial W_2} = \frac{\log_e L_2 - \log_e L_1}{(L_2 - L_1)(t_2 - t_1)},$$

$$\frac{\partial \mathrm{E}}{\partial L_1} = \frac{\{(-1/L_1)(L_2 - L_1) + (\log_e L_2 - \log_e L_1)\}\{W_2 - W_1\}}{(L_2 - L_1)^2 (t_2 - t_1)},$$

$$\frac{\partial \mathrm{E}}{\partial L_2} = \frac{\{(1/L_2)(L_2 - L_1) - (\log_e L_2 - \log_e L_1)\}\{W_2 - W_1\}}{(L_2 - L_1)^2 (t_2 - t_1)},$$

$$\frac{\partial^2 \mathrm{E}}{\partial W_1^2} = \frac{\partial^2 \mathrm{E}}{\partial W_2^2} = 0,$$

$$\frac{\partial^2 \mathrm{E}}{\partial L_1^2} = \frac{\{(1/L_1^2)(L_2 - L_1)^2 - (2/L_1)(L_2 - L_1) + 2(\log_e L_2 - \log_e L_1)\}\{W_2 - W_1\}}{(L_2 - L_1)^3 (t_2 - t_1)}$$

$$\frac{\partial^2 \mathrm{E}}{\partial L_2^2} = \frac{\{(-1/L_2^2)(L_2 - L_1)^2 - (2/L_2)(L_2 - L_1) + 2(\log_e L_2 - \log_e L^1)\}\{W_2 - W_1\}}{(L_2 - L_1)^3 (t_2 - t_1)}$$

$$\frac{\partial^2 \mathrm{E}}{\partial W_1 \partial L_1} = \frac{(1/L_1)(L_2 - L_1) - (\log_e L_2 - \log_e L_1)}{(L_2 - L_1)^2 (t_2 - t_1)},$$

$$\frac{\partial^2 \mathrm{E}}{\partial W_2 \partial L_2} = \frac{(1/L_2)(L_2 - L_1) - (\log_e L_2 - \log_e L_1)}{(L_2 - L_1)^2 (t_2 - t_1)}.$$

All the W_i and L_i on the right-hand sides of the above formulae are to be read as sample mean values of W and L, respectively.

Thus, valid estimates (exact in the case of mean relative growth rate, but approximate in the case of mean unit leaf rate) can be obtained of the two rate parameters in a plant growth analysis, and of the variances of these estimates. Because the results for mean relative growth rate are exact, there is no doubt that this non-pairing method should be used. Results for mean unit leaf rate are, however, still approximate although there is still the biological advantage of no pairing. We must now compare the two methods for estimating mean unit leaf rate from a statistical viewpoint.

The statistical validity of the methods for estimating mean unit leaf rate: a simulation study

The underlying whole plant dry weights and leaf areas for this simulation study are shown in Table 2.1; they are mean values taken from actual data (*Helianthus annuus*, page 43) to be biologically reasonable. Three mean unit leaf rate values were then calculated from these basic data, using equation (2.22): between harvests 1 and 2, between harvests 2 and 3, and between harvests 1 and 3; the results are also shown in Table 2.1.

Table 2.1 Underlying harvest times, whole plant dry weights, leaf areas, and calculated mean unit leaf rates for the simulation study of the estimators of mean unit leaf rate.

Harvest number	1	2	3
Time, day	2	4	6
Whole plant dry weight, g	0.03	0.07	0.102
Foliage area, m^2	0.00065	0.00158	0.00295
Unit leaf rate (harvests 1–2: 2–3) $(\overline{E}_u)$ g m^{-2} day^{-1}	19.09		7.28
Unit leaf rate (harvest 1–3) $(\overline{E}_u)$ g m^{-2} day^{-1}		11.84	

For each of the three harvest intervals, 20 replicate observations were generated, using random normal deviates. The deviates were added to the logarithmic values of the weights and areas in Table 2.1, and these deviates were adjusted as required to simulate data with standard deviations of the logarithmically transformed weights and areas of 0.2, 0.4, and 0.6, corresponding to coefficients of variation of 20.2%, 41.7%, and 65.8%, respectively. These coefficients of variation cover the usual range of variability of biological data, but missing out a coefficient of variation of 10% which would be considered 'good' biological data; however, data having a high coefficient of variation are likely to create most problems. No attempt was made to build in a positive correlation which invariably exists between W and L_A.

The whole study comprised nine different experiments – every combination of the three standard deviations and the three harvest intervals – and in each experiment about $N = 50$ individual data sets, of 20 replicates each, were generated. The scheme for a single one of the 9 experiments is shown in Table 2.2. For each of the N generated data sets an estimate of $\overline{E}$, $\hat{E}_i$, and the variance of this estimate, $\mathcal{V}(\hat{E}_i)$, were calculated, and at the end of the experiment $\hat{E}_m$ and $\mathcal{V}(\hat{E}_m)$ were calculated as shown at the bottom of Table 2.2.

The entire study was done twice: once using random pairing for the calculation of each $\hat{E}_i$ and its variance (equations (2.29) and (2.31)), and again using the new non-pairing method (equations (2.34) and (2.35)).

The prime purpose of the simulation study was to compare $\hat{E}_m$ (Table 2.2) with $\overline{E}_u$ (Table 2.1) for each of the 9 experiments. This was done by a t-test of the form

$$t_{[N-1]} = \frac{\hat{E}_m - \overline{E}_u}{\sqrt{\{\mathcal{V}(\hat{E}_m)/N\}}} \qquad (2.36)$$

and results for the random pairing method are given in Table 2.3, while those for the non-pairing method appear in Table 2.4.

Table 2.2   Scheme of a single experiment in the simulation study of the estimators mean unit leaf rate. See page 33 for details of the notation involved.

Data set	Estimate of $\overline{E}$ $\hat{\overline{E}}$	Variance of estimate of $\overline{E}$ $\mathcal{V}(\hat{\overline{E}})$
1	$\hat{\overline{E}}_1$	$\mathcal{V}(\hat{\overline{E}}_1)$
.	.	.
.	.	.
.	.	.
N	$\hat{\overline{E}}_N$	$\mathcal{V}(\hat{\overline{E}}_N)$
Mean value of the $\hat{\overline{E}}_j$	$\hat{\overline{E}}_m$	–
Variance of the N estimates of $\overline{E}$	$\mathcal{V}(\hat{\overline{E}}_m)$	–

$$\hat{\overline{E}}_m = (1/N).\sum_{j=1}^{N}\hat{\overline{E}}_j \qquad \mathcal{V}(\hat{\overline{E}}_m) = \{1/(N-1)\}.\sum_{j=1}^{N}(\hat{\overline{E}}_j - \hat{\overline{E}}_m)^2$$

Table 2.3   Results of the simulation study using the random pairing method for estimating mean unit leaf rate. For each experiment, $\hat{\overline{E}}_m$ is shown, $100(\hat{\overline{E}}_m - \overline{E}_u)/\overline{E}_u$ is given in parentheses, and Student's t-value (calculated according to equation (2.36)) is shown below with its significance level: * – $P < 0.05$,   ** – $P < 0.01$,   *** – $P < 0.001$.

Standard deviation \\ Harvest interval	1–2	2–3	1–3
0.2	19.72 (3.3) 3.28**	7.57 (4.0) 1.89	12.21 (3.1) 3.88***
0.4	21.48 (12.5) 5.04***	8.27 (13.6) 2.66*	13.31 (12.4) 6.19***
0.6	24.71 (29.4) 5.98***	9.51 (30.6) 3.05**	15.36 (29.7) 7.44***

It is immediately apparent that, although all $\hat{\overline{E}}_m$ values are positively biased, the results for the non-pairing method are greatly superior to those for the pairing method. Six out of the nine results for the former method yield non-significant differences between $\hat{\overline{E}}_m$ and $\overline{E}_u$, whereas there is only one non-significant difference shown in the latter method; moreover, there are only two highly significant differences ($P < 0.01$) shown in the non-pairing method, but no less than seven in the random pairing method. The larger t-values given by the random pairing method are brought about by

Table 2.4 Results of the simulation study using the non-pairing method for estimating mean unit leaf rate. Details as for Table 2.3.

Standard deviation \ Harvest interval	1–2	2–3	1–3
0.2	19.41 (7.1) 1.35	7.85 (1.0) 0.39	12.21 (2.4) 2.08
0.4	20.32 (6.4) 1.75	8.78 (20.6) 1.96	13.35 (12.8) 5.38***
0.6	22.04 (15.5) 2.63*	11.47 (57.6) 4.42***	15.11 (27.6) 1.91

the smaller variances of the estimates rather than larger deviations between $\hat{\bar{E}}_m$ and $\bar{E}_u$, when compared to the non-pairing method. In both methods there is a tendency for larger significant differences to occur as the standard deviation of the data increases, but this is no more pronounced in one method than the other. In the random pairing method, there is no evidence of an effect of duration of harvest interval nor of an effect of the magnitude of the underlying mean unit leaf rate, $\bar{E}_u$, on the size or significance of the deviations between $\hat{\bar{E}}_m$ and $\bar{E}_u$. In the non-pairing method, however, there may be some indication of a negative correlation between the magnitudes of the underlying unit leaf rates and the deviations between $\hat{\bar{E}}_m$ and $\bar{E}_u$. The overall conclusion is that the estimator of the non-pairing method appears to be well-behaved for data having a coefficient of variation up to about 30%.

It is impossible, without knowing how the true value of $\bar{E}$ varies from plant to plant, to evaluate the variances of the estimates of $\mathscr{E}(\bar{E})$ in a rigorous manner. However, in the long run, the average of the $\mathscr{V}(\hat{\bar{E}}_j)$ should equal $\mathscr{V}(\hat{\bar{E}}_m)$, and this null hypothesis may be assessed by the sign test. Firstly, the number of $\mathscr{V}(\hat{\bar{E}}_j)$ greater than $\mathscr{V}(\hat{\bar{E}}_m)$ are recorded, N_g; then, if the null hypothesis is true, N_g is binomially distributed with mean $N/2$ and variance $N/4$. Because N is relatively large (about 50 in each experiment), we then have

$$z = \frac{N_g - N/2}{\sqrt{(N/4)}} \tag{2.37}$$

and z may be assessed against a table of standard normal deviates.

The results are remarkably clear cut, and disturbing (Table 2.5). In the random pairing method three results, each associated with data of the highest variability, were non-significant; while in the non-pairing method

Table 2.5 Results of the simulation study for the assessment of the variances of the estimates of $\hat{\bar{E}}$. The *z*-values of the sign test (calculated according to equation (2.37)) are shown, together with their significance levels (asterisk code as in Table 2.3). For each experiment the upper value is for the random pairing method, and the lower value is for the non-pairing method.

Standard deviation \ Harvest interval	1–2	2–3	1–3
0.2	5.7***	5.7***	5.4***
	1.0	1.0	−4.1***
0.4	4.5***	2.5***	3.9***
	−6.3***	−6.6***	−2.4***
0.6	0.8	0.8	0.3
	5.8***	−6.6***	−6.6***

only two results were not significant, and these were found in the cases of lowest data variability and short harvest interval. All the other six or seven results, respectively, were highly significant. Moreover, there was a distinct, but different, bias of the variance estimate associated with each method: positive in the case of random pairing, and negative in the case of non-pairing. The conclusion, reached by this simulation study, is that $\mathscr{V}(\hat{\bar{E}})$ is likely to be too large in the random pairing method, and too small in the non-pairing method; on the other hand, $\hat{\bar{E}}$ is likely to be near $\mathscr{E}(\bar{E})$ in the non-pairing method, and so the use of this latter method is to be preferred.

Classical growth analysis: applications

In this section we consider a variety of topics, starting with aspects of methodology of a more practical nature than those discussed hitherto, and ending with a comparison of the growth of four species, using our own data.

Difficulties in the interpretation of rate trends

In a growth analysis experiment where several harvests have been taken, it would be expected that the various ratios and rates, derived from the data, would show trends with time, but superimposed on these trends are fluctuations due to both sampling variation and environmental fluctuations. Sampling variation of the ratios does not appear to be much more serious in its effects than is the variability of the underlying data, but in the case of the rates sampling variation can be very large, obscuring not only real fluctuations but long term trends also. The variability of calculated rates increases as the harvest frequency increases because the sampling variation has a greater effect (page 30).

There is usually a negative correlation between adjacent calculated rate values; this is because the middle harvest of three is used in the calculation of two adjacent rate values, and if the central harvest size values are small, this will have the effect of lowering the rate calculated over the first time interval but raising that over the second interval. This effect becomes more pronounced as the deviation of observed weight and area values from the underlying smooth trends become larger in relation to the difference between harvests, as when the interval between harvests is short.

An attempt has been made to solve this problem by weighting the observations (McIntyre & Williams, 1949), and the weighting factor becomes progressively more important as harvest intervals become shorter. Consider two plants, which at time t_1 have dry weights W_{11} and W_{12}; plant 1 is harvested, but plant 2 is allowed to grow on until time t_2 when it is harvested and weighs W_{22}. The mean relative growth rate can only be calculated as

$$\overline{\mathrm{R}} = \frac{\log_e W_{22} - \log_e W_{11}}{t_2 - t_1}$$

but this takes no account of the differences between the plants at the first harvest; if $W_{12} - W_{11}$ is relatively large in comparison with $W_{22} - W_{11}$ then the calculated mean relative growth rate will be inaccurate. Similar arguments apply to the usual practical situation in which there are N replicate plants per harvest instead of just 1, and weighting attempts to take into account the original differences between samples. However, the method has been scarcely used, and we will not discuss it further here.

A simulation experiment to find a suitable harvest interval for realistic rate estimation

Underlying relationship curves for plant weight and leaf area with time were defined as those Richards functions that described these attributes in our *Helianthus annuus* experiment (Table 2.6). These curves were chosen purely for convenience, and to give biologically realistic underlying trends. Using these two Richards functions for $w(t)$ and $l_A(t)$ (growth functions are always fitted to data in logarithmic form; page 100), relative growth and unit leaf rate functions were obtained using equation (2.1) and an appropriately re-arranged version of (2.5). Since underlying trends only were required of the fitted functions, no account was needed to be taken of data variability; so $W = \exp\{w(t)\}$ and $L_A = \exp\{l_A(t)\}$ directly, and equations (2.1) and (2.5) were used without modification. The resulting curves obtained for relative growth and unit leaf rates are shown in Fig. 2.3 and Fig. 2.4, respectively.

Table 2.6 Underlying relationship harvest values, together with harvest means of the simulated data and their standard deviations for the simulation experiment on the effect of harvest interval on the variability of mean relative growth and unit leaf rates.

	Plant dry weight			Leaf area		
	Underlying relationship	Simulated data		Underlying relationship	Simulated data	
t	w	$\bar{w}$	s_w	l_A	$\bar{l}_A$	s_{lA}
1	−3.29	−3.27	0.22	1.61	1.62	0.20
3	−2.91	−2.92	0.24	2.08	2.10	0.30
5	−2.53	−2.52	0.30	2.54	2.57	0.11
7	−2.14	−2.15	0.26	3.01	3.01	0.21
9	−1.76	−1.71	0.24	3.47	3.52	0.27
11	−1.38	−1.29	0.27	3.91	3.94	0.20
13	−1.01	−1.05	0.21	4.34	4.33	0.16
15	−0.63	−0.61	0.22	4.74	4.72	0.20
17	−0.27	−0.26	0.18	5.10	5.11	0.29
19	0.09	0.07	0.20	5.40	5.39	0.19
21	0.44	0.53	0.25	5.64	5.61	0.19
23	0.77	0.74	0.19	5.80	5.86	0.18
25	1.08	1.07	0.14	5.92	5.89	0.16
27	1.36	1.37	0.28	5.99	5.97	0.22
29	1.61	1.63	0.29	6.03	6.07	0.30
31	1.81	1.82	0.20	6.05	5.98	0.21
33	1.98	1.99	0.24	6.07	6.05	0.21

Simulated plant weight and leaf area data were obtained by adding random normal deviates, adjusted to give a standard deviation of 0.2, to the underlying $w(t)$ and $l_A(t)$ relationship values at distinct 'harvest times'; harvests were designated as occurring every two days, and ten replicate weight and area observations were obtained at each harvest. The level of variability selected for this simulation experiment is equivalent to a coefficient of variation of 20.2%, which is the lowest of the three variabilities used in the simulation study for examining the estimators of unit leaf rate (pages 32–36) but represents an average for our experimental data described in this book. The means and standard deviations of the simulated data are compared with the underlying relationship values in Table 2.6.

Estimates of mean relative growth and unit leaf rates were obtained, together with their variances, using equations (2.32–2.35); from the variances, 95% confidence intervals for the estimated rates were also obtained, for example for a relative growth rate

$$\hat{\bar{\mathrm{R}}} - t_{[N-1]} \cdot \sqrt{\{\mathscr{V}(\bar{\mathrm{R}})/N\}} \text{ to } \hat{\bar{\mathrm{R}}} + t_{[N-1]} \cdot \sqrt{\{\mathscr{V}(\bar{\mathrm{R}})/N\}}$$

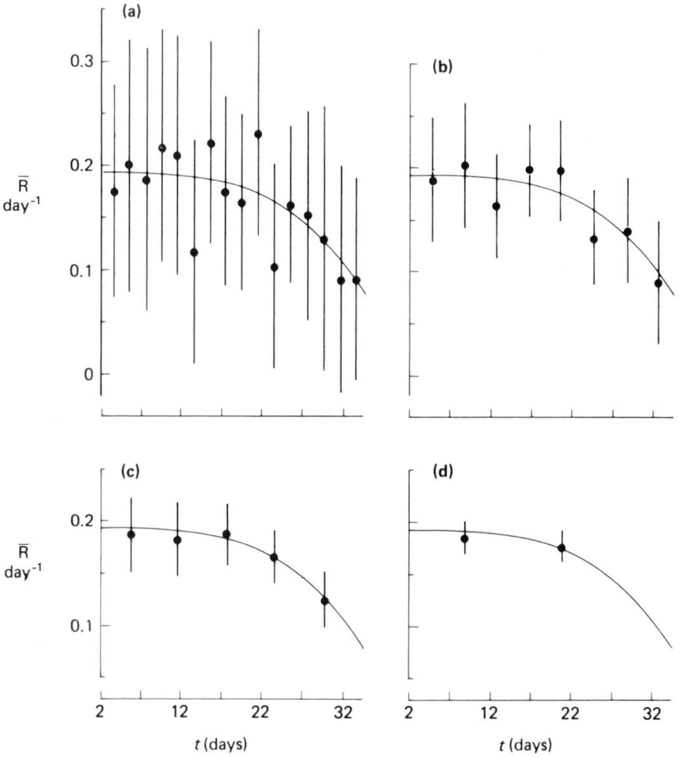

Fig. 2.3 Relative growth rates calculated for varying harvest intervals (see text for details) compared with the underlying relative growth rate curve. Bars respresent ± students-t × standard error of $\overline{R}$.

These calculations were repeated for four different sub-sets of the total simulated data:

(a) all harvests (17 in all),
(b) alternate harvests only (9),
(c) every third harvest only (6),
(d) harvests on 'days' 1, 13, and 27 only (3).

Results: relative growth rate

The effects of an increasing harvest interval on the estimates of mean relative growth rate and their variances (in the form of the 95% confidence interval) are shown in Fig. 2.3.* When all harvests were used (Fig. 2.3a), the

* Strictly, a histogram is the correct form of graphical presentation of mean rate-values over an interval of time where the horizontal axis is time itself; but such graphs could become very confused when confidence intervals are included as well, and so we have avoided using histograms in this book.

variability of the estimated rates is very high, and the estimates themselves oscillate greatly about the underlying relationship. The estimated mean relative growth rate over the interval from day 11 to day 13 is low because the simulated mean plant size at day 11 was larger than the underlying value, whereas that for day 13 was smaller (Table 2.6). Consequently, this rate is negatively correlated with the next estimated value (13 to 15 days) which is too large, for similar reasons. Corresponding evidence can be found to explain other oscillations, indicating that harvests are so close together that the discrepancies between the simulated data and the underlying relationship are large in relation to differences between harvests.

The problems found above become less obvious when only alternate harvests are considered (Fig. 2.3b), although the estimated rate over the interval 9 to 13 days is conspicuously low. Also, the estimated mean relative growth rates over the intervals 17 to 21 days and 21 to 25 days exhibit a strong negative correlation with each other.

When only 6 harvests are taken, the fluctuations of the estimated mean relative growth rates about the underlying relationship are much less evident, and the variability of the estimates is considerably reduced (Fig. 2.3c). This is taken a stage further when only three harvests are considered. However, these only yield two estimated mean relative growth rates which can provide only a very limited indication of the underlying relationship (Fig. 2.3d). It therefore appears that the former sub-set of data, consisting of harvests taken every 6 days, strikes a compromise between depicting the underlying trend of relative growth rate and having reasonably small variability of the estimated mean relative growth rates. The favourability of the 6-day harvest interval in this instance is a happy coincidence with the weekly regime commonly used.

The above results are supported by theory. In equation (2.33) it will be seen that $\mathscr{V}(\overline{\mathrm{R}})$ is inversely proportional to the square of the inter-harvest duration. The variance of $\overline{\mathrm{R}}$ is also directly proportional to the variance of $\log_e W$, so arranging for suitable harvest intervals must be done in conjunction with some knowledge of the coefficient of variation of the plant size attributes.

Results: unit leaf rate

Similar conclusions can be drawn for the estimated mean unit leaf rates (Fig. 2.4) as for the corresponding relative growth rates. Reduction in the number of harvests used in calculation, i.e., increase in the time between harvests, results in a decrease in variability and a reduction in the oscillations around the underlying relationship as the negative correlation effect becomes less. The results for the data sub-set (c), consisting of harvests every 6 days, can again be suggested as the most realistic (Fig. 2.4c),

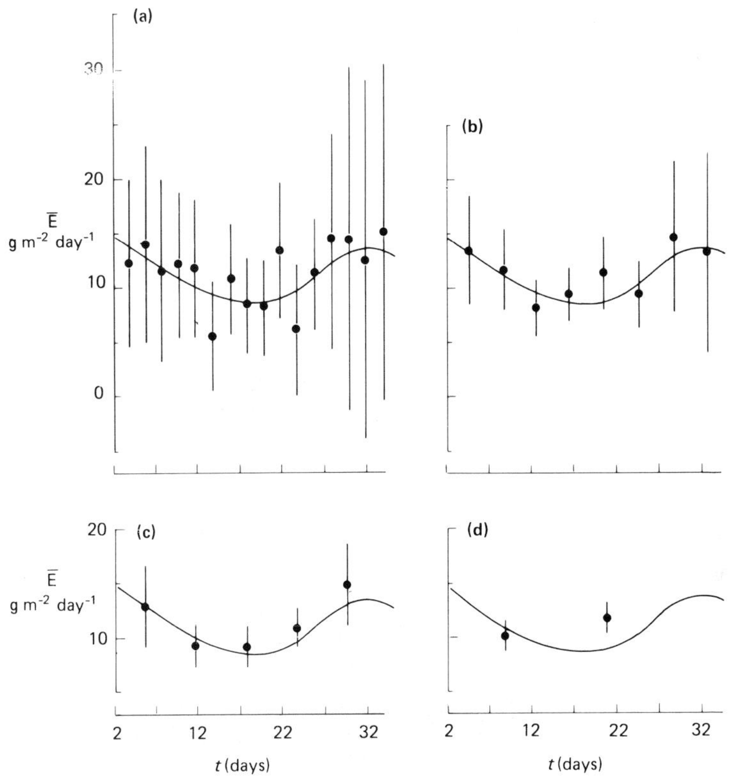

Fig. 2.4 Unit leaf rate calculated for varying harvest intervals (see text for details) compared with the underlying unit leaf rate curve. Bars represent $\pm$ students-t $\times$ standard error of $\overline{E}$.

although this data sub-set gives no indication of the final downward trend of the underlying relationship; however, the results from data sub-set (b) oscillate too much to support their preference as an alternative.

It therefore appears that for data of this type, with a coefficient of variation of about 20% and 10 replicates (variability of a sample estimate is inversely proportional to the square root of the number of replicates in the sample), harvests taken at about 6-day intervals provide the most realistic rate results. This strategy will be adopted in the presentation of our results in the section after next.

A variety of uses for whole plant growth analysis

In this section, we shall not distinguish between growth analysis studies pursued by the classical or the functional methods, as the aims and results associated with both methods are broadly similar; it is the utility of the principles of whole plant growth analysis that will be our concern here.

Early workers used growth analysis methods primarily to give a preliminary description of the growth of individual species (page 8). More recently, the techniques have been used to assess the effects of particular environmental factors on growth, and the interaction of these factors on the ontogeny of growth (e.g. Rajan, Betteridge & Blackman, 1971; Hughes & Cockshull, 1972; Voldeng & Blackman, 1973; Rajan & Blackman, 1975; Gifford, 1977). Besides the earlier work already mentioned, classical methods have also been used to compare species performance (e.g. Bradshaw, Lodge, Jowett & Chadwick, 1958; Blackman & Black, 1959; Jefferies & Willis, 1964; Myerscough & Whitehead, 1967; Hutchinson, 1967, 1968; Pollard & Wareing, 1968; Rorison, 1968; Higgs & James, 1969). Most of the above investigations were concerned with the comparison of a limited number of species often, but not always, grown under identical conditions. On the other hand, Jarvis & Jarvis (1964) collected a large body of information about growth rates of woody species from a variety of sources, and were able to draw realistic conclusions from the comparisons.

Hitherto, however, the ultimate in species comparison must be the work of Grime & Hunt (1975), who surveyed no less than 132 species in the British flora over their first few weeks from germination in a series of growth cabinet experiments using sand culture. Small groups of species were used in each experiment, and the conditions between experiments were as near identical as possible. Grime and Hunt used the functional method, developed by Hunt & Parsons (1974), to estimate the maximum relative growth rate at the beginning of growth, R_{max}, and the mean relative growth rate over the duration of each experiment (in the very many cases where the logarithmically transformed plant dry weight data were adequately described by a straight line, the two relative growth rates were identical), and compared these for the different species. These authors give a convincing argument in support of the comparison of rates between experiments; they did, however, always grow the species in what they termed 'potentially productive environments' but, bearing in mind the range of original habitats and variety of life strategies (Grime, 1979), they proposed a number of interesting and realistic species groupings in terms of R_{max}. Grime and Hunt's work was concerned with the characterization of each species by a single estimate of relative growth rate R_{max}, but it is important to remember that all growth analysis quantities show changes as the plants grow and develop (ontogenetic 'drift'); thus, it is of considerable interest to compare the ontogenetic drifts of these quantities between species over a period of time.

Experimental data: a comparative study of four species

The species and the experiments

The four experiments, one for each species, to be introduced here are used to provide illustrative material throughout the book. The species used were *Betula pubescens* (birch) – a woody dicotyledon; *Helianthus annuus* (sunflower) – an herbaceous dicotyledon; *Triticum aestivum* (wheat) – a monocotyledon; and *Zea mays* (maize) – a monocotyledon having the C_4 photosynthetic mechanism with associated anatomical modifications. The experiments were conducted under similar glasshouse conditions during spring and early summer, and the wheat and maize experiments ran concurrently. The plants were grown in John Innes (J.I.) No. 2 potting compost in 75 mm plastic pots; but owing to the much longer duration of the birch experiment these plants were grown in 100 mm plastic pots, and after 4 weeks the pots were each placed on top of a 125 mm pot which was also filled with J.I. No. 2 potting compost. This was an attempt to minimize the effects of 'pot binding', as roots could grow out through the holes in the bottom of the smaller pot into the additional volume of soil in the larger pot. Repotting was dismissed as an alternative because this could only be done effectively when the soil volume in the smaller pot was already full of roots implying that pot binding would have already commenced. Remaining details concerning the experiments are given in Table 2.7.

The results of classical growth analyses, performed by the methods given on pages 24 and 30–32, are presented as a series of graphs aimed at facilitating comparisons between the four species.

Results: leaf weight, stem weight, and root weight ratios

Trends in these ratios show some fundamental differences in the species' growth strategies (Fig. 2.5). In the sunflower, the stem plays an increasingly more important role throughout the growth period examined, reaching nearly 50 % of the total plant dry weight by the end of the experiment, and is significantly larger than in the other species for almost the entire growth period (Fig. 2.5b). In the other species examined, the stem has a more minor role to play, although there is an indication of an increase in stem weight ratio at the end of the wheat experiment; this is associated with the switch to reproductive development and associated stem elongation. Both the grass species (wheat and maize) have very small stems until the reproductive phase commences. The stem weight ratio in birch is intermediate between that in the grasses and the sunflower but, in marked contrast to the other three species, the stem weight ratio in birch declines during part of the experimental period. This is despite the fact that the stem soon becomes a prominent feature of the young birch plant and, in fact, results for the entire

Table 2.7 Details of the four species experiments used as illustrative material.

Species	First harvest (days from germination)	Harvest intervals	Total number of harvests h	Number of replicates N	Duration of experiment (days)
Betula pubescens (birch)	12	3- & 4-day intervals	24 (8 for whole plant growth analysis)	10	80 (35 for whole plant growth analysis)
Helianthus annuus (sunflower)	3	2- & 3-day intervals	16	9	35
Triticum aestivum (wheat)	4 (6 for classical growth analysis)	10 daily, then alternate days	18	6	29
Zea mays (maize)	6	7 daily, then alternate days	16	6	30

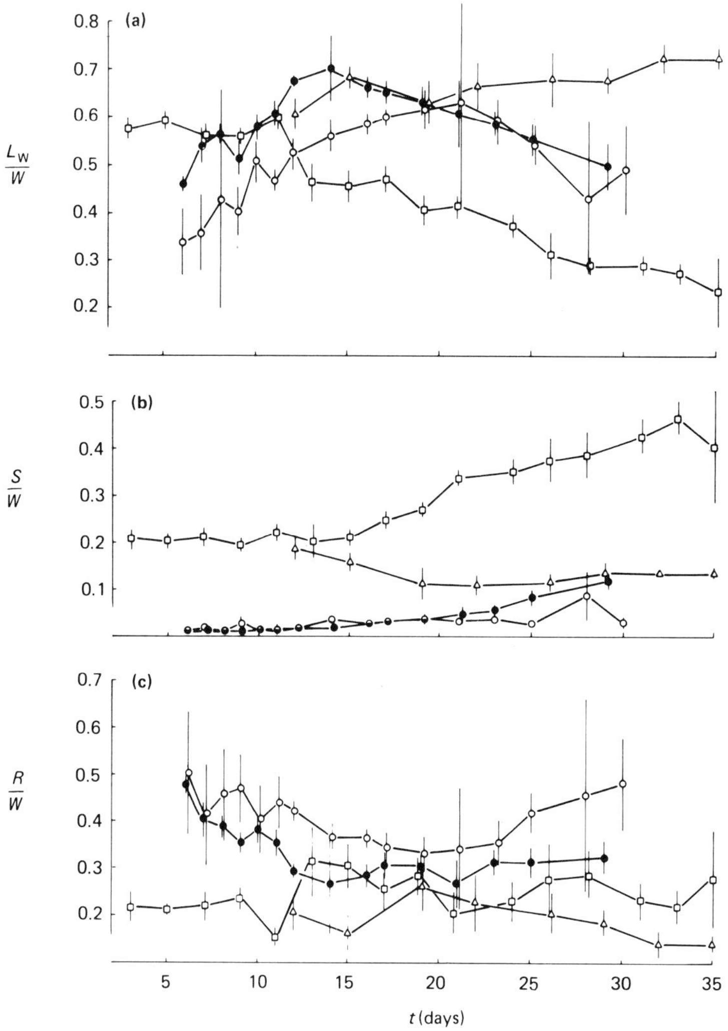

Fig. 2.5 Classical growth analysis comparison of wheat (●), maize (o), sunflower (□) and birch (△): (**a**) leaf weight ratio, (**b**) stem weight ratio and (**c**) root weight ratio. Growth measurements with 95% confidence intervals.

80-day period show a steady increase in stem weight ratio after 35 days (Venus, 1978).

Although the clearest species differences are discernable in the stem weight ratio, there are interesting differences in terms of both the leaf weight and root weight ratios (Fig. 2.5a,c). In sunflower, leaf weight ratio declines throughout the experiment, reflecting the increase in stem weight ratio discussed above. The leaf weight ratio trends of wheat and maize are interesting in that both appear curvilinear; wheat, however, reaches a larger maximum considerably before (7 days) that attained by maize, and for the later part of the experiment the two species have very similar leaf weight ratios (this also applies to their stem and root weight ratios). Root weight ratio estimates are more variable than those of the other ratios, and significant differences between the species therefore appear less frequently. The higher variability of the root weight ratio estimates may be a reflection of the fact that greatly differing amounts of the root system are retrieved from the soil from one plant to another.

Results: leaf area ratios and specific leaf areas

Turning to the area ratios (Fig. 2.6) it is immediately apparent that significant differences between the four species are not common in the leaf area ratio estimates (Fig. 2.6a). Sunflower has slightly lower values than the

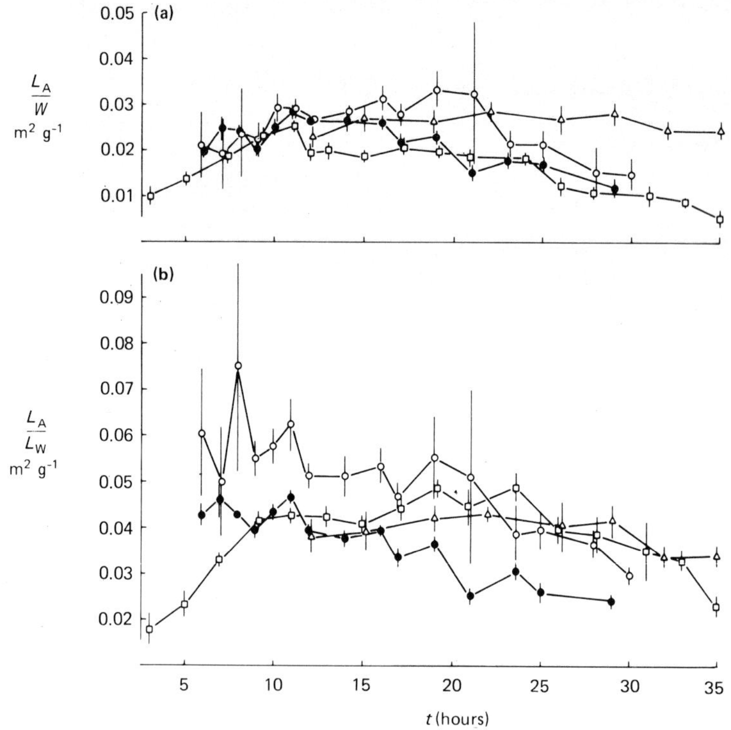

Fig. 2.6 Classicial growth analysis comparison of wheat (•), maize (o), sunflower (□) and birch (△): **(a)** leaf area ratio and **(b)** specific leaf area. Growth measurements with 95% confidence intervals.

two grass species for the latter part of the experiment, whereas the birch leaf area ratio is on the whole marginally higher.

Among the specific leaf area trends, however, species differences are more apparent (Fig. 2.6b). At first glance, there appears to be an overall decreasing trend in specific leaf area with age; but closer examination reveals that this decline is preceded by a rise in both sunflower and birch, and this is more pronounced in the former species. When the plants are young, maize has a significantly larger specific leaf area than the other three species – until day 17 – whereas from day 17 wheat has a significantly lower specific leaf area than the other species.

When leaf area ratio (Fig. 2.6a) and leaf weight ratio (Fig. 2.5a) graphs are compared, we find similarities during the second half of the experiment. During the earlier part of the experiment, differences between species in respect of leaf weight ratio tend to be the opposite to those of specific leaf area (Fig. 2.6b); thus, leaf area ratios between the species are very similar.

Results: relative growth and unit leaf rates

Relative growth and unit leaf rates have been estimated using harvest intervals as near as possible to 6 days – the interval suggested as optimal for data with a coefficient of variation of around 20% by the investigation described on pages 37–41, and the results are shown in Fig. 2.7. For each species, relative growth rate rises at first before showing the more usual decline with time over almost the whole of ontogeny (Fig. 2.7a). The maximum relative growth rate is highest in maize, but there are no significant differences between the four species in this respect; however, the maximum is attained markedly later in birch than in the three herbaceous species. The rate of decline appears to be greater in the two grass species than in the two dicotyledon species.

Unit leaf rate trends (Fig. 2.7b) for wheat, maize, and birch follow the same time course of an increase preceding a slow decline; here, the maximum for wheat may be just significantly higher than for birch (the same may be true of maize, but the difference is obscured by greater variability in this species). In contrast, unit leaf rate of sunflower appears to increase throughout the experimental period, which is somewhat unusual.

In order to avoid unnecessary repetition, the biological significance of these trends will be discussed after the presentation of results of the analysis of the same data by the functional method (page 62).

Consequences of the interdependence of unit leaf rate and leaf area ratio

Equation (2.8) expresses the interdependence of unit leaf rate and leaf area ratio at an instant of time; for convenience, put $x = \rho/(L_A/W)$. Over an

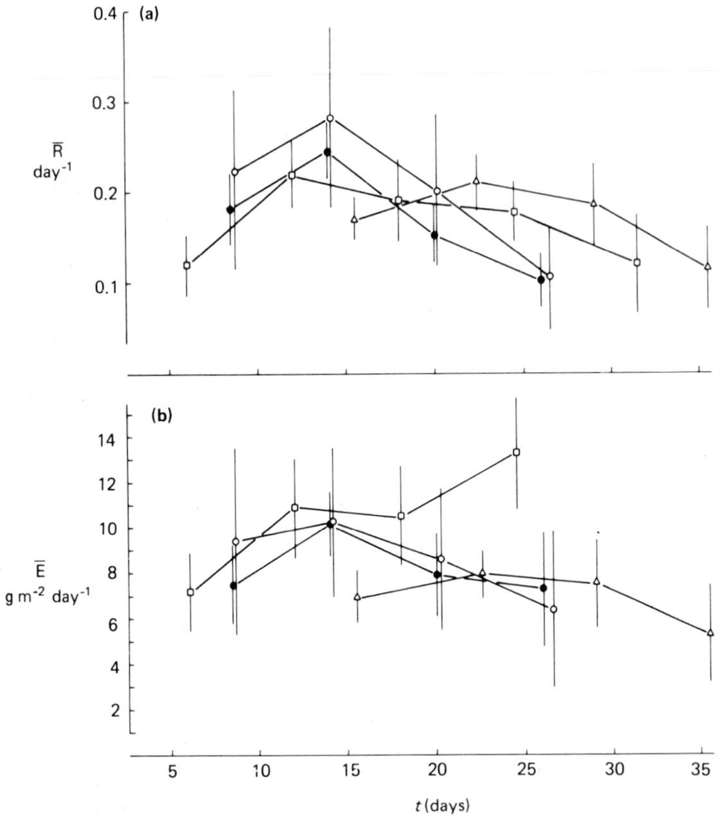

Fig. 2.7 Classical growth analysis comparison of wheat (●), maize (o), sunflower (□) and birch (Δ): (**a**) relative growth rate and (**b**) unit leaf rate. Growth rates with 95% confidence intervals.

interval of time, between t_1 and t_2, we have

$$\bar{E} = \pi - \bar{x} \tag{2.38}$$

where $\pi = (t_2 - t_1)^{-1} \int_{t_1}^{t_2} \pi \, dt$ and $\bar{x} = (t_2 - t_1)^{-1} \int_{t_1}^{t_2} x \, dt$.

The relationships between all the quantities in both equations (2.8) and (2.38) are shown for a simple situation, a plant growing in a controlled environment over a 24-hour period, in Fig. 2.8. In this hypothetical example, the photoperiod is 12 hours, the temperature is constant throughout, and the irradiance during the light period is not too high (say about 100 W m^{-2} of photosynthetically active radiation) so that the photosynthetic rate per unit leaf area stays constant over the photoperiod.

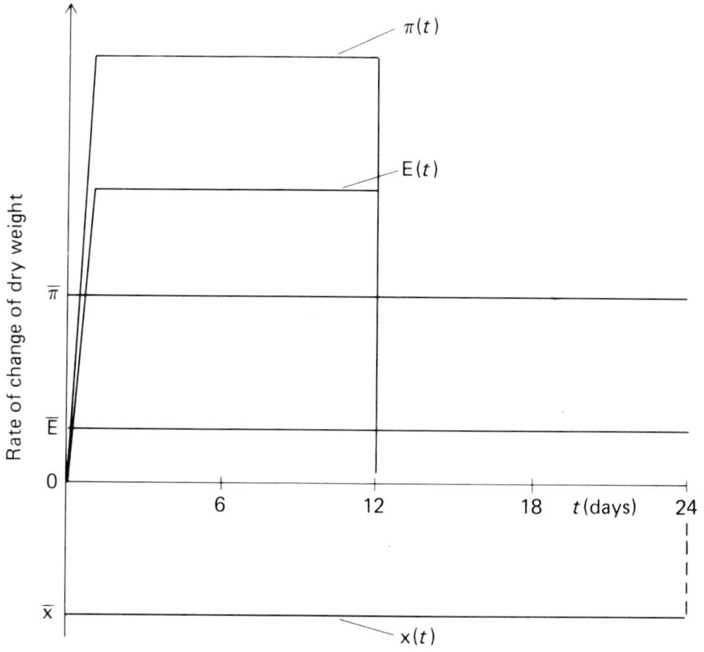

Fig. 2.8 A hypothetical curve of rate of change of dry weight for a plant growing in a constant environment with a 12 hour photoperiod (see text for details).

We also assume that, over the 24-hour period, there is no change in respiration rate per unit plant dry weight nor in leaf area ratio; hence $x(t)$ is represented by a horizontal line, which is also equal to $\bar{x}$. The interpretation of π is also straightforward, but it is important to remember that it represents mean photosynthetic rate per unit leaf area over the whole 24 hours, and not just over the photoperiod. Hence, the magnitude of π is slightly less than half of the constant value of π because photosynthetic induction at the start of the light period takes an appreciable time.

In a normal, uncontrolled, environment, variations in the magnitude of π could be rapid, following the fluctuations in light level; there would also be slower changes in ρ as temperature varied but, again, the slowest change would be in leaf area ratio. Hence, approximately, we have

$$\bar{x} = \frac{1}{(L_A/W)(t_2 - t_1)} \int_{t_1}^{t_2} \rho \, dt$$

Considering a growing plant over an extended period of time, it is clear that $\bar{\rho}$ decreases because the metabolically active parts of the plant continuously decrease in proportion to the whole. Leaf area ratio may rise at first (Fig. 2.6a), but its more prolonged trend is downwards. Hence, $\bar{x}$ may fall rapidly in a very young plant, but mostly the decline in $\bar{x}$ should be small; $\bar{x}$ may even rise slowly, depending on the relative magnitudes of the rates of change of $\bar{\rho}$ and L_A/W. Thus, changes in $\bar{E}$ should be largely a reflection of changes in $\bar{\pi}$ during most of the plant's ontogeny, but at the beginning $\bar{E}$ would be expected to rise faster than $\bar{\pi}$ if leaf area ratio were increasing.

Another application of equation (2.38) is in relation to the so-called sun and shade species of plants. At low light intensities, shade plants are reported to have both a higher net photosynthetic rate and leaf area ratio than sun plants (Bohning and Burnside, 1956; Wassink, Richardson and Pieters, 1956; Loach, 1967; Larcher, 1969; Blackman and Wilson, 1951b; Grime, 1965). Part of the difference between the two groups of species in respect of net photosynthetic rate may be due to the fact that shade plants are reputed to have lower dark respiration rates than sun plants (Grime, 1965; Loach, 1967), but whether there is a difference in gross photosynthetic rates between sun and shade plants at low light intensities is unknown.

In equation (2.38), if we suppose that $\bar{\pi}$ is the same for both sun and shade species at low light intensities, then either a lower $\bar{\rho}$, or a higher L_A/W, or both, in shade plants would ensure that their unit leaf rates were higher than those of sun plants. Indeed, both $\bar{\pi}$ and $\bar{\rho}$ could be the same for both groups of plants at low light intensities, then a higher leaf area ratio in the shade plants would be sufficient to give them a higher unit leaf rate under these conditions than that of sun plants.

The functional approach: methods

Classical growth analysis methods were originally designed to be independent of underlying mathematical functions because of very considerable difficulties of function fitting in practice at that time. The last 12 years or so, since the advent of the 'computer revolution', have seen something of a takeover by curve fitting methods in the field of growth analysis, and there is an increasing number of projects undertaken via the functional approach. The two problems associated with the methodology of the functional approach are the selection of suitable functions and, often, their mode of use. No attempt will be made at a comprehensive coverage of the functional approach to whole plant growth analysis here, as this will be the subject of a forthcoming book (Hunt, 1982). Also, much of the remainder of this book is concerned with the analysis of plant growth in terms of fitted functions, although not along the lines of the present chapter. However, a brief consideration of the functional approach to whole plant growth analysis will be given for the sake of completeness.

The principles of the functional method are not as straightforward as they might appear at first sight. Specifically, we have at time t:

$$\left.\begin{aligned} \mathrm{R} &= w'(t) \\ \frac{L_{\mathrm{W}}}{W} &= \exp\{l_{\mathrm{W}}(t) - w(t)\} \\ \frac{L_{\mathrm{A}}}{W} &= \exp\{l_{\mathrm{A}}(t) - w(t)\} \\ \frac{L_{\mathrm{A}}}{L_{\mathrm{W}}} &= \exp\{l_{\mathrm{A}}(t) - l_{\mathrm{W}}(t)\} \\ \mathrm{E} &= \frac{w'(t)}{\exp\{l_{\mathrm{A}}(t) - w(t)\}} \end{aligned}\right\} \qquad (2.39)$$

where, for example, $w(t)$ is the logarithm of total plant dry weight as a defined function of time. If we are able to acquire the functions $w(t)$, $l_{\mathrm{W}}(t)$, and $l_{\mathrm{A}}(t)$, then it would seem to be a simple matter to apply the relationships in (2.39). However, to be useful in practice, confidence intervals are required for the curves of the functions in (2.39). Further, because rates and ratios of variates are involved, even their estimated values at a particular time are not given as simply as implied by (2.39).

Hitherto, only two main function types have been used in the functional approach to whole plant growth analysis. We shall consider these in turn, and deal with the methods of obtaining estimates of relative growth rates and their variances at particular times. Thereafter, the more general problem of obtaining estimates of the ratios and unit leaf rates, and their variances, at particular times, will be discussed.

Polynomial exponential functions

A polynomial exponential function is a polynomial function of the natural logarithm of a growth attribute in relation to time, i.e.

$$w = \alpha + \beta_1 t + \beta_2 t^2 + \ldots \qquad (2.40)$$

or

$$W = \exp(\alpha + \beta_1 t + \beta_2 t^2 + \ldots)$$

Since the first form of (2.40) is always used in growth analysis, the term 'polynomial' will suffice (see footnote on page 9).

Polynomials have, until very recently, reigned supreme in the functional approach to whole plant growth analysis. This is not because of any biological reason; indeed, no kind of mechanistic model in biology proposed so far gives rise to a polynomial, as far as we are aware, except the exponential (first degree polynomial). However, polynomials are the simplest kind of mathematical function; they are very flexible and, most important, they can be fitted to data and confidence intervals established by the straightforward and exact methods of linear regression. Thus the only remaining problem is, which degree of polynomial should be fitted to a particular set of data?

Users of polynomials in growth analysis have become polarized in their approach to the problem of 'which polynomial?': on the one hand are those who maintain that a particular degree of polynomial should be used all or most of the time (Hughes & Freeman, 1967; Hurd, 1977), and on the other hand there are the advocates of a statistically objective selection procedure (Nicholls & Calder, 1973; Hunt & Parsons, 1974; Elias & Causton, 1976). Both approaches have their merits and disadvantages, and these pertain to the derived functions (equations (2.39)) rather than to those fitted to the primary growth data (equations (2.40)). The curves of functions derived from low-degree polynomials probably over-simplify the situation; for example, the relative growth rate trend with time obtained from a second-degree polynomial fitted to data is linear. Relative growth rate curves derived from higher-order polynomials exhibit various forms of curvilinearity, but often the results are not biologically satisfactory. On the other hand, a unit leaf rate trend with time derived from second-degree polynomials fitted to whole plant dry weight and leaf area data is curvilinear, and may exhibit two turning points (Causton, 1967); analogously, a unit leaf rate curve derived from higher-order polynomials may show such a complexity of form that it is wholly unacceptable from a biological viewpoint. Similar problems arise with the ratios but the problems are not as acute. It is rare to be able to strike a balance between over-simplicity and over-complexity whichever strategy is adopted in the selection of polynomial functions, and the whole problem exists in the first place because this family of functions has no biological basis, only mathematical tractability. Statistically, a polynomial can always be found to adequately describe a particular set of data; the derived functions are then also statistically valid, but may be meaningless biologically.

Mathematically, the corresponding relative growth rate is given by a polynomial of one degree less than that fitted to data:

$$\mathrm{R} = \beta_1 + 2\beta_2 t + \ \ldots \tag{2.41}$$

and, because polynomials are linear in their parameters, the estimate of the expected value of R at time t, $\hat{\mathrm{R}}$, is also given by (2.41). An estimate of the variance of R is given by

$$\mathscr{V}(\hat{R}) = \sum_{i=1}^{n} i^2 t^{2(i-1)} \mathscr{V}(\beta_i) + \sum_{i=1}^{n} \sum_{j=1}^{n} ijt^{(i+j-2)} \mathscr{C}(\beta_i, \beta_j) \quad i \neq j \tag{2.42}$$

where n is the degree of polynomial.

The Richards function

Venus & Causton (1979a) have introduced the Richards function into whole plant growth analysis, and compared its performance on two sets of data with that of polynomial exponentials. They found that the Richards function gave trends which were more sensible from a biological viewpoint, and that there was rarely any statistically significant differences between the two methods, as judged by the overlap of confidence intervals. The latter were, however, sometimes wider in the case of the Richards function than for polynomials, but occasionally this tendency was reversed. Functions derived from low order polynomials tend to have narrow confidence intervals, due to the changing number of variance and covariance terms in equation (2.42); the Richards function, however, has a constant number of parameters (four). Confidence intervals for the Richards function have to be obtained by approximate methods (Venus & Causton, 1979c).

The biologically improved trends given by the Richards function derived curves in whole plant growth analysis is doubtless due to the fact that the function is based on a biologically realistic model (page 90), and that the function is a bounded monotonic one. Thus, Venus and Causton recommend that the Richards function be used in preference to polynomials wherever possible. Occasionally, if a set of data do not extend sufficiently near to the asymptote, a fit may be unobtainable; hitherto, such instances have been uncommon in our experience (they could be anticipated), but time will tell!

The statistical treatment of the Richards function is comprehensively discussed in Chapter 4, and includes a section on the relative growth rate form of this function.

Splined functions

One of the aims in applying a single mathematical function to describe a set of growth data is to smooth out irregularities. The fluctuations of actual data points about the fitted curve may be due to sampling variability and/or environmental fluctuations; also, there may be systematic deviations from the curve over one or more parts of its range, owing to inadequacies of the model. The greater the number of parameters a function contains, the more it will tend to follow deviations and fluctuations in the data but, as discussed for polynomials above, parameter number in a single function can very rapidly become excessive for a sensible description of the growth analysis rates and ratios.

Hunt & Parsons (1977) suggested dividing a set of data into two or more segments and fitting a function to each of these. The example they presented, dividing the data set into two segments on a subjective basis, seemed promising, and subsequently they have developed this idea further, using the idea of splined functions (Cox, 1972).

The main advantage of spline functions is that they have the flexibility to describe lengthy and complicated trends where we do not know the true underlying relationship(s), but where a smooth function is required. Spline functions are also important where it is obvious that different parts of the data set are behaving differently, i.e. not following the same function.

A spline function is a chain of separate polynomial functions, each of degree n. Although discrete, neighbouring functions meet at so-called knots, and here they fulfil continuity conditions both in the functions themselves and in their first $n-1$ derivatives (Wold, 1974). Polynomials of any degree can be splined together but, when applied to logarithmically transformed plant growth data, the cubic is the minimum degree of polynomial that can be considered. This is because the first derivative (relative growth rate) must be free to change smoothly with time, in other words there must be continuity in the second derivative as well as the first. Splines of lower order cannot possess this feature.

The application of spline functions to plant growth data is very recent, presumably because of the complexity of the mathematics. However, Parsons & Hunt (1980) have now re-analysed the classical data of Kreusler, Prehn & Hornberger (1879) using β-splines and have also applied spline functions to two artificial sets of data. The results presented indicate the usefulness of the spline functions in analysing extended data sets. The work also brings together the literature relating to spline functions and provides guide-lines for the use of β-splines, supporting Wold's (1974) four main rules of thumb in the use of the functions: (1) have as few knots as possible; (2) have not more than one maximum or minimum and one inflexion per interval (this is imposed anyway when using cubics); (3) have maxima and minima centred in the intervals, and (4) have knots close to inflexions.

Ratios

Ratios are calculated, at time t, by the system of equations (2.11–2.14) which exemplify the procedure for leaf area ratio; where $\mathscr{E}(l_A)$ and $\mathscr{E}(w)$ are replaced by $\hat{l}_A$ and $\hat{w}$, obtained from the fitted $l_A(t)$ and $w(t)$ relationships, respectively, and $\mathscr{V}(l_A)$ and $\mathscr{V}(w)$ are also obtained from the fitted relationships (see Chapter 3 for details relating to polynomials, and Chapter 4 for the Richards function). The covariance term presents a difficulty which is hard to resolve satisfactorily. Hunt & Parsons (1974) calculated this term as

$$\mathscr{C}(l_A, w) = 1/(Nh) \, . \sum_{i=1}^{Nh} (\hat{l}_{Ai} - \bar{l}_A)(\hat{w}_i - \bar{w}) \qquad (2.43)$$

which gives a single covariance for the entire length of the fitted curve, that is, over the whole time interval covered by the data. However, we believe that, while the variances, $\mathscr{V}(l_A)$ and $\mathscr{V}(w)$, are adequately estimated in the function fitting process, the covariance is a property of the plants themselves, and is not reflected in the two fitted functions because they are fitted to the two sets of data separately. Thus, on this line of argument, the only way in which the covariance can be estimated is from the basic data themselves. Doing this alone, however, results in variances being obtained from the fitted functions, and the covariance from the raw data. In order to rationalize this anomaly, the following procedure is suggested.

Firstly, calculate the correlation coefficient, r, between the leaf areas and whole plant dry weight:

$$r = \frac{\sum_{i=1}^{h} \sum_{j=1}^{N} (l_{Aij} - \bar{l}_{Ai})(w_{ij} - \bar{w}_j)}{\left\{ \sum_{j=1}^{h} \sum_{j=1}^{N} (l_{Aij} - \bar{l}_{Ai})^2 \, . \sum_{i=1}^{h} \sum_{j=1}^{N} (w_{ij} - \bar{w}_i)^2 \right\}^{\frac{1}{2}}} \qquad (2.44)$$

where l_{Aij} and w_{ij} represent the jth replicate leaf area and whole plant dry weight (logarithmically transformed), respectively, at the ith harvest, and $\bar{l}_{Ai}$ and $\bar{w}_i$ represent the corresponding mean values of the N replicates at the ith harvest. Then

$$\mathscr{C}(l_A, w) = r\{\mathscr{V}(l_A) \, . \, \mathscr{V}(w)\}^{\frac{1}{2}}$$

In this way, the calculated covariance may be regarded as a 'fitted' covariance, since the variances of the fitted functions are used in its calculation, and the correlation coefficient calculated from the raw data provides the underlying measure of association.

Unit leaf rate

Unit leaf rate, at time t, is estimated from the relationship $\mathrm{E} = \mathrm{R}/(L_A/W)$. Although L_A/W is approximately log-normally distributed, R is approximately normally distributed and, because of the quotient relationship, estimates of the expected value and variance of unit leaf rate at time t are calculated by approximation formulae (Kendall and Stuart, 1977; and appendix, p. 261). Applying these formulae to the present situation gives

$$\begin{aligned} \mathscr{E}(\mathrm{E}) = {} & \mathscr{E}(\mathrm{R})/\mathscr{E}(L_A/W) + \mathscr{E}(\mathrm{R}) \, . \, \mathscr{V}(L_A/W)/\{\mathscr{E}(L_A/W)\}^3 \\ & - \mathscr{C}(\mathrm{R}, L_A/W)/\{\mathscr{E}(L_A/W)\}^2 \end{aligned} \qquad (2.45)$$

$$\mathscr{V}(\mathrm{E}) = \mathscr{V}(\mathrm{R})/\{\mathscr{E}(L_A/W)\}^2 + \{\mathscr{E}(\mathrm{R})\}^2 . \mathscr{V}(L_A/W)/\{(L_A/W)\}^4 - 2.\mathscr{E}(\mathrm{R}).\mathscr{C}(\mathrm{R}, L_A/W)/\{\mathscr{E}(L_A/W)\}^3 \quad (2.46)$$

Estimates of the quantities $\mathscr{E}$(R), $\mathscr{E}(L_A/W)$, $\mathscr{V}$(R), and $\mathscr{V}(L_A/W)$ having already been obtained, only the covariance term remains and, again, it presents a problem.

At the commencement of a discussion on possible magnitudes of $\mathscr{C}(\mathrm{R}, L_A/W)$, the nature of R estimated from a fitted function must be clearly understood. In the classical method, the meaning of $\overline{\mathrm{R}}$ is clear – it is the mean relative growth rate over a specified interval of time (p. 25). In the functional method, R at time t might appear to be an instantaneous value, but this is not so because the trend of the derived $\mathrm{R}(t)$ curve is smooth over a prolonged period of time; there are no oscillations in the curve, neither does $\mathrm{R}(t)$ assume negative values during periods of darkness. In reality, the $\mathrm{R}(t)$ curve is a smoothed $\overline{\mathrm{R}}(t)$ curve but, in this sense, $\overline{\mathrm{R}}$ cannot be defined over a time period of less than one day. So we can interpret a derived $\mathrm{R}(t)$ curve as a smoothed set of consecutive $\overline{\mathrm{R}}$-values, each one derived on the basis of a 24-hour period. Similar remarks apply to E.

Now, it is immediately obvious that the covariance between leaf area ratio and true instantaneous values of R would be zero, because the latter can oscillate quite independently of the former; but the situation is not nearly as clear cut concerning correlation between a mean relative growth rate and leaf area ratio. In fact, there is evidence (Parsons & Hunt, 1980, and pers. comm.) that the correlation between mean relative growth rate and leaf area ratio is high and positive, but we should point out that this implies a relatively constant unit leaf rate because of the relationship

$$\mathrm{R} \simeq \overline{\mathrm{E}} . \frac{L_A}{W}$$

An estimate of $\mathscr{C}(\mathrm{R}, L_A/W)$ can be found by using the relationship

$$\mathscr{C}(\mathrm{R}, L_A/W) = r\{\mathscr{V}(\mathrm{R}).\mathscr{V}(L_A/W)\}^{\frac{1}{2}}$$

Altering the value of r only affects the estimate of $\mathscr{E}$(E) in the third or fourth significant figure, but $\mathscr{V}$(E) can be considerably affected (Venus & Causton, 1979a). In the absence of a good universal method of estimating the true correlation between relative growth rate and leaf area ratio the correlation coefficient of zero, which gives intermediate values of $\mathscr{V}$(E), will be used in this book.

The functional approach: applications

In the functional methods of whole plant growth analysis, there is no problem about inter-harvest duration, provided there are a sufficient number of harvests compared with the numbers of parameters in the functions used. Further, the uses to which the results may be put are not changed by the method of analysis, even though the results given by the two methods for the same set of experimental data would be slightly different. An examination of these differences will now be made by using the same four species experimental data, already analysed by the classical methods, in a functional growth analysis.

Experimental data: a comparative study of four species (continued)

Richards functions were used to describe the growth of foliage, stem, root, and whole plant dry weights, and of leaf area for the three herbaceous species. However, owing to lack of curvature of the weight and area data of birch, the Richards function could not be used for this species, and polynomials had to be employed. Using the test criterion of Elias & Causton (1976), the appropriate polynomial for the data up to day 35 was, in every case, linear (i.e. the exponential function).

Results: leaf weight, stem weight, and root weight ratios

The weight ratios are summarized in Fig. 2.9, and differences between the species are more obvious than those given by the classical analysis. This is markedly so for the leaf and root weight ratios (Fig. 2.9a,c and Fig. 2.5a,c), but less so for the stem weight ratio (Fig. 2.9b) because species differences shown by the classical method were quite clear cut (Fig. 2.5b). In both trends and values, the weight ratios given by the two methods are very similar.

Results: leaf area ratios and specific leaf areas

The few species differences in leaf area ratio demonstrated by the classical method (Fig. 2.6a) are shown with greater clarity by the functional method in Fig. 2.10a. The same is true for specific leaf areas (Fig. 2.10b and Fig. 2.6b), but close examination reveals some discrepancies between the results given by the two methods, although the overall picture is very similar in both cases.

Results: relative growth and unit leaf rates

Relative growth rate trends by the functional method (Fig. 2.11a) show some marked differences from those exhibited by the classical method (Fig. 2.7a). In particular, the initial rise, shown by the latter procedure, is not shown by the functional method; in fact, the Richards function model (page 90) only allows for a declining relative growth rate as a plant develops, and a polynomial, fitted to raw data, below the third degree cannot show a change in direction of relative growth rate. Further, because an exponential function has been used to describe the growth of birch, the relative growth

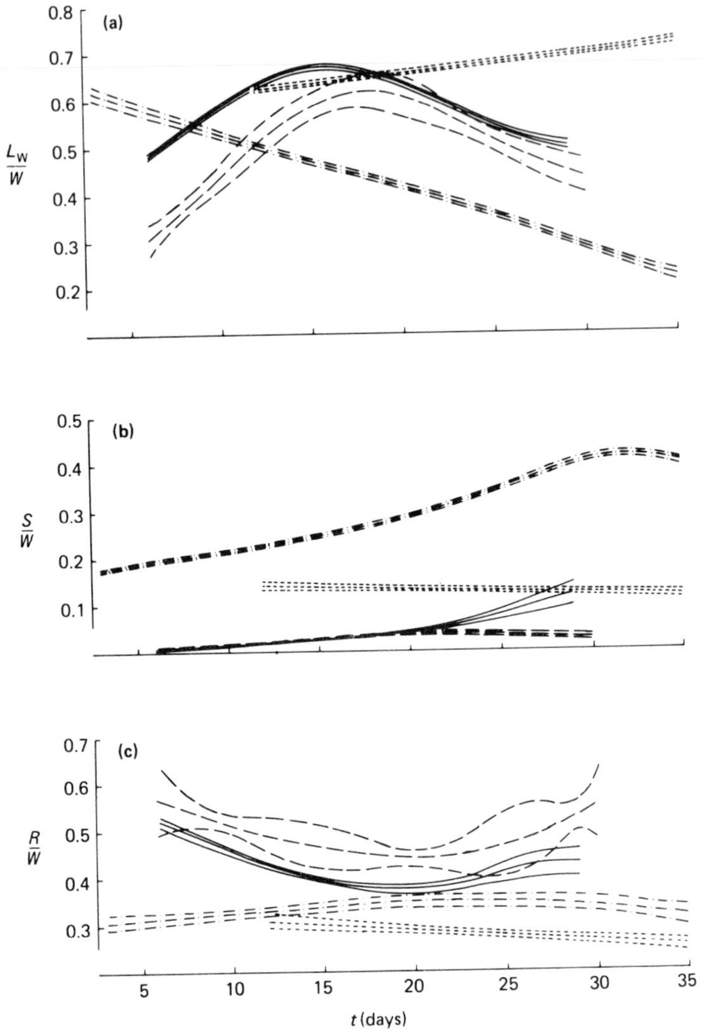

Fig. 2.9 Functional growth analysis comparison of wheat (——), maize (— —), sunflower (— · —) and birch (-----): (**a**) leaf weight ratio, (**b**) stem weight ratio and (**c**) root weight ratio. Ratio curves with 95% confidence bands.

rate is constrained to be constant. Apart from these differences, the magnitudes of the relative growth rates over their decline with time are similar for both methods.

The discrepancies between the two methods are even more apparent for unit leaf rate. Apart from birch, which shows a very gradual (almost linear,

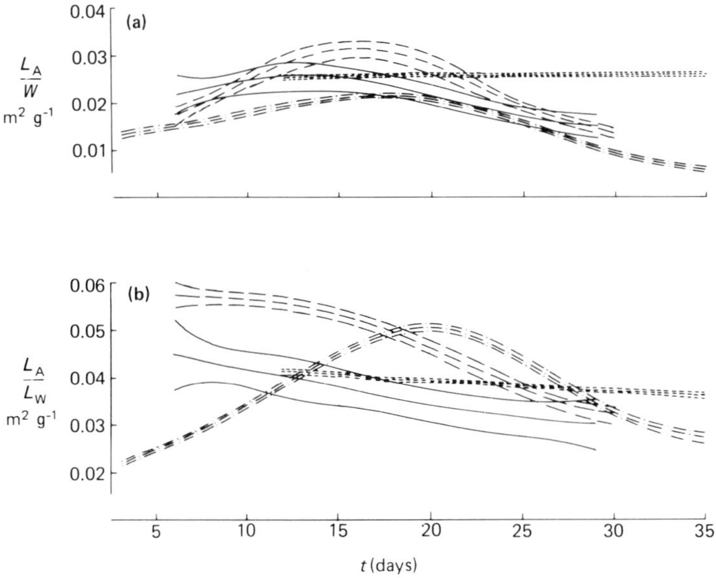

Fig. 2.10 Functional growth analysis comparison of wheat (——), maize (— —), sunflower (— · —) and birch (----): (**a**) leaf area ratio and (**b**) specific leaf area. Ratio curves with 95% confidence bands.

but actually exponential) decline with time, the trends in unit leaf rate are from high to low, with a reversal in the middle to a greater or lesser degree, by the functional method (Fig. 2.11b). This is in considerable contrast from the results given by the classical method (Fig. 2.7b), especially in the early stages of growth. Later, the differences in the results between the two methods are not as great, but the confidence intervals can become extremely large by the functional approach – larger than in the classical method. The best result, in terms of comparability between the two methods, is shown by wheat.

Discussion: the functional and the classical methods

Two features of the functional methods, in respect of these experiments, are prominent. Firstly, the methods give estimates of the ratios and relative growth rates having smaller variances than those yielded by the classical methods, although the variances of the estimates of unit leaf rate are comparable between the two methods. Secondly, the similarity in trends with time shown by the two methods decreases in the sequence, weight ratios, area ratios, rates.

Few would doubt the opinion that results from the functional methods should be viewed against the classical methods, rather than the other way

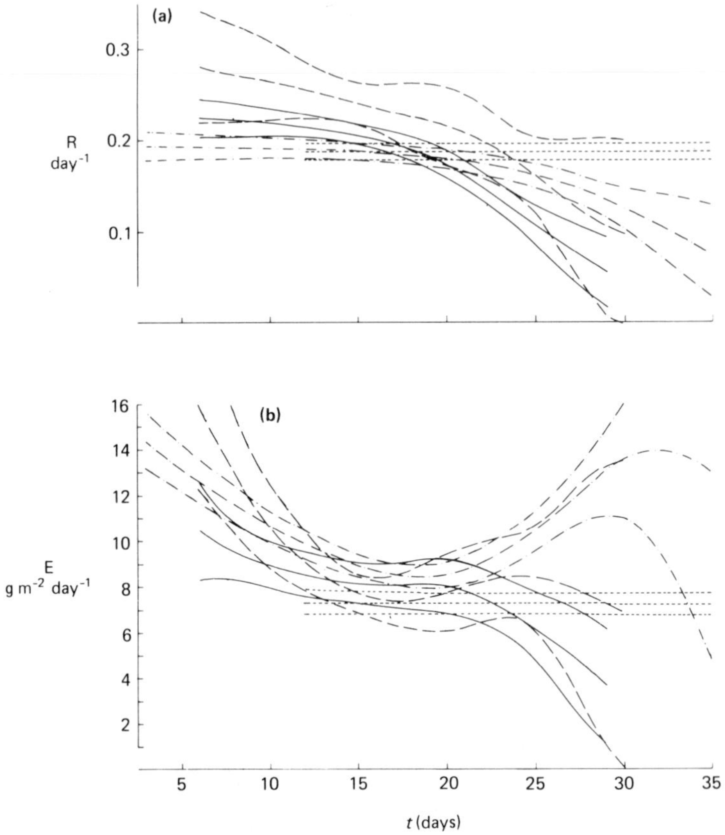

Fig. 2.11 Functional growth analysis comparison of wheat (----), maize (——), sunflower (—·—) and birch (----): (**a**) relative growth rate and (**b**) unit leaf rate. Rate curves with 95% confidence bands.

round, since the latter transform the basic data without any inbuilt assumptions (one exists in the case of unit leaf rate, but has been shown to have only minor consequences if it is wrong, page 27). In the present experiments, the trends shown by the rates and ratios of an individual species by the classical methods are clear, even for the rates which is often not the case in many experiments. No doubt adoption of the 'optimal' harvest interval, indicated by the simulation study, has helped here, but the functional methods have not helped to clarify the results to any great degree.

On the other hand, the confidence intervals given by the functional methods are mostly smaller than those given by the classical methods, and this assists in species comparison. Obviously, the reduction in the sizes of the confidence intervals is a consequence of the extra assumptions made when fitting a curve to data, and the question that must be asked is what degree of reduction, consequent upon fitting a curve, is acceptable to still give biologically valid comparisons?

The reduction in size of confidence intervals by fitting a function is brought about because, at any one time, information from the whole curve is being utilized, and not just that available at the one particular harvest. This implies that the best possible model should be used to describe the basic data and, in particular, strenuous efforts should be made to avoid the situation in which there are systematic deviations of data points from the fitted function, as opposed to irregular fluctuations. Statistically, there is usually no difficulty in finding a function that can be shown to fit the data well, but such a function may not necessarily satisfy biological criteria. This means that although a polynomial can always be found to adequately describe a set of data from a statistical viewpoint, the growth analyst must feel happier with a model which has a more biological basis (page 90), however slight; hence our advocacy of the Richards function in preference to polynomials.

Assistance in the selection of a suitable model may be provided by relative growth rate trends estimated by the classical method. For these experiments, it must be admitted that the Richards function was not a suitable choice for any of the species because of the initial rise in relative growth rate in every case (Fig. 2.7a). However, even a third degree polynomial, which is the simplest polynomial function showing a change in direction of relative growth rate, would not be satisfactory (except, possibly, for birch) because the maximum is symmetrically placed in the relative growth rate form of this function, but this does not appear to be the case for the actual results. Hence, polynomials of at least the fourth degree would have been required for this experiment. The probable reason for this initial rise in relative growth rate is because the plants were harvested as soon as they were large enough to handle, i.e. very young: growth rate is negative in the very young seedling until photosynthetic gain has risen sufficiently to exceed dark respiratory loss in a 24-hour period, and it may be that the initial harvests included the last stage of this phase with relative growth rate still rising from its originally negative value. Most growth experiments commence after this phase, and in these cases the Richards function should be a good model. Where more than one distinct phase of growth is covered by a set of experimental data, splined functions could be useful; we present an alternative approach in Chapter 7.

The greater stringency of biological criteria as against statistical ones, mentioned above, is well shown in the case of birch. The relative growth rate trend, estimated by the classical method, indicates very clearly that at least a third degree polynomial should be fitted to the whole plant dry weight data;

but the statistical test showed that a first degree polynomial was appropriate. The reason for the discrepancy is not far to seek: the statistical test was conducted on the growth data themselves, whereas the biological criterion rests on a manifestation of growth rates, and it takes a large change in a growth rate to be evident on a graph of dry weight against time. Thus the growth analyst would prefer to use a fitted function that statistically is not the best available, because on biological grounds an alternative function is preferable. This conflict of ideals can be overcome if the biologist first defines the range of polynomials that he considers to be reasonable; for example, eliminating first and second order curves which, on relative growth rate criteria, are inappropriate. The statistical goodness of fit tests are then applied only to the actual contenders and hence the biologist has no *a priori* reason to reject the statistical results.

Discussion: biological implications of the results

The stem weight ratios show readily interpretable trends. The most obvious (and most easily observed without any analysis) is the increasing stem weight ratio of the sunflower. This plant has no basal rosette of leaves, and its growth strategy of rapid increase in stem height reflects its very high demand for light, so that it can potentially overgrow any competing species in its vicinity. Most grasses have very little stem growth during their vegetative phase; stem elongation is consequent upon the switch to the reproductive phase, in which the cells of the sub-apical meristem rapidly divide and elongate, and the lateral structures initiated on the apical meristem itself are floral rather than foliar. Evidently, the reproductive phase has commenced in wheat, but perhaps not in maize, by the end of the experiment (Fig. 2.9b). The birch result is interesting in that there is a slight decline, followed by a constant phase. It would seem that between about 20 to 35 days after germination the proportion of the plant in stem is the lowest attained in birch, because the later results of the experiment, referred to on page 43, show a continuing rise in stem weight ratio. Indeed in the long term, stem weight ratio in a woody species must continually rise.

Birch also maintains a high leaf weight ratio. This is because the whole experimental duration was extremely short compared with the life span of birch, but was an appreciable proportion of the life span of the other three species, which are annuals. As a plant develops, its leaf weight ratio, high to begin with to maximize the photosynthetic capacity of the establishing seedling, falls as vegetative growth proceeds with greater emphasis on the stem and root; but by 35 days after germination, a birch seedling is scarcely established. Birch is a small seeded pioneer species, and the high leaf weight ratio, leading to a high leaf area ratio (Fig. 2.6a), is doubtless necessary to raise the chance of an individual's survival in the early stages, especially as unit leaf rate is rather low. By contrast, sunflower is a large seeded plant, and the necessity of maintaining a high leaf weight ratio is considerably less for a high chance of survival.

Maize has the highest root weight ratio of the four species. This may be correlated with the fact that maize is a C_4 species, and these species are thought to have evolved under semi-arid conditions: hence the requirement for a large root system.

The comments made about leaf weight ratio also apply to a lesser extent to leaf area ratio, but differences in leaf weight ratio between the species are much less evident in the composite leaf area ratio due to differences in specific leaf area. For example, owing to the higher specific leaf area in sunflower, the very low leaf weight ratio is not reflected in leaf area ratio being outstandingly low. Maize starts with a higher specific leaf area than any of the other three species, but finishes the experiment quite low. Conversely, leaf weight ratio in this species is very low at the beginning, but rises to values intermediate between the extremes of birch and sunflower, with the result that leaf area ratio is not greatly different from the other species throughout the experiment.

Unit leaf rate is highest in sunflower, and this species is already known to have high unit leaf rates under suitable conditions (Warren Wilson, 1966b). Maize might have been expected to have the highest unit leaf rate, since it is a C_4 species, but it must be remembered that C_4 plants only show a high photosynthetic rate under high light intensities at high temperatures. Some high maximum temperatures were recorded in the glasshouse during the experiment, and minima were normally about 18°C; but it may be that average radiation levels (which were not measured) were insufficient to promote a high unit leaf rate. It is of interest to note that Warren Wilson (1966c) also did not find a particularly high unit leaf rate in maize in comparison with other species. In contrast, but in accordance with previous observations of woody species (Coombe, 1960; Jarvis & Jarvis, 1964; Grime & Hunt, unpublished), the unit leaf rate of birch was markedly lower than in the other three species.

On the whole, maize shows the highest relative growth rate of the four species, due to a combination of a moderate unit leaf rate combined with a high leaf area ratio. The low unit leaf rate of birch is offset by a high leaf area ratio, giving an average relative growth rate for this species as compared with the herbaceous ones. Conversely, the high unit leaf rate in sunflower, combined with a low leaf area ratio, gives this species a very mediocre relative growth rate.

In conclusion, then, it is evident that besides providing material for illustrating the more statistical aspects of this work, the experiments have demonstrated how whole plant growth analysis can highlight different growth strategies between species. Results such as these can then be pursued further: by the plant physiologist who will wish to examine why and how these different growth patterns are caused, and by the ecologist who will enquire how these different growth strategies enable the species to find and maintain their niche in the ecosystem.

3

Linear regression theory

Fitting a mathematical function to a set of data and analysis of the fitted function, in statistical parlance, has usually been called regression. Two broad categories of regression may be distinguished – linear and non-linear – and the distinction lies in whether or not the function to be fitted is linear or non-linear in its parameters. For example, a polynomial is linear in its parameters whereas the exponential function, $y = ae^{bx}$ where a and b are the parameters, is not. The term 'model', instead of 'regression', is gaining ground: one would nowadays tend to refer to a polynomial function as a linear model, and to the exponential function as a non-linear model, giving rise to linear and non-linear estimations, respectively; but we shall retain the word regression here, since this term will already be familiar to all readers. Although the underlying principles and methods are the same in each case, the implementation of linear regression methodology is explicit and exact, whereas non-linear regression relies on various approximate and indirect methods of implementation.

The reason for devoting a whole chapter to an exposition of standard underlying theory of linear regression is because of the central place occupied in this book by the use of fitted mathematical functions to describe quantitative aspects of plant growth. The Richards function, which we use to describe individual leaf growth (Chapters 4 and 5) and also whole plant and Level 1a (Table 1.1, page 3) component growth in whole plant growth analysis (Chapter 2), gives rise to a non-linear regression situation. Before the availability of electronic computers, a rigorous non-linear regression analysis was not practicable except in the simplest cases, and functions non-linear in their parameters were usually dealt with by transforming the data, and hence the function, to a linear form. However, such a transformation may invalidate the estimation procedure because the error distribution assumptions inbuilt to the regression method may not conform to the transformed data, and as it is now feasible to analyse a function which is non-linear in its parameters directly without transformation to linearity, there is no reason for not doing so. Over several years, we have come to realize that the best possible estimates of the parameters of a function (statistically speaking) are necessary in order to have confidence in one's biological interpretation of the parameter estimates; bad estimates by unrigorous methods are worse than useless, because they can lead to erroneous conclusions. It has already been mentioned that the principles of non-linear and linear regression methods are the same, and we feel that it is much easier to discuss the former as an extension to the treatment of the latter – an extension not only of concepts but also of a system of notation which, once established, can be used unchanged in the remainder of the book. Also, there are problems of a different kind relating to the allometric functions of Chapter 6, but the starting point for discussion of these problems is also basic linear regression theory. Biological readers are unlikely to have this necessary background in regression analysis in sufficient detail, and this may also apply to some mathematical readers who are not primarily statisticians. Three other useful sources of reference are: Acton (1959), the popular book by Draper & Smith (1966), and Sprent's (1969) excellent monograph.

The present chapter is intended as a source of reference and, as such, has been written *in extensio*. After a general introduction to regression principles, the method of least squares is introduced and applied to fitting the straight line; then the sampling distributions of the parameter estimates are derived. Least squares will be used in succeeding chapters where it is conceptually easy to do so, but there are situations, particularly in Chapter 6, where the method of maximum likelihood is conceptually the simplest approach. One of the purposes of the present chapter is to thoroughly introduce this lesser known (to non-statisticians) technique as it applies to linear regression. Having demonstrated the relationships between the two methods, the chapter will end by showing the extension, using least squares, to polynomial regression in general.

Regression in general

Consider a number of individual (data) points on an x-y plane, (x_i, y_i) $i = 1, \ldots, h$, in which the quantity represented by the x-axis is not subject to random fluctuations but that the quantity represented by the y-axis is subject to random fluctuations. We further suppose that the values of the x_i are all different. In biological terms, the y-axis can represent some attribute of an organism (e.g. the dry weight of a whole plant) and the x-axis may represent some factor external to the organism (e.g. time); thus we may imagine the dry weight of a single plant to be measured on each of h occasions. In statistical terms, a whole population of y-values is inferred to exist at each value of x_i selected for the measurement of y, so there are h populations (Fig. 3.1a). The actual y_i-values occurring in the data set are thus sample values from the underlying populations, and in the present situation there is only one sample observation of y from each of the h populations (Fig. 3.1b). Situations in which several observations, N, of y at

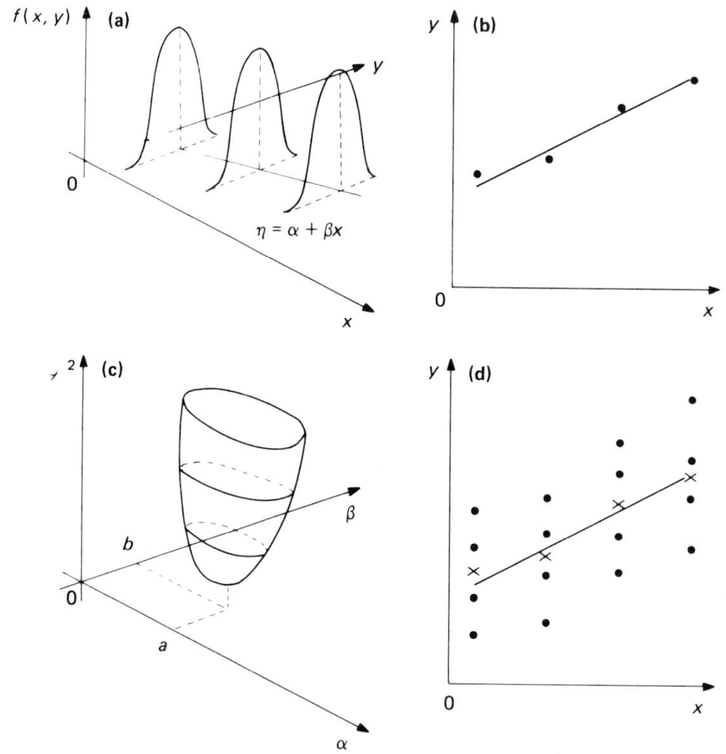

Fig. 3.1 (**a**) Population structure for straight-line regression. (**b**) Regression with a single observation from each population. (**c**) Geometric representation of least squares applied to straight-line regression. (**d**) Regression with several replicate observations from each population.

each x-value are made; $(x_{ij}, y_{ij})\ i = 1, \ldots, h, j = 1, \ldots, N$, and all x_{ij} equal to one another for any one i; are not only possible, but are desirable as they enable a more extensive regression analysis to be undertaken (Fig. 3.1d). A third situation exists which is a blend of the first two; several y-values at each x may be taken, but these are then combined into a single y-value at each x, the mean $\left(\bar{y}_i = (1/N) . \sum_{j=1}^{N} y_{ij}\right)$. Hence, in this third case, we have data points on the x-y plane defined as $(\bar{x}_i, \bar{y}_i)\ i = 1, \ldots, h$. This third situation should not normally be employed, except where required for special purposes such as Elias & Causton's (1976) test for the most appropriate polynomial in whole plant growth analysis, since if replicate y-values at each x are available, they should be used as such. The principles of regression analysis, however, remain unchanged for the three cases.

Now suppose we wish to fit a function of the form

$$\eta = f(\beta_k : x) \qquad k = 1, \ldots, p \tag{3.1}$$

which contains p parameters, β_k, to a set of data; the symbol η corresponds to the 'true' value of the variable y for any x, as given by $f(\beta_k : x)$, as opposed to any observed value of y corresponding to an x-value appearing in the original data. Fitting a particular function to a particular set of data implies that the expected y-values (η-values) of the h populations lie exactly on the curve of the function. Selection of an appropriate function (i.e. a particular form of (3.1)) must be done before any regression analysis can be started, and in what follows we assume that this has been done.

Fitting the selected function to a particular set of data implies calculating estimates of the p parameters; the results will be relevant for the particular set of data in hand, but we would hope that these calculated quantities would be efficient and unbiased estimates of the underlying parameters of the function which, as we have seen, are directly related to the underlying populations sampled. Thus, in regression analysis we have the situation that the parameters of the selected mathematical function, β_k, are intimately related to the underlying populations of y-values sampled at each x – the parameters of the function may therefore be regarded as population parameters. As a set of actual observations are sample values from those populations, the calculated values of the p parameters of the function for that set of observations constitute estimates of the function's population parameters: the calculated values are sample statistics, just as (say) $\bar{x}$ is a sample statistic (sample mean) estimating the population parameter μ (the population mean) in the case of a univariate normally distributed population from which a single sample has been taken.

We can go further. A sample statistic can only be regarded as a constant for the one particular sample from which it is calculated; a different sample from the same population would yield a different value of the sample statistic, but both values would be valid *estimates* of the underlying population parameter which is a constant for that population. Thus, sample statistics are themselves subject to random variation, and a single one obtained from a single sample without some idea of its variability is not very helpful. The latter is measured as the standard error of the sample statistic (although often quoted as the variance of the sample statistic, which is the square of the standard error, in more theoretical work), and from this a confidence interval for the sample statistic may be defined. An exactly parallel situation obtains in regression analysis for the parameters of the fitted mathematical function: point estimates of their expected values are obtained together with their variances or standard errors.

Returning to equation (3.1), we consider that this function rigidly describes the relationship between x and y for the underlying populations of y-values at each x. From a particular sample set of data values, an estimate of the function is given by

$$\hat{\eta} = f(\hat{\beta}_k : x) \tag{3.2}$$

where the p $\hat{\beta}_k$-values are estimates of the underlying β_k population parameters, calculated from the sample data, and $\hat{\eta}$ is a y-value for any given x using the functional relationship specified by f and the estimated parameters, $\hat{\beta}_k$. The 'hats' thus specify estimates, and the hat notation is conventionally associated with the maximum likelihood method of estimation. A closely allied technique in regression analysis is the method of least squares, which we shall discuss first in what follows, and here the convention is to use corresponding Roman letters without hats to represent estimates; thus, (3.2) can be written as

$$\hat{y} = f(b_k : x) \tag{3.3}$$

The hat must, however, be retained over the y in order to distinguish $\hat{y}$, calculated from the fitted function (3.3) for a given x, from y as an item of data corresponding to that same x-value.

It would be relevant to our present purpose to consider regression analysis solely in terms of the method of maximum likelihood. However, at the outset, the analogous method of least squares is so easy in concept that we shall start with it. The principle of least squares will first be given, and then applied to the regression analysis of a straight line; then, still using the latter example, the transition to maximum likelihood methodology will be made.

The principle of least squares

Consider a set of h x-y points on a plane with the properties alluded to earlier, namely, the x_i known without random error and the y_i being sample values from underlying populations, one at each x. Further consider that the expected value of each population would lie on a function defined in (3.1). The y-values of the data points themselves, being samples, do not lie on the curve of (3.1); they will deviate from it to a greater or lesser extent. Thus, deviations of the data points from the curve are parallel to the y-axis only (Fig. 3.1b). We now require the best estimate of (3.1), and this will be provided by (3.3). The criterion of least squares gives the best fitting curve of the form (3.3) when the sum of the squares of the deviations of the data from the curve is a minimum. Now we have shown that deviations in this context exist solely in the y-direction, and so the sum of the squares of the deviations of the data from the curve is given by

$$\mathcal{S}^2 = \sum_{i=1}^{h} (y_i - \hat{y}_i)^2$$

or, using (3.3),

$$\mathscr{S}^2 = \sum_{i=1}^{h} \{y_i - f(b_k : x_i)\}^2 \tag{3.4}$$

At this point point it must be clearly understood that for a particular set of data the x_i and y_i in equation (3.4) are constant; hence, changes in $\mathscr{S}^2$ can only be occasioned by changes in the parameters, b_k, of the estimated function. Therefore, to minimize $\mathscr{S}^2$ we must choose the values of the p parameters to ensure that this is done, and this is what is meant by fitting a curve to data.

To minimize $\mathscr{S}^2$, we partially differentiate (3.4) with respect to each b, set all the partial derivatives to zero, and solve the resulting simultaneous equations for the p b_k. It is only at this stage that the distinction between linear and non-linear regression enters; a linear regression situation gives a set of linear equations at this stage which can be solved explicitly, whereas a non-linear regression situation gives a set of non-linear equations which do not, in general, have explicit solutions.

Polynomial regression: the straight line

Least squares estimation

In this situation we employ the model that the expected value of every population of y-values, corresponding to each x_i, lies on the straight line

$$\eta = \alpha + \beta x \tag{3.5}$$

Although not stated in the preceding section, three other criteria have to be incorporated into the model before (3.4) can be used. These are: (i) that all the populations are normally distributed; (ii) that all the populations have the same variance, σ^2; (iii) that each of the observations in the sample data set is independent of all the others. We may note here that criterion (i) is unnecessary insofar as obtaining estimates of the parameters is concerned, but as an assumption concerning the distributions of the populations has to be made before variances of these estimates can be calculated, it is convenient to make the assumption at the outset as criterion (i). Criteria (ii) and (iii) can be altered, and such changes give rise to different least squares formulations; for the moment, however, we shall adhere to the simplest situation.

To estimate the straight line given in (3.5) we require estimates of the constants, and the equation of the estimated line is written as

$$\hat{y} = a + bx \tag{3.6}$$

Substituting into (3.4), and assuming summation over all data points so that the limits on the Σ sign need not be shown, we have

$$\mathscr{S}^2 = \Sigma(y_i - a - bx_i)^2 \tag{3.7}$$

Now, partially differentiate the right-hand side of (3.7) successively with respect to a and b:

$$\frac{\partial \mathscr{S}^2}{\partial a} = -2\Sigma(y_i - a - bx_i) \tag{3.8}$$

$$\frac{\partial \mathscr{S}^2}{\partial b} = -2\Sigma(y_i - a - bx_i)x_i \tag{3.9}$$

then equate to zero, eliminate the brackets and rearrange to give

$$ah + b\Sigma x_i = \Sigma y_i \tag{3.10}$$

$$a\Sigma x_i + b\Sigma x_i^2 = \Sigma x_i y_i \tag{3.11}$$

Dividing through (3.10) by h gives

$$a = \bar{y} - b\bar{x} \tag{3.12}$$

where $\bar{x} = (1/h).\ \Sigma x_i$ and $\bar{y} = (1/h).\ \Sigma y_i$. Substituting into (3.11):

$$(\bar{y} - b\bar{x}).\Sigma x_i + b\Sigma x_i^2 = \Sigma x_i y_i$$

i.e.

$$\bar{y}\Sigma x_i - b\bar{x}\Sigma x_i + b\Sigma x_i^2 = \Sigma x_i y_i$$

i.e.

$$\frac{\Sigma x_i \Sigma y_i}{h} + b\left(\Sigma x_i^2 - \frac{\Sigma^2 x_i}{h}\right) = \Sigma x_i y_i$$

$$b = \frac{\Sigma x_i y_i - (\Sigma x_i \Sigma y_i)/h}{\Sigma x_i^2 - (\Sigma^2 x_i)/h} \tag{3.13}$$

which is equavalent to

$$b = \frac{\Sigma(x_i - \bar{x})(y_i - \bar{y})}{\Sigma(x_i - \bar{x})^2} \tag{3.14}$$

Equations (3.12) and (3.13) or (3.14) are, of course, the well-known results for calculating the best fitting straight line to a set of data by least squares.

An alternative, and more generally useful, approach to the solution of equations (3.10) and (3.11) is through matrix algebra: the pair of equations (3.10 and 3.11) is equivalent to

$$\begin{bmatrix} h & \Sigma x_i \\ \Sigma x_i & \Sigma x_i^2 \end{bmatrix} \begin{bmatrix} a \\ b \end{bmatrix} = \begin{bmatrix} \Sigma y_i \\ \Sigma x_i y_i \end{bmatrix} \tag{3.15}$$

The inverse of the square matrix on the left-hand side of (3.15) is

$$\frac{1}{h\Sigma x_i^2 - \Sigma^2 x_i}\begin{bmatrix} \Sigma x_i^2 & -\Sigma x_i \\ -\Sigma x_i & h \end{bmatrix}$$

and so

$$\begin{bmatrix} a \\ b \end{bmatrix} = \frac{1}{h\Sigma x_i^2 - \Sigma^2 x_i}\begin{bmatrix} \Sigma x_i^2 & -\Sigma x_i \\ -\Sigma x_i & h \end{bmatrix}\begin{bmatrix} \Sigma y_i \\ \Sigma x_i y_i \end{bmatrix}$$

On multiplying out the right-hand side, we find that

$$b = \frac{h\Sigma x_i y_i - \Sigma x_i \Sigma y_i}{h\Sigma x_i^2 - \Sigma^2 x_i}$$

which, after division of both numerator and denominator by h, leads to the result (3.13). The expression for a from the above matrix equation can be shown, after some algebra, to be the same as (3.12).

After obtaining estimates of the parameters, it is desirable to find a more convenient expression for the deviations sum of squares than (3.7). Substituting for a in (3.7), we have

$$\begin{aligned}\mathcal{S}^2 &= \Sigma\{y_i - (\bar{y} - b\bar{x}) - bx_i\}^2 \\ &= \Sigma\{(y_i - \bar{y}) - b(x_i - \bar{x})\}^2 \\ &= \Sigma(y_i - \bar{y})^2 - 2b\Sigma(x_i - \bar{x})(y_i - \bar{y}) + b^2\Sigma(x_i - \bar{x})^2\end{aligned}$$

On substituting for b in the third term above, using (3.14), we have

$$\mathcal{S}^2 = \Sigma(y_i - \bar{y})^2 - 2b\Sigma(x_i - \bar{x})(y_i - \bar{y}) + \frac{b\Sigma(x_i - \bar{x})(y_i - \bar{y})\cdot\Sigma(x_i - \bar{x})^2}{\Sigma(x_t - \bar{x})^2}$$

and so

$$\mathcal{S}^2 = \Sigma(y_i - \bar{y})^2 - b\Sigma(x_i - \bar{x})(y_i - \bar{y}) \tag{3.16}$$

which, again, is a well-known result.

Geometrical interpretation of the least squares estimation of the straight line

Figure 3.1c shows the geometrical representation of equation (3.7). The three-dimensional figure is a paraboloid, the height of the minimum corresponds to the minimized sum of squares, and the values of α and β at which this minimum occurs are a and b, respectively. On any plane intersecting the paraboloid and parallel to the α-β axis, the surface of the paraboloid traces out an ellipse. Such an ellipse may be called a contour ellipse, and three of these are shown in Fig. 3.1c.

For straight line estimation, we require three dimensions for the geometrical description of the process – one for each of the two parameters, and one for the residual sum of squares. For estimating a polynomial of degree n (page 84), we require $(n+1)$ dimensions for the geometrical interpretation of the least squares equation, for example (3.52), and the paraboloid becomes a hyper-paraboloid. Similarly, in the case of a second degree polynomial the contour ellipses are three-dimensional contour ellipsoids, and for higher order polynomials they are hyper-ellipsoids.

The sampling distributions of the parameters estimates

The derivations below essentially follow those presented by Bulmer (1965). The assumptions, or criteria, already given concerning the structure of our data (page 70) are equivalent to the statement

$$y_i = (\alpha + \beta x_i) + \varepsilon_i \tag{3.17}$$

in which an observation of y, y_i, is the sum of two components – a value dependent on the corresponding value of x, x_i, and a deviation from this regression value, ε_i, consequent upon y_i being an observation from a whole population of y-values. The ε_i are thus normally distributed with a mean of zero and a variance of σ^2.

The variance of b

The numerator of b (equation (3.14)) is $\Sigma(x_i - \bar{x})(y_i - \bar{y})$.

Now $$\Sigma(x_i - \bar{x})(y_i - \bar{y}) = \Sigma(x_i - \bar{x})y_i - \Sigma(x_i - \bar{x})\bar{y}$$

i.e. $$\Sigma(x_i - \bar{x})(y_i - \bar{y}) = \Sigma(x_i - \bar{x})y_i \tag{3.18}$$

since $\bar{y}$ is a constant for the set of data in hand, and $\Sigma(x_i - \bar{x}) = 0$. Now, from (3.17),

$$\begin{aligned}\Sigma(x_i - \bar{x})y_i &= \Sigma(x_i - \bar{x})(\alpha + \beta x_i + \varepsilon_i)\\ &= \alpha\Sigma(x_i - \bar{x}) + \beta\Sigma(x_i - \bar{x})x_i + \Sigma(x_i - \bar{x})\varepsilon_i\\ &= \beta\Sigma(x_i - \bar{x})^2 + \Sigma(x_i - \bar{x})\varepsilon_i\end{aligned}$$

since $\alpha\Sigma(x_i - \bar{x}) = 0$ and $\Sigma(x_i - \bar{x})x_i = \Sigma(x_i - \bar{x})^2$ by arguments analogous to those involved in the derivation of (3.18). Hence,

$$b = \frac{\Sigma(x_i - \bar{x})(y_i - \bar{y})}{\Sigma(x_i - \bar{x})^2} = \frac{\beta\Sigma(x_i - \bar{x})^2 + \Sigma(x_i - \bar{x})\varepsilon_i}{\Sigma(x_i - \bar{x})^2}$$

so $$b = \beta + \frac{\Sigma(x_i - \bar{x})\varepsilon_i}{\Sigma(x_i - \bar{x})^2} \tag{3.19}$$

Because, for any set of data, $\Sigma(x_i - \bar{x})^2$ is a constant and any $(x_i - \bar{x})\varepsilon_i$ is a normal deviate, b is normally distributed, since it is the sum of h independent normal deviates. Also from (3.19), $\mathscr{E}(b) = \beta$, since $\mathscr{E}(\varepsilon_i) = 0$ and so $\mathscr{E}\{\Sigma(x_i - \bar{x})\varepsilon_i\} = 0$.

The variance of the estimate of β, i.e. the variance of b is given by

$$\mathscr{V}(b) = \mathscr{E}\{(b-\beta)^2\} \tag{3.20}$$

Substituting for b, using (3.19), gives

$$\mathscr{V}(b) = \mathscr{E}\left\{\left(\frac{\Sigma(x_i-\bar{x})\varepsilon_i}{\Sigma(x_i-x)^2}\right)^2\right\}$$

$$= \mathscr{E}\left\{\frac{\Sigma^2(x_i-\bar{x})\varepsilon_i}{\Sigma^2(x_i-\bar{x})^2}\right\}$$

$$= \mathscr{E}\left\{\frac{\Sigma(x_i-\bar{x})^2\varepsilon_i^2+2\Sigma(x_i-\bar{x})(x_j-\bar{x})\varepsilon_i\varepsilon_j}{\Sigma^2(x_i-\bar{x})^2}\right\} \qquad i \neq j$$

The second term in the numerator is equal to zero since $\mathscr{E}(\varepsilon_i\varepsilon_j) = 0$, and so

$$\mathscr{V}(b) = \frac{\mathscr{E}(\varepsilon_i^2)}{\Sigma(x_i-\bar{x})^2} = \frac{\sigma^2}{\Sigma(x_i-\bar{x})^2} \tag{3.21}$$

The variance of a

From (3.12) and the results of a standard theorem on the linear combination of variates, we have

$$\mathscr{E}(a) = \mathscr{E}(\bar{y}) - \bar{x}\,.\,\mathscr{E}(b) \tag{3.22}$$

$$\mathscr{V}(a) = \mathscr{V}(\bar{y}) + \bar{x}^2\,.\,\mathscr{V}(b) + 2x\,.\,\mathscr{C}(\bar{y}, b) \tag{3.23}$$

Now, $\bar{y} = (1/h)\,.\,\Sigma y_i = \alpha + \beta\bar{x} + (1/h)\,.\,\Sigma\varepsilon_i$, and so $\bar{y}$ is normally distributed with mean of $\alpha + \beta\bar{x}$ and variance σ^2/h. The covariance of $\bar{y}$ and b is

$$\mathscr{E}\left\{\frac{\Sigma\varepsilon_i}{h}\cdot\frac{\Sigma(x_i-\bar{x})\varepsilon_i}{\Sigma(x_i-\bar{x})^2}\right\}$$

and is zero because $\mathscr{E}(\varepsilon_i) = 0$. Hence, from (3.22), $\mathscr{E}(a) = \alpha + \beta\bar{x} - \beta\bar{x} = \alpha$, and from (3.23)

$$\mathscr{V}(a) = \sigma^2\left\{\frac{1}{h} + \frac{\bar{x}^2}{\Sigma(x_i-\bar{x})^2}\right\} \tag{3.24}$$

The covariance of a and b

From the relationship between $\bar{y}$ and $\bar{x}$, namely $\bar{y} = a + b\bar{x}$, we have an analogous result to (3.23):

$$\mathscr{V}(\bar{y}) = \mathscr{V}(a) + \bar{x}^2\,.\,\mathscr{V}(b) + 2\bar{x}\,.\,\mathscr{C}(a, b)$$

$$\mathscr{C}(a, b) = \frac{\mathscr{V}(\bar{y}) - \mathscr{V}(a) - \bar{x}^2\,.\,\mathscr{V}(b)}{2\bar{x}} \tag{3.25}$$

On substituting for the variances on the right-hand side of (3.60), we immediately obtain

$$\mathscr{C}(a, b) = -\frac{\sigma^2 \bar{x}}{\Sigma (x_i - \bar{x})^2} \tag{3.26}$$

Maximum likelihood estimation

Using the same model for the data as before (page 70), we start by specifying the *likelihood* of all the observations. The likelihood of a single observation, y_i, is given by the normal density function as

$$\frac{1}{\sigma\sqrt{(2\pi)}}.\exp\left[-\frac{1}{2}\left\{\frac{y_i - (\alpha + \beta x_i)}{\sigma}\right\}^2\right]$$

Because all the h observations are independent of one another, we may use the multiplication law of probability to give the likelihood of all the observations: it is the product of all the individual likelihoods, and is given by

$$L = \frac{1}{\{\sigma\sqrt{(2\pi)}\}^h}.\prod_{i=1}^{h} \exp\left[-\frac{1}{2}\left\{\frac{y_i - \alpha - \beta x_i}{\sigma}\right\}^2\right] \tag{3.27}$$

Values of the parameters α, β, and σ^2 are now chosen to maximize L, but it is more convenient to work with $\log_e L$; this is quite valid since $\log_e L$ is an increasing function of L and the former will have a maximum at the same values of the parameters as the latter. Hence, from (3.27), we have

$$\log_e L = -\frac{h}{2}.\log_e \sigma^2 - \frac{h}{2}.\log_e 2\pi - \frac{1}{2\sigma^2}.\sum_{i=1}^{h} (y_i - \alpha - \beta x_i)^2 \tag{3.28}$$

Again, we shall assume summation over all the h observations, and so no longer quote the limits of summation. Differentiating (3.28) successively with respect to α, β, and σ^2 gives

$$\frac{\partial(\log_e L)}{\partial \alpha} = \frac{1}{\sigma^2}.\sum (y_i - \alpha - \beta x_i) \tag{3.29}$$

$$\frac{\partial(\log_e L)}{\partial \beta} = \frac{1}{\sigma^2}.\sum (y_i - \alpha - \beta x_i)x_i \tag{3.30}$$

$$\frac{\partial(\log_e L)}{\partial \sigma^2} = -\frac{h}{2\sigma^2} + \frac{1}{2\sigma^4}.\sum (y_i - \alpha - \beta x_i)^2 \tag{3.31}$$

On equating each of the right-hand sides of (3.29) and (3.30) to zero, we obtain identical expressions to equations (3.10) and (3.11) after removal of brackets and slight rearrangement, which are the least squares equations, and so the maximum likelihood estimates of α and β are, respectively,

$$\hat{\alpha} = \bar{y} - \hat{\beta}\bar{x} \tag{3.32}$$

$$\hat{\beta} = \frac{\Sigma(x_i - \bar{x})(y_i - \bar{y})}{\Sigma(x_i - \bar{x})^2} \tag{3.33}$$

as before. On equating the right-hand side of (3.31) to zero and rearranging, we have

$$\hat{\sigma}^2 = \frac{1}{h} \cdot \sum (y_i - \hat{\alpha} - \hat{\beta}x_i)^2 \tag{3.34}$$

that is, the estimate of the variance of the underlying populations is given by the deviations sum of squares divided by the number of observations; the hats over the parameters denote maximum likelihood estimates. It can be shown, however, that this is a biased estimate of σ^2, and it can be further shown that an unbiased estimate can be obtained by dividing by $h-2$ instead of by h. Thus, in least squares symbols, we may write the estimate of σ^2 as

$$s^2 = \frac{1}{h-2} \cdot \sum (y_i - a - bx_i)^2 \tag{3.35}$$

Two points can be made at this stage. Firstly, maximum likelihood estimates may sometimes be biased, as in the case of $\hat{\sigma}^2$. Secondly, apart from this bias, the method of maximum likelihood gives us directly an estimate of σ^2; the same result could be proved for least squares only by using advanced concepts of matrix algebra.

The variances and covariance of the parameter estimates

The variances and covariance of the parameter estimates given in (3.21), (3.24), and (3.26) were derived from first principles, and are exact. For other, more complicated, functions than the straight line it is not usually possible to derive the variances and covariances of the parameter estimates from first principles, and more indirect methods are called for. We shall now show how the method of maximum likelihood can be extended to obtain the variances and covariance of the parameter estimates for the straight line.

The starting point is the result of a theorem, known as the Cramer-Rao theorem, that the variance of an unbiased estimator of a parameter (say θ) cannot be less than

$$-\left[\mathscr{E}\left\{\frac{\partial^2(\log_e L)}{\partial\theta^2}\right\}\right]^{-1}$$

When, as at present, we have a function of more than one parameter there are variances and covariances to be treated simultaneously. If there are p parameters, the symmetric matrix of second order derivatives of the form

$$\begin{bmatrix} \dfrac{\partial^2(\log_e L)}{\partial\theta_1^2} & \dfrac{\partial^2(\log_e L)}{\partial\theta_1\partial\theta_2} & \cdots\cdots & \dfrac{\partial^2(\log_e L)}{\partial\theta_2\partial\theta_p} \\ \dfrac{\partial^2(\log_e L)}{\partial\theta_1\partial\theta_2} & \dfrac{\partial^2(\log_e L)}{\partial\theta_2^2} & \cdots\cdots & \dfrac{\partial^2(\log_e L)}{\partial\theta_1\partial\theta_p} \\ \vdots & \vdots & & \vdots \\ \dfrac{\partial^2(\log_e L)}{\partial\theta_1\partial\theta_p} & \dfrac{\partial^2(\log_e L)}{\partial\theta_2\partial\theta_p} & \cdots\cdots & \dfrac{\partial^2(\log_e L)}{\partial\theta_p^2} \end{bmatrix}$$

is known as the information matrix, and is usually denoted by **I** (but risk of confusion with the unit matrix); then the negative of the expected values of the elements in the leading diagonal of the inverse of **I** give the minimum possible variances, and the other elements give the minimum possible covariances. Thus, $-\mathbf{I}^{-1}$ can be regarded as the variance–covariance matrix of the parameter estimates.

For the straight line regression, the second order derivatives, obtained from the first derivatives (3.29) and (3.30), are:

$$\frac{\partial^2(\log_e L)}{\partial\alpha^2} = -\frac{h}{\sigma^2} \tag{3.36}$$

$$\frac{\partial^2(\log_e L)}{\partial\alpha\partial\beta} = -\frac{1}{\sigma^2}.\Sigma x_1 \tag{3.37}$$

$$\frac{\partial^2(\log_e L)}{\partial\beta^2} = -\frac{1}{\sigma^2}.\Sigma x_i^2 \tag{3.38}$$

and so the information matrix is

$$\mathbf{I} = -\frac{1}{\sigma^2}\begin{bmatrix} \mathrm{h} & \Sigma x_i \\ \Sigma x_i & \Sigma x_i^2 \end{bmatrix} \tag{3.39}$$

The determinant of **I** is given by

$$|\mathbf{I}| = \frac{h.\Sigma x_i^2 - \Sigma^2 x_i}{\sigma^4}$$

so the variance–covariance matrix of the parameters is

$$-\mathbf{I}^{-1} = \frac{\hat{\sigma}^2}{h.\Sigma x_i^2 - \Sigma^2 x_i}\begin{bmatrix} \Sigma x_i^2 & -\Sigma x_i \\ -\Sigma x_i & h \end{bmatrix} \tag{3.40}$$

and dividing both numerator and denominator of the right-hand side by h, we have

$$-\mathbf{I}^{-1} = \frac{\hat{\sigma}^2/h}{\Sigma(x_i-\bar{x})^2}\begin{bmatrix} \Sigma x_i^2 & -\Sigma x_i \\ -\Sigma x_i & h \end{bmatrix} \qquad (3.41)$$

Notice that in (3.40) and (3.41) the negative sign preceding all the elements has disappeared, and that $\hat{\sigma}^2$ has replaced σ^2. These changes are in accord with the result of Cramer-Rao's theorem, in that it is the negative of the elements of $\mathbf{I}^{-1}$ that are required, and also that expected values of these elements are obtained by using expected values (i.e. the maximum likelihood estimates of the expected values) of any parameters contained in the elements.

From (3.41) we see immediately that $\mathscr{V}(\hat{\beta}) = \hat{\sigma}^2/\Sigma(x_i-\bar{x})^2$, and $\mathscr{C}(\hat{\alpha}, \hat{\beta}) = -\hat{\sigma}^2\,\bar{x}/\Sigma(x_i-\bar{x})^2$. The variance of $\hat{\alpha}$ from (3.41) is given by

$$\mathscr{V}(\hat{\alpha}) = \frac{\hat{\sigma}^2 . \Sigma x_i^2}{h . \Sigma(x_i-\bar{x})^2}$$

$$= \frac{\hat{\sigma}^2\{h . \Sigma(x_i-\bar{x})^2 + \Sigma^2 x_i\}}{h^2 . \Sigma(x_i-\bar{x})^2}$$

$$= \hat{\sigma}^2\left\{\frac{1}{h} + \frac{\bar{x}^2}{\Sigma(x_i-\bar{x})^2}\right\}$$

where the second line above is obtained through the identity $\Sigma(x_i-\bar{x})^2 = \Sigma x_i^2 - (1/h) . \Sigma^2 x_i$. Comparison of the above results for the variances and covariance of the parameter estimates of the straight line with those obtained from first principles ((3.21), (3.24), and (3.26)) show exact correspondence between the two. Hence, for the straight line, the maximum likelihood solutions for the variances and covariance of the parameter estimates are exact.

Finally, reverting to least squares, we see that if second order derivatives, $\partial^2 \mathscr{S}^2/\partial a^2$, $\partial^2 \mathscr{S}^2/(\partial a \partial b)$, and $\partial^2 \mathscr{S}^2/\partial b^2$, are obtained from equations (3.8) and (3.9), and these derivatives are set up as a matrix analogous to (3.39), then if the constant 2 is disregarded and the matrix multiplied by the scalar $-1/\sigma^2$ we again have the information matrix (3.39). This again demonstrates the similarity between the least squares and the maximum likelihood methods.

Several replicate y-values at each x

Although in relation to the analysis of plant growth data several y-values at each x is the normal situation (i.e. several replicate growth measurements at each harvest), we need only to sketch the outline of the mathematics here since the procedure is merely an extension of that already given for the single y-value at each x situation. The method of maximum likelihood will be used.

Consider the situation in which there are h distinct x-values (as before) with N replicate y-values at each: thus, we have Nh observations in total. The logarithm of the likelihood of these observations is

$$\log_e L = -\frac{Nh}{2}.\log_e \sigma^2 - \frac{Nh}{2}.\log_e 2\pi - \frac{1}{2\sigma^2}.\sum_{i=1}^{h}\sum_{j=1}^{N}(y_{ij}-\alpha-\beta x_{ij})^2 \tag{3.42}$$

Differentiating successively with respect to α, β, and σ^2 gives

$$\frac{\partial(\log_e L)}{\partial\alpha} = \frac{1}{\sigma^2}.\sum_{i=1}^{h}\sum_{j=1}^{N}(y_{ij}-\alpha-\beta x_{ij}) \tag{3.43}$$

$$\frac{\partial(\log_e L)}{\partial\beta} = \frac{1}{\sigma^2}.\sum_{i=1}^{h}\sum_{j=1}^{N}(y_{ij}-\alpha-\beta x_{ij})x_{ij} \tag{3.44}$$

$$\frac{\partial(\log_e L)}{\partial\sigma^2} = \frac{Nh}{2\sigma^2} + \frac{1}{2\sigma^4}.\sum_{i=1}^{h}\sum_{j=1}^{N}(y_{ij}-\alpha-\beta\alpha_{ij})^2 \tag{3.45}$$

From (3.43) and (3.44) the same procedure as before yields

$$\hat{\alpha} = \bar{y} - \hat{\beta}\bar{x} \tag{3.46}$$

$$\hat{\beta} = \frac{\sum_{i=1}^{h}\sum_{j=1}^{N}(x_{ij}-\bar{x})(y_{ij}-\bar{y})}{\sum_{i=1}^{h}\sum_{j=1}^{N}(x_{ij}-\bar{x})^2} \tag{3.47}$$

where $\bar{x} = (Nh)^{-1}.\sum_{i=1}^{h}\sum_{j=1}^{N} x_{ij}$ and $\bar{y} = (Nh)^{-1}.\sum_{i=1}^{h}\sum_{j=1}^{N} y_{ij}$.

Equating (3.45) to zero gives

$$\hat{\sigma}^2 = \frac{1}{Nh}.\sum_{i=1}^{h}\sum_{j=1}^{N}(y_{ij}-\hat{\alpha}-\hat{\beta}x_{ij})^2 \tag{3.48}$$

which, without the initial multiplier $(Nh)^{-1}$ on the right-hand side is the deviations sum of squares:

$$\mathcal{S}^2 = \sum_{i=1}^{h}\sum_{j=1}^{N}(y_{ij}-\bar{y})^2 - \hat{\beta}.\sum_{i=1}^{h}\sum_{j=1}^{N}(x_{ij}-\bar{x})(y_{ij}-\bar{y}) \tag{3.49}$$

analogous to (3.16). In both (3.16) and (3.49), the first of the two terms on the right-hand sides is the total sum of squares of deviations of the y_{ij} about the overall y-mean, $\bar{y}$. The second term is the sum of squares due to regression, since this term subtracted from the first gives the deviations from regression sum of squares, $\mathcal{S}^2$.

Test of linearity

Without any regression analysis at all, the differences between the means of the y-values between the groups (defined by the h x-values) can be assessed by a single classification analysis of variance (Table 3.1). The total sum of squares (of deviations about the overall y-mean) is given by $\sum_{i=1}^{h} \sum_{j=1}^{N} (y_{ij} - \bar{y})^2$. The sum of squares between groups is given by $N . \sum_{i=1}^{j} (\bar{y}_i - \bar{y})^2$ where $\bar{y}_i = (1/N) . \sum_{j=1}^{N} y_{ij}$, $i = 1, \ldots, h$; the multiplier N is necessary to make the magnitude of this sum of squares term, comprising only h mean values, comparable with the total and within groups sums of squares, each comprising Nh individual quantities. The within groups sum of squares is then the difference between the total and the between groups sum of squares, and the within groups mean square, M_W, is an estimate of σ^2. The differences of the between group $\bar{y}_i$-values can then be assessed as $F = M_B/M_W$.

Because of the regression analysis, however, the between groups sum of squares can be partitioned into a regression component and a deviations component, the latter is N times the sum of squares of the deviations of the $\bar{y}_i$-values from the regression line. It will be observed that the deviations sum of squares and the within groups sum of squares add to the total deviations sum of squares given in (3.49), as would be expected. In other words, $\mathscr{S}^2$ can be resolved into two components: one (within groups) due to the intrinsic variability of the y-values themselves, and the other (deviations) due to the variability of the h $\bar{y}_i$-values about the regression line.

Now, if the underlying population means of y, η_i, lay exactly on the straight line, the deviations sum of squares would be merely a reflection of the variability of the y-values themselves, and so the deviations mean square and the within groups mean square would both be valid estimates of σ^2. Hence the F-ratio, M_D/M_W, would be approximately unity. Significant deviations of the data from linearity, however, add an extra component of variability to the deviations mean square, and the significance of these deviations can be assessed by the F-ratio.

This test is, therefore, a direct test of an assumption in the model, namely, that the underlying population means, η_i, lie on a straight line. Thus, it is the best test to use under most circumstances, but it can only be applied if there are several replicate y-values at each x.

Polynomial regression: general

The extension of the methods of least squares and maximum likelihood to polynomial curve fitting is straightforward, and we shall illustrate it with the second degree polynomial, using the method of least squares. The

Table 3.1 Breakdown of the sums of squares, and scheme of testing for linearity, for N replicate y-observations at each of h x-values.

	Sum of squares	Degrees of freedom	Mean Square	F
Between groups	$N \cdot \sum_{i=1}^{h} (\bar{y}_i - \bar{y})$	$h-1$	M_B	$\frac{M_B}{M_W}$
Regression	$\beta \cdot \sum_{i=1}^{h} \sum_{j=1}^{N} (x_{ij} - \bar{x})(y_{ij} - \bar{y})$	1	M_R	$\frac{M_R}{M_W}$
Deviations	$N \cdot \sum_{i=1}^{h} (\bar{y}_i - \bar{y})^2 - \beta \cdot \sum_{i=1}^{h} \sum_{j=1}^{N} (x_{ij} - \bar{x})(y_{ij} - \bar{y})$	$h-2$	M_D	$\frac{M_D}{M_W}$
Within groups	$\sum_{i=1}^{h} \sum_{j=1}^{N} (y_{ij} - \bar{y})^2 - N \cdot \sum_{i=1}^{h} (\bar{y}_i - \bar{y})^2$	$h(N-1)$	M_W	–
TOTAL	$\sum_{i=1}^{h} \sum_{j=1}^{N} (y_{ij} - \bar{y})^2$	$Nh-1$	–	–

analogies between this method and maximum likelihood are exactly the same as for the straight line. The model here is

$$\eta = \alpha + \beta_1 x + \beta_2 x^2 \tag{3.50}$$

which is estimated by

$$\hat{y} = a + b_1 x + b_2 x^2 \tag{3.51}$$

We shall assume only one y-value at each x, using the same notation as previously, and the limits of the summations will not be quoted; the extension to several replicate y-values is the same as for the straight line. The model assumptions are also the same as before (page 70), except that now the means of the underlying populations, η_i, are related to the x_i by relationship (3.50).

The expression to be minimized is

$$\mathscr{S}^2 = \Sigma(y_i - a - b_1 x_i - b_2 x_i^2)^2 \tag{3.52}$$

Differentiating the right-hand side with respect to each parameter in turn, gives

$$\frac{\partial \mathscr{S}^2}{\partial a} = -2\Sigma(y_i - a - b_1 x_i - b_2 x_i^2) \tag{3.53}$$

$$\frac{\partial \mathscr{S}^2}{\partial b_1} = -2\Sigma(y_i - a - b_1 x_i - b_2 x_i^2)x_i \tag{3.54}$$

$$\frac{\partial \mathscr{S}^2}{\partial b_2} = -2\Sigma(y_i - a - b_1 x_i - b_2 x_i^2)x_i^2 \tag{3.55}$$

Equating to zero, eliminating the brackets, and rearranging gives the following set of linear equations:

$$ah + b_1 \Sigma x_i + b_2 \Sigma x_i^2 = \sum y_i \tag{3.56}$$

$$a\Sigma x_i + b_1 \Sigma x_i^2 + b_2 \Sigma x_i^3 = \Sigma x_i y_i \tag{3.57}$$

$$a\Sigma x_i^2 + b_1 \Sigma x_i^3 + b_2 \Sigma x_i^4 = \Sigma x_i^2 y_i \tag{3.58}$$

Dividing through equation (3.56) by h, we have

$$a = \bar{y} - b_1 \bar{x} - b_2 \overline{x^2} \tag{3.59}$$

where $\overline{x^2} = (1/h).\Sigma x_i^2$, which is analogous to the straight line case. The b_i may be found by solving the two equations (3.57) and (3.58) after substituting for a from (3.59) giving, after rearrangement

$$b_1\Sigma(x_i-\bar{x})^2+b_2\Sigma(x_i-\bar{x})(x_i^2-\overline{x^2})=\Sigma(x_i-\bar{x})(y_i-\bar{y})$$

$$b_1\Sigma(x_i-\bar{x})(x_i^2-\overline{x^2})+b_2\Sigma(x_i^2-\overline{x^2})^2=\Sigma(x_i^2-\overline{x^2})(y_i-\bar{y})$$

hence

$$\begin{bmatrix}\Sigma(x_i-\bar{x})^2 & \Sigma(x_i-\bar{x})(x_i^2-\overline{x^2})\\ \Sigma(x_i-\bar{x})(x_i^2-\overline{x^2}) & \Sigma(x_i^2-\overline{x^2})^2\end{bmatrix}\begin{bmatrix}b_1\\ b_2\end{bmatrix}=\begin{bmatrix}\Sigma(x_i-\bar{x})(y_i-\bar{y})\\ \Sigma(x_i^2-\overline{x^2})(y_i-\bar{y})\end{bmatrix}$$

and so, putting $D=\Sigma(x_i-\bar{x})^2.\Sigma(x_i^2-\overline{x^2})^2-\Sigma^2(x_i-\bar{x})(x_i^2-\overline{x^2})$, the determinant of the square matrix on the left-hand side, we obtain the solution:

$$\begin{bmatrix}b_1\\ b_2\end{bmatrix}=\frac{1}{D}\begin{bmatrix}\Sigma(x_i^2-\overline{x^2})^2 & -\Sigma(x_i-\bar{x})(x_i^2-\overline{x^2})\\ -\Sigma(x_i-\bar{x})(x_i^2-\overline{x^2}) & \Sigma(x_i-\bar{x})^2\end{bmatrix}$$

$$\times\begin{bmatrix}\Sigma(x_i-\bar{x})(y_i-\bar{y})\\ \Sigma(x_i^2-\overline{x^2})(y_i-\bar{y})\end{bmatrix} \qquad (3.60)$$

The sum of squares of deviations from the regression curve (3.52) can be shown to be equivalent to

$$\mathscr{S}^2=\Sigma(y_i-\bar{y})^2-\{b_1\Sigma(x_i-\bar{x})(y_i-\bar{y})+b_2\Sigma(x_i^2-\overline{x^2})(y_i-\bar{y})\} \qquad (3.61)$$

which is analogous to the straight line result (3.16). The last two terms on the right-hand side, within the braces, together constitute the sum of squares due to regression. Further, $\mathscr{S}^2/(h-3)$ is an unbiased estimate of σ^2: one extra degree of freedom is removed from the denominator, relative to the straight line case, because an extra parameter, β_2, has been estimated.

The variances and covariances of the parameter estimates

First, we obtain the second derivatives of (3.52) from (3.53) to (3.55):

$$\frac{\partial^2\mathscr{S}^2}{\partial a^2}=2h \qquad \frac{\partial^2\mathscr{S}^2}{\partial a\partial b_1}=2\Sigma x_i \qquad \frac{\partial^2\mathscr{S}^2}{\partial a\partial b_2}=2\Sigma x_i^2$$

$$\frac{\partial^2\mathscr{S}^2}{\partial b_1^2}=2\Sigma x_i^2 \qquad \frac{\partial^2\mathscr{S}^2}{\partial b_1\partial b_2}=2\Sigma x_i^3 \qquad \frac{\partial^2\mathscr{S}^2}{\partial b_2^2}=2\Sigma x_i^4$$

Then, disregarding the constant 2, the information matrix is

$$\mathbf{I}=-\frac{1}{\sigma^2}\begin{bmatrix}h & \Sigma x_i & \Sigma x_i^2\\ \Sigma x_i & \Sigma x_i^2 & \Sigma x_i^3\\ \Sigma x_i^2 & \Sigma x_i^3 & \Sigma x_i^4\end{bmatrix} \qquad (3.62)$$

and $-\mathbf{I}^{-1}$ is the variance–covariance matrix of the three parameter estimates. The actual variances and covariances are lengthy to quote; in any case there is no need to, since nearly all polynomial regression is now carried out by computer routines.

The variance of $\hat{y}$ is given by

$$\mathscr{V}(\hat{y}) = \mathscr{V}(a) + x^2.\mathscr{V}(b_1) + x^4.\mathscr{V}(b_2) + 2x.\mathscr{C}(a, b_1) + 2x^2.\mathscr{C}(a, b_2) + 2x^3.\mathscr{C}(b_1, b_2) \quad (3.63)$$

and all the variances and covariances on the right-hand side are obtained from $-\mathbf{I}^{-1}$.

Higher order polynomials

Comparison of the following homologous pairs of equations: (3.6) with (3.51), (3.7) with (3.52), (3.8) and (3.9) with (3.53), (3.54) and (3.55), (3.16) with (3.61), and (3.39) with (3.62); will show how the theory of linear regression is extended to higher order polynomials. The resulting expressions for the estimates of the parameters and their variances and covariances rapidly become very complicated with increasing degree of polynomial, and so they are never quoted explicitly in algebraic terms. The variance of $\hat{y}$ can, however, be succinctly written, and is given by

$$\mathscr{V}(\hat{y}) = \sum_{i=0}^{n} x^{2i}.\mathscr{V}(b_i) + 2.\sum_{i=0}^{n}\sum_{j=0}^{i-1} x^{(i+j)}.\mathscr{C}(b_i, b_j) \quad (3.64)$$

where n is the degree of polynomial, and $b_0 \equiv a$.

Although some of the covariances are negative, all the variances are necessarily positive; and so (3.64) indicates that, in general, $\mathscr{V}(\hat{y})$ increases with degree of polynomial fitted to a particular set of data. This means that there is a danger of overfitting; that is, fitting a polynomial of too high a degree to a set of data. The ultimate in over-fitting is a polynomial of degree $h-1$, the curve of which would go through every datum point; but less extreme over-fitting can occur, and this is the subject of the final section of this chapter.

Testing the appropriateness of a function

Where there are several replicate y-values at each x, the adequacy of any function linear in its parameters (e.g. all polynomials) can be tested by a procedure analogous to the test of linearity presented earlier (page 80); the only change in the scheme shown in Table 3.1 will be the regression sum of squares, which is unique for a given function.

If there is only one y-value for each x in the data, a test of adequacy of a function is not possible; all that can be done is to fit the particular function in question, then fit an alternative and see whether the alternative function fits the data *significantly* better than the original. The word 'significant' is important here, because the more parameters a function has, in general, the better the fit to a given set of data; hence, the important question is, does a function with a larger number of parameters fit a set of data significantly better than another with fewer parameters? For example, when assessing the most appropriate polynomial to describe a particular set of data, one fits a straight line, then a second degree polynomial, and so on. At each stage the coefficient of the highest power of x is tested, using the null hypothesis that the coefficient in question is zero. If the null hypothesis is accepted for the coefficient of x^i, then a polynomial of degree $i-1$ is appropriate; if the null hypothesis is rejected, then a polynomial of degree $i+1$ is fitted and the same test carried out on the coefficient of $x^{(i+1)}$.

Where several replicate y-values at each x are available, the direct test is the correct one to use. However, where polynomials are used purely empirically, as in whole plant growth analysis, complications may arise simply because the model being fitted is unsuitable. At present, the polynomial is firmly entrenched in whole plant growth analysis methodology even though this model is unsuitable and, as a result of this, Elias & Causton (1976) found that the sensitivity of the direct test is dependent upon the variablility of the data – data of low variability required ridiculously high degrees of polynomial function (up to the 6th degree) according to the direct test. The indirect test is, however, not affected by data variability, especially if the fittings and the test are carried out on $(\bar{x}_i, \bar{y}_i)$-values, and this obviated the difficulties encountered by Elias and Causton in using the direct test.

Therefore, when polynomials are being used empirically, it may be advantageous to test for the appropriate degree of polynomial in both ways (direct and indirect tests), and examine the results in relation to both the source and the appearance of the data. This may seem to be undesirably subjective, but we believe some subjectivity is unavoidable in this particular artificial situation of empiricism.

4

Single leaf growth and the Richards function: methodology

Determinate growth and sigmoid functions

Viewed as a whole, plants are organisms of indeterminate growth, whereas animals usually have a determinate growth pattern. An animal grows to a certain maximum size then stops when adulthood is reached, and the same can be said for the animal's main organs; most plants, on the other hand, do not have a definite maximum size, although vegetative and reproductive phases (corresponding to juvenile and adult phases of animals) do exist. An adult animal at its maximum size lives for a greater or lesser period of time, whereas many plants continue growing through their reproductive phase until senescence ensues. A pattern of growth showing a definite and prolonged maximum is known as determinate, whereas the opposite pattern, exhibited by most plants, is indeterminate.

Quite obviously, perennial and biennial plants are of indeterminate growth, but the situation is more variable in the case of annuals depending upon the type of inflorescence formed. Where the inflorescence is terminal, for example in temperate cereals, the pattern of growth can be determinate to a high degree of approximation; usually a definite number of leaves and internodes is produced first, followed by a complete change of the shoot apex to reproductive structures. The number of tillers can, of course, vary, but uniculm (non-tillering) cultivars have now been bred which would eliminate that source of indeterminism. Root growth is always indeterminate, but since fairly precise relationships normally exist between the Level 1 components of a plant under constant conditions (Chapter 6), the amount of indeterminism here is restricted. Thus, a uniculm annual grass plant would have a relatively determinate growth pattern. In contrast, an annual species with axillary inflorescences (e.g. the annual *Veronica* (speedwell) species) would be unlikely to have a determinate growth pattern, even if a plant consisted of only one main stem. Here, the shoot apex does not change from a vegetative to a reproductive structure, and an indefinite number of new leaves is added on during the reproductive phase thus making growth indeterminate.

For any pattern of biological growth, however, a mathematical function giving rise to a sigmoid curve (i.e. a function that is bounded by two horizontal asymptotes and having everywhere a positive first derivative) can empirically describe growth, since even indeterminate growth will cease at some stage. But if it is desired to use such a function (which itself can be based on a simple but meaningful biological model) less empirically, then the approach must be more rigorous and, in particular, a sigmoid growth function should be applied only to those organisms or parts of organisms showing a determinate pattern of growth. In the higher plant, only individual leaves are likely to show such a pattern during the vegetative phase, with the addition of individual flowers and fruits in the reproductive phase. In our original classification of component and entity systems (Table 1.1, p. 3), the highest level containing organs showing determinate growth are the components of Level 3. Even here, however, individual internodes of dicotyledon species would be unlikely to have a determinate pattern because of secondary growth; in monocotyledons, internodes may well have a determinate growth pattern.

Sigmoid functions

A sigmoid curve may be either symmetric or asymmetric according to whether the point of inflexion is or is not midway between the two asymptotes. Before proceeding to a full consideration of the Richards function, we shall discuss three other sigmoid functions which have been proposed as growth functions in order to highlight their merits and limitations in this respect.

The logistic function

The logistic is the best known sigmoid function, and may be written in the form

$$f(t) = A(1 + e^{(\beta - \kappa t)})^{-1} \tag{4.1}$$

with asymptotes at $f(t) = 0$ and $f(t) = A$. Note that Greek and Roman 'A's are indistinguishable, and that in equation (4.1) and similar we are using Greek symbols for the parameters of a function, in accord with the statistical idea that they are population parameters (Chapter 3). Of the other two function parameters, κ is a 'rate' parameter – a high value indicating a rapid rise of the function between the two asymptotes, and vice versa – and β/κ defines the value of t at the point of inflexion. Thus, the parameters β and κ are not biologically independent of one another.

The logistic function has been employed many times in the past as a growth function (Chapter 1): originally because it is the function of an autocatalytic chemical reaction, and growth was considered to be controlled by one such 'master' reaction (Robertson, 1923); and lately because the logistic function may be considered to be defined by the differential equation

$$\frac{dW}{dt} = \frac{\kappa W}{A}(A - W) \qquad (4.2)$$

implying that relative growth rate is a declining linear function of size. Equation (4.2) shows that the two parameters A and κ are biologically correlated, since the gradient of the line is κ/A. Hence, all three parameters in (4.1) are correlated and, because the statistical estimation of the logistic function involves non-linear regression, complications can arise when fitting the function to data.

The Gompertz function

This is a non-symmetric sigmoid function which may be written in the form

$$f(t) = A.\exp(-e^{(\beta - \kappa t)}) \qquad (4.3)$$

where the parameters have the same general meaning as in the logistic function. The asymptotes are again at $f(t) = 0$ and $f(t) = A$, but the value of $f(t)$ at the point of inflexion is A/e instead of $A/2$.

The Gompertz function, (Gompertz, 1825) has been found to be more appropriate in biological work than any other sigmoid function. Amer & Williams (1957) considered that the asymmetry of the Gompertz function was more appropriate to leaf growth data than the symmetry of the logistic; the same view was also taken by Hackett & Rawson (1974). On the animal side, Laird, Tyler & Barton (1965), and Laird (1965) demonstrated the usefulness of the function in analysing the pre-natal growth of the Guinea Pig. Much earlier though, Medawar (1940) went further, when he deduced from theoretical considerations that the growth of the embryo chicken heart should follow the Gompertz function – a true mechanistic growth model.

The Gompertz function may be defined by the differential equation

$$\frac{dW}{dt} = \kappa W(\log_e A - \log_e W) \qquad (4.4)$$

which shows relative growth rate to be a declining linear function of the logarithm of size. Also, there is a particularly simple relationship between relative growth rate and the time in this function, i.e.

$$R = \kappa e^{(\beta - \kappa t)} \qquad (4.5)$$

an exponential decline.

Pearl and Reed's generalized logistic function

A major disadvantage of the two preceding functions is that each has an unalterable shape. This means that the point of inflexion occurs at a given distance between the two asymptotes regardless of the parameter values, although change of the latter will alter the position of the upper asymptote, the position of the point of inflexion with respect to the t-axis, and the rate at which the curve rises from one asymptote to the other. Clearly it would be desirable to introduce one or more extra parameters to allow the sigmoid function to change its shape according to the requirements of the data.

The logistic function may be generalized by extending the function of t in equation (4.1) which, in its most general form may be written as

$$f(t) = A(1 + e^{\phi(t)})^{-1} \tag{4.6}$$

For the logistic, we have $\phi(t) = \beta - \kappa t$, a linear function, in which $d\{\phi(t)\}/dt$ is negative and equal to $-\kappa$. Provided that $d\{\phi(t)\}/dt$ is everywhere negative, equation (4.6) can represent an asymmetric sigmoid curve. The simplest forms of $\phi(t)$, mathematically speaking, are the odd degree polynomials in which, with suitable combinations of the parameters, $d\{\phi(t)\}/dt$ may be negative throughout. Pearl & Reed (1923) used a third degree polynomial for $\phi(t)$ to describe the growth of the fruit of *Cucurbita pepo* (pumpkin); the fit of the function to the data was good, but this would be expected as the function has five parameters.

However, it appears that this generalized logistic function has been not, or little, used subsequently. Although very flexible, having five parameters or more, it has no simple form of $R = F(W)$, as has the logistic or the Gompertz functions. This means that this generalized logistic function cannot be the basis of a simple growth model, and so it is very much more difficult, if not impossible, to interpret the parameters biologically in a direct and simple manner. In fact,

$$R = -\phi'(t).(A - W)/A \tag{4.7}$$

Doubtless a form of $\phi(t)$ could be found to make (4.7) a simple function of W, but almost certainly without sufficient generalization to make the exercise worth while. Further, with the polynomial generalization of $\phi(t)$, the point of inflexion must always lie between $f(t) = 0$ and $f(t) = A/2$; it cannot lie in the upper half of the 0 to A range. Since the Richards function (page 89) (i) has only four parameters, (ii) has a simple form of $R = F(W)$, and (iii) may have the point of inflexion anywhere between 0 and A by the adjustment of only one of its four parameters, there would seem to be no incentive to pursue Pearl and Reed's function further.

The Richards function

Since in this book the Richards function is largely applied to single leaf growth, L will be used instead of W as the symbol for the growth attribute.

The function is defined by the differential equation

$$\frac{\mathrm{d}L}{\mathrm{d}t} = \frac{\kappa L}{\nu A^{\nu}}(A^{\nu} - L^{\nu}) \tag{4.8}$$

with initial condition $0 < L < A$ when $t = 0$, where L is the value of a size attribute at time t, and A, κ, ν are parameters; A, $\kappa > 0$, and $-1 \leqslant \nu < \infty$ $\nu \neq 0$. The variables in (4.8) are separable so, on rearranging and making the substitution $u = L^{\nu}$ giving $\mathrm{d}u/u = \nu.\mathrm{d}L/L$, equation (4.8) becomes

$$\int \frac{\mathrm{d}u}{u(A-u)} = \int \frac{\kappa \mathrm{d}t}{A^{\nu}}$$

The left hand side can be split into partial fractions and integrated to yield

$$L = A(1 \pm \mathrm{e}^{(\beta - \kappa t)})^{-1/\nu} \tag{4.9}$$

The negative alternative is used when ν is negative, and *vice versa*.

Derivations of the Richards function

By a generalization of the logistic

If $\nu = 1$, equation (4.8) becomes

$$\frac{\mathrm{d}L}{\mathrm{d}t} = \frac{\kappa L}{A}(A - L)$$

and (4.9) becomes

$$L = A(1 + e^{(\beta - \kappa t)})^{-1}$$

which are identical with the logistic function (c.f. equations (4.2) and (4.1), respectively). Equation (4.2) can be written as

$$\frac{1}{L}.\frac{\mathrm{d}L}{\mathrm{d}t} = \kappa\{1 - (L/A)\}$$

which implies that the relative growth rate declines linearly with increasing size, L. Similarly, equation (4.8) can be rewritten in the form

$$\frac{1}{L}.\frac{\mathrm{d}L}{\mathrm{d}t} = \frac{\kappa}{\nu}\left\{1 - \left(\frac{L}{A}\right)^{\nu}\right\} \tag{4.10}$$

showing that relative growth rate declines linearly with some power, ν, of increasing size. Thus the Richards function is another generalized logistic but, unlike Pearl & Reed's function, it is based on a simple biological model embodied in equation (4.10). The criterion of biological simplicity here is that the relative growth rate of the attribute concerned declines in a mathematically simple manner with increasing size of the attribute, but there is sufficient flexibility in the Richards function to allow for varying durations of initial, nearly constant, relative growth rates (i.e. approximations to exponential growth).

Before leaving this sub-section, it should be pointed out that Nelder's (1961, 1962) generalizations of the logistic function are essentially re-parameterizations of the Richards function.

From the Von Bertalanffy function

Richards (1959) first derived his function from one developed by Von Bertalanffy (1941, 1957) which was based on theoretical considerations of animal growth. For details, the reader is referred to the above papers, but for completeness we will note the salient points here, using their original notation.

Essentially, Von Bertalanffy's function is defined by

$$\frac{dW}{dt} = \eta W^m - \kappa W \tag{4.11}$$

where η and κ are constants representing the rates of anabolism and catabolism, respectively, and m is a positive constant theoretically lying in the range $\frac{2}{3} \leqslant m \leqslant 1$. Dividing (4.11) throughout by W, we have

$$\frac{1}{W} \cdot \frac{dW}{dt} = \eta W^{(m-1)} - \kappa \tag{4.12}$$

Finally, putting $\eta/\kappa = A^{(1-m)}$ and $(1-m)\kappa = k$, after some rearrangement (4.12) becomes

$$\frac{1}{W} \cdot \frac{dW}{dt} = \frac{k}{1-m} \left\{ \left(\frac{A}{W} \right)^{(1-m)} - 1 \right\} \tag{4.13}$$

which is the same form as (4.10) with $k = \kappa$ and $m = \nu + 1$.

Richards then pointed out that if m took values greater than unity ($\nu > 0$), then for a limited growth situation η and κ would have to be negative. As these quantities were defined as metabolic rates, the latter situation was regarded as impossible. He was further of the opinion that when m was assessed from growth data instead of being assigned a value from theoretical considerations, values of m in excess of unity would be found to be the rule. It is not known whether Richards was referring specifically to plant growth data or not, but in our experience positive values of ν ($m > 1$) are almost invariably found when the Richards function is fitted to plant growth data of any kind (see appendix tables).

However, it is of interest to note that when the Richards function was employed to describe the accumulation of net photosynthate with time in an individual leaf, negative values of ν, and particularly in the range $-\frac{1}{3} \leqslant \nu < 0$ (corresponding to $\frac{2}{3} \leqslant m < 1$), were normally found (Hadley, 1978).

Properties of the Richards function

The interesting variations in the geometric properties of the Richards function are consequent upon changes in ν. The diagrams in Fig. 4.1 therefore have fixed values of the other parameters and parameter combinations: A, κ/ν, and $A(1 \pm e^{\beta})^{-1/\nu}$; the second representing a theoretical initial relative growth rate, the first representing the upper asymptote of L, and the last representing the value of L, L_0, when $t = 0$ (i.e. the intercept of the $L(t)$ curve on the L-axis.

The R(L) form

This is the relative growth rate as a function of size form of the Richards function (equation (4.10)), and curves of the function are shown in Fig. 4.1a,b,c,d for ν-values of -0.5, 0.5, 2, and 10 respectively. For positive values of ν the curve intersects the R-axis at $\mathrm{R} = \kappa/\nu$, although in one sense this result is spurious since $\mathrm{R}\,(= (1/L)(\mathrm{d}L/\mathrm{d}t))$ is undefined when $L = 0$; thus κ/ν is a limiting value. As ν increases in value there is an increasing duration (as measured by growth of L) in which R is almost constant, implying approximate exponential growth. When ν is negative, the $\mathrm{R}(L)$ curve is asymptotic to the R-axis.

A mean relative growth rate (weighted as described by Richards, 1959) over the whole duration of growth from $L = 0$ to $L = A$ is given by

$$\overline{\mathrm{R}} = \frac{1}{A - 0}\int_0^A \frac{\kappa}{\nu}\left\{1 - \left(\frac{L}{A}\right)^{\nu}\right\}\mathrm{d}L$$

i.e.

$$\overline{\mathrm{R}} = \frac{\kappa}{\nu A^{(\nu+1)}}\int_0^A (A^{\nu} - L^{\nu})\,\mathrm{d}L \tag{4.14}$$

which integrates to

$$\overline{\mathrm{R}} = \frac{\kappa}{\nu + 1} \tag{4.15}$$

As will be seen later, this is a most useful quantity.

For $\nu = -1$ and $\nu = 1$ the Richards function reduces to two special cases – the monomolecular and the logistic functions, respectively. For $\nu = 0$, equation (4.10) becomes indeterminate but, as will be shown below, this value of ν represents a limiting situation – the Gompertz function. Curves of $\mathrm{R}(L)$ for the monomolecular, logistic, and Gompertz functions are shown in Fig. 4.2a,b,c.

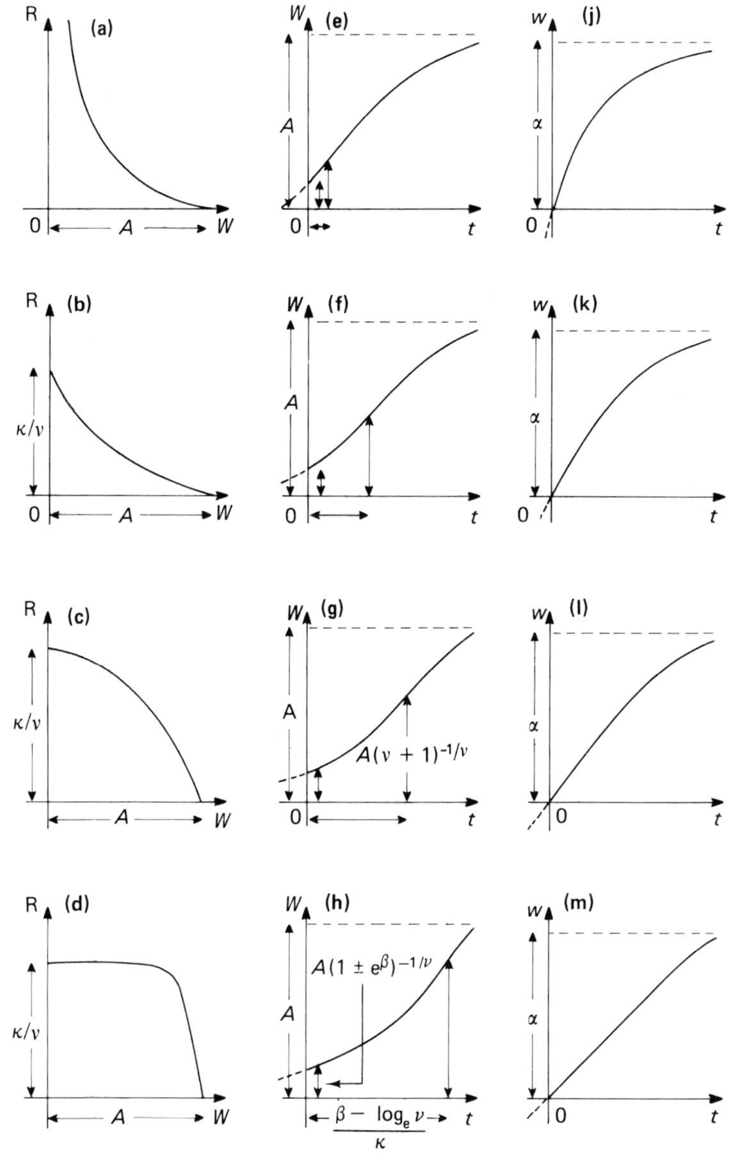

Fig. 4.1 The Richards function for four values of ν: $\nu = -0.5$ (**a**, **e**, **j**), $\nu = 0.5$ (**b**, **f**, **k**), $\nu = 2$ (**c**, **g**, **l**) and $\nu = 10$ (**d**, **h**, **m**). Curves of relative growth rate as a function of dry weight (**a**, **b**, **c**, **d**), dry weight as a function of time (**e**, **f**, **g**, **h**) and $\log_e$ dry weight as a function of time (**j**, **k**, **l**, **m**). In the central column of graphs the intercept and point of inflexion are indicated. The fact that the logarithmic forms pass through the origin is fortuitous. (After Causton, Elias & Hadley, 1978.)

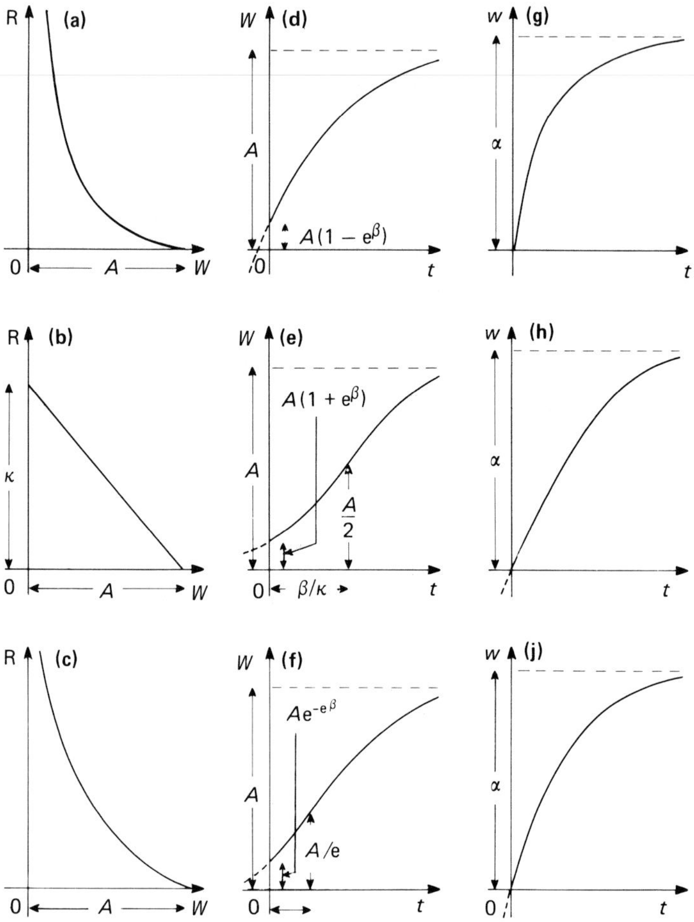

Fig. 4.2 The monomolecular function (**a**, **d**, **g**), the logistic function (**b**, **e**, **h**) and the Gompertz function (**c**, **f**, **j**). Curves of relative growth rate as a function of time (**d**, **e**, **f**) and $\log_e$ dry weight as a function of time (**g**, **h**, **j**). Other notes as for Fig. 4.1. (After Causton, Elias & Hadley, 1978.)

The $L(t)$ form

For $\nu > 0$ both the lines $L = 0$ and $L = A$ are asymptotes, but for $\nu < 0$ there is no lower asymptote, as $L \to -\infty$ when $t \to -\infty$. Evidently the introduction of the negative sign into equation (4.9) gives an element of artificiality to the function for negative ν and, perhaps, makes it less suitable as a growth function. When $t = 0$, $L_0 = A(1 \pm e^{\beta})^{-1/\nu}$, so that

$$\beta = \log_e\{\pm(A/L_0{}^{\nu} \mp 1\} \tag{4.16}$$

Successive differentiation of (4.9) gives

$$\frac{dL}{dt} = \pm \frac{A\kappa e^{(\beta - \kappa t)}(1 \pm e^{(\beta - \kappa t)})^{-(1/\nu + 1)}}{\nu} \tag{4.17}$$

and

$$\frac{d^2L}{dt^2} = \frac{\pm A\kappa^2 e^{(\beta - \kappa t)}}{\nu^2(1 \pm e^{(\beta - \kappa t)})^{(1/\nu + 2)}} \left\{ \pm e^{(\beta - \kappa t)} - \nu \right\} \tag{4.18}$$

On equating (4.18) to zero, we find that the point of inflexion occurs when

$$t_I = \frac{\beta - \log_e(\pm \nu)}{\kappa} \tag{4.19}$$

and, on substituting (4.19) for t in (4.9), we find that L at the point of inflexion, L_I, as a fraction of A is given by

$$L_I/A = (\nu + 1)^{-1/\nu} \tag{4.20}$$

This confirms that the position of the point of inflexion, as measured by L_I/A, depends solely on the parameter ν; relationship (4.20) is illustrated in Fig. 4.3. On putting $\nu = 1$ in (4.20) we have $L_I/A = 0.5$, which is appropriate for the logistic function; similarly, by putting $\nu = -1$ we have $L_I/A = 0$, corresponding to the monomolecular function which has no point of inflexion.

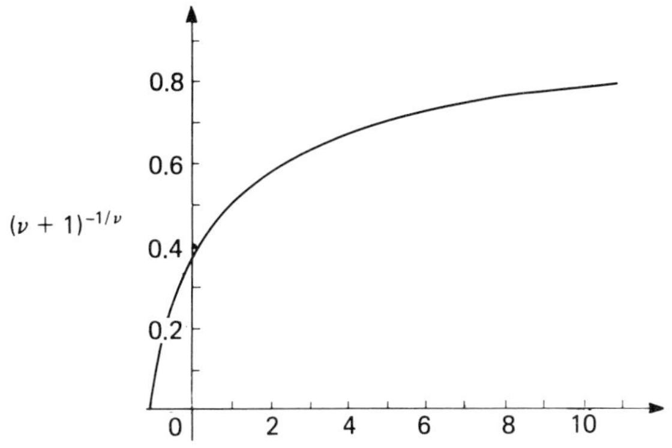

Fig. 4.3 The relationship between ν and the point of inflexion (see text for details).

The limiting case of $\nu = 0$ representing the Gompertz function can now be proved. Expanding the right-hand side of (4.20) by the binomial theorem, we have for small non-zero ν

$$(1+v)^{-1/v} = 1 + (-1/v) + \frac{(-1/v)(-1/v-1)}{2!}v^2$$
$$+\frac{(-1/v)(-1/v-1)(-1/v-2)}{3!}v^3$$
$$+\frac{(-1/v)(-1/v-1)(-1/v-2)(-1/v-3)}{4!}v^4 + \;....$$

i.e.

$$(1+v)^{-1/v} = \frac{1+v}{2!} - \frac{1+3v+2v^2}{3!} + \frac{1+6v+11v^2+6v^3}{4!} - ... \qquad (4.21)$$

Putting $v = 0$ in the right-hand side of (4.21) gives

$$\frac{1}{2!} - \frac{1}{3!} + \frac{1}{4!} - \;.... = e^{-1}$$

As will be shown later (page 99), L_I/A for the Gompertz function is precisely e^{-1}, and so the result is proved.

Curves of equation (4.9) are shown in Fig. 4.1e,f,g,h for the same values of v as previously, and curves of the monomolecular, logistic, and Gompertz functions are given in Fig. 4.2d,e,f.

The $l(t)$ form

The logarithmic form of equation (4.9) is

$$l = \alpha - (1/v).\log_e(1 \pm e^{(\beta - \kappa t)}) \qquad (4.22)$$

and there is only one horizontal asymptote ($l = \alpha$). For $v > 0$, when $t \to -\infty$ $W \to -\infty$; for $v < 0$, as $t \to \beta/\kappa$ $W \to -\infty$, so here the curve is asymptotic to the vertical line $t = \beta/\kappa$. Curves of equation (4.22) are shown in Fig. 4.1i,j,k,l for the same values of v as previously, and curves of the logarithmic form of the monomolecular, logistic, and Gompertz functions are given in Fig. 4.2g,h,i.

The $R(t)$ form

Differentiation of (4.22) gives directly the relative growth rate as a function of time:

$$R = \pm\frac{\kappa e^{\beta}}{v(e^{\kappa t} \pm e^{\beta})} \qquad (4.23)$$

The curves given by (4.23) are rather different in form according to whether v is positive or negative (Fig. 4.4), and so the two sets of properties will be considered separately.

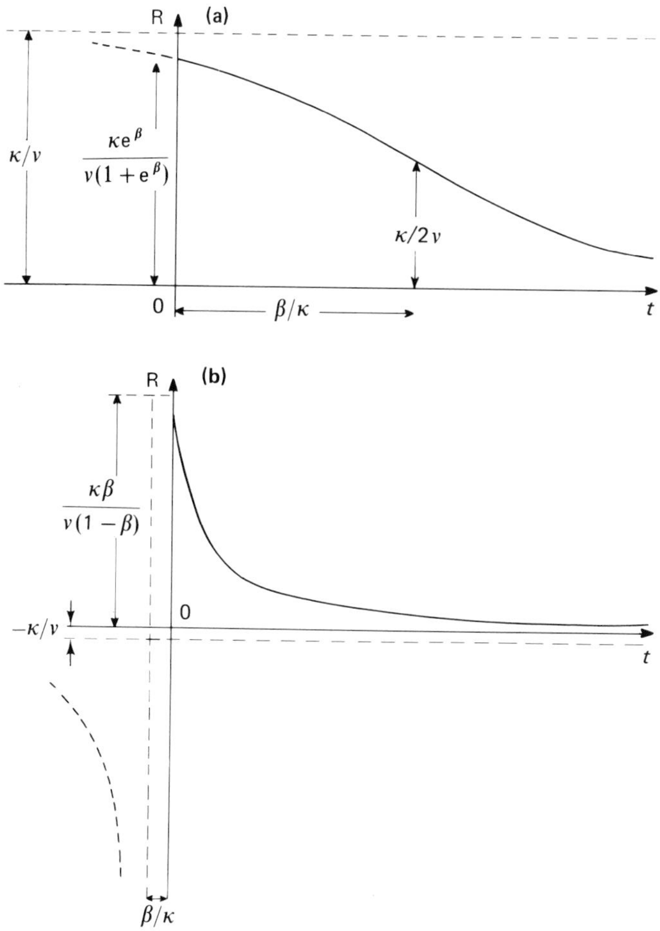

Fig. 4.4 Relative growth rate as a function of time for the Richards function with (**a**) $\nu > 0$ and (**b**) $\nu < 0$.

For $\nu > 0$, inspection of (4.23) shows that the lines $R = 0$ and $R = \kappa/\nu$ are asymptotes. Differentiation of (4.23) gives

$$\frac{dR}{dt} = -\frac{\kappa^2 e^{(\beta + \kappa t)}}{\nu(e^{\kappa t} + e^{\beta})^2} \tag{4.24}$$

which is always negative. Differentiating (4.24) gives

$$\frac{d^2R}{dt^2} = \frac{\kappa^3 e^{(\beta + \kappa t)}}{\nu(e^{\kappa t} + e^{\beta})^3}(e^{\kappa t} - e^{\beta}) \tag{4.25}$$

Expression (4.25) is negative when $t < \beta/\kappa$ and positive when $t > \beta/\kappa$. Thus there is a point of inflexion in R(t) curve at $t = \beta/\kappa$. Substituting this value of t into (4.23) gives R at the point of inflexion, R_I, as

$$R_I = \frac{\kappa}{2\nu} \tag{4.26}$$

Thus, whatever the value of ν, relationship (4.23) represents a 'reverse' sigmoid curve (continuous negative gradient) which is, moreover, symmetrical with its point of inflexion lying midway between the two asymptotes.

For $\nu < 0$, the lines $R = 0$ and $R = \kappa/\nu$ are, again, asymptotes, but we now also have a vertical asymptote at $t = \beta/\kappa$, which is evident by rearranging (4.23) to give

$$t = \frac{\beta + \log_e\left(1 - \dfrac{\kappa}{\nu R}\right)}{\kappa} \tag{4.27}$$

and letting $R \to \pm\infty$. Differentiation of (4.23) gives

$$\frac{dR}{dt} = \frac{\kappa^2 e^{(\beta + \kappa t)}}{\nu(e^{\kappa t} - e^{\beta})^2} \tag{4.28}$$

which, again, is negative throughout. Differentiating again gives

$$\frac{d^2R}{dt^2} = -\frac{\kappa^2 e^{(\beta + \kappa t)}(e^{\kappa t} + e^{\beta})}{\nu(e^{\kappa t} - e^{\beta})^3} \tag{4.29}$$

which again shows a change in the direction of the change in gradient at $t = \beta/\kappa$, but this does not correspond to a point of inflexion because the R(t) curve is asymptotic at this point. Finally, it can be deduced from (4.27) that R cannot lie in the range $\kappa/\nu \leqslant R \leqslant 0$; hence the curve of (4.23) where $\nu < 0$ is as shown in Fig. 4.4b.

The Gompertz function

As already described, the Gompertz function is a limiting case of the Richards function; it is, therefore, defined by somewhat different equations and this fact, together with that of the Gompertz being an important function from a biological viewpoint in its own right, requires that this function be reviewed separately.

The R(L) form

This is given by

$$R = \kappa(\log_e A - \log_e L) \tag{4.30}$$

showing that the relative growth rate is a declining linear function of $\log_e L$. The curve is asymptotic to the line $L = 0$. A weighted mean relative growth rate over the whole duration of growth from $L = 0$ to $L = A$ is defined as

$$\overline{\mathrm{R}} = \frac{1}{A-0} \int_0^A \kappa(\log_e A - \log_e L)\mathrm{d}L$$

i.e.

$$\overline{\mathrm{R}} = \frac{\kappa}{A} \int_0^A (\log_e A - \log_e L)\mathrm{d}L \tag{4.31}$$

which reduces to

$$\overline{\mathrm{R}} = \kappa \tag{4.32}$$

This result is identical to the general result for the Richards function (equation (4.15)) with $v = 0$.

The $L(t)$ form

This is given by

$$L = A\,.\exp\left(-\mathrm{e}^{(\beta - \kappa t)}\right) \tag{4.33}$$

which is derived from (4.30) by making the substitution $u = \log_e A - \log_e L$. The lines $L = 0$ and $L = A$ are asymptotes, and when $t = 0$ $L_0 = A\,.\exp(-\mathrm{e}^{\beta})$ giving

$$\beta = \log_e\{-\log_e(L_0/A)\} \tag{4.34}$$

Successive differentiation of (4.33) gives

$$\frac{\mathrm{d}L}{\mathrm{d}t} = A\kappa\mathrm{e}^{(\beta - \kappa t)}\,.\exp\left(-\mathrm{e}^{(\beta - \kappa t)}\right) \tag{4.35}$$

and

$$\frac{\mathrm{d}^2 L}{\mathrm{d}t^2} = A\kappa^2\mathrm{e}^{(\beta - \kappa t)}\,.\exp\left(-\mathrm{e}^{(\beta - \kappa t)}\right)\left(\mathrm{e}^{(\beta - \kappa t)} - 1\right) \tag{4.36}$$

On equating (4.36) to zero, we find that the point of inflexion occurs when

$$t_I = \beta/\kappa \tag{4.37}$$

and, on substitution of this value of t into (4.33), we find that

$$L_I/A = \mathrm{e}^{-1} \simeq 0.3679 \tag{4.38}$$

Thus the Gompertz curve is markedly asymmetric, with the point of inflexion occurring well before growth is half completed.

The l(t) form

The logarithmic form of equation (4.33) is

$$l = \alpha - e^{(\beta - \kappa t)} \tag{4.39}$$

and there is only one asymptote ($l = \alpha$).

The R(t) form

Differentiation of (4.39) gives relative growth rate as a function of time, and is

$$\mathrm{R} = \kappa e^{(\beta - \kappa t)} \tag{4.40}$$

which shows that R declines exponentially with time. Thus, there is no point of inflexion in the curve, and the latter resembles the upper right-hand portion of the curve in Fig. 4.4b.

Estimating the Richards function

The assumptions in the estimation procedure to be used initially are:

(a) that at each harvest time, t_i, which should be known without error, there is a normally distributed population of l-values (equivalently, a log-normally distributed population of L-values);

(b) the mean of each population, λ_i, is given by the Richards function

$$\lambda_i = \alpha - \frac{1}{\nu} . \log_e(1 \pm e^{(\beta + \kappa t_i)}) \tag{4.41}$$

(note that $K = -\kappa$ for the convenience of eliminating a negative sign, and that Greek and Roman 'K's cannot be distinguished);

(c) the variance of each population of l-values is the same, and equal to σ^2;

(d) the sample observations at each t_i are independent of one another.

Because of the necessity of destructive sampling for dry weight determinations, different plants are used at each harvest; in practice, this applies to leaf areas too, as these measurements are made on the same plants as are used for dry weight determinations. Adherence of the data to assumption (d) is thereby assured. The other assumptions in the model will be examined in the next part of this chapter.

As in the case of linear regression, the least squares and maximum likelihood methods give identical results; we shall employ the former method here. Also, for economy of notation, we shall assume only one replicate l-value at each t_i; the extension to N replicate l-values at each harvest is straightforward and is the same, in principle, as for linear regression. Thus, (4.41) is estimated as

$$\hat{l}_i = a - \frac{1}{n}.\ \log_e(1 \pm e^{(b+Kt_i)}) \tag{4.42}$$

and the expression to be minimized over h harvest times is

$$\mathscr{S}^2 = \sum_{i=1}^{h} \left\{ l_i - a + \frac{1}{n}.\ \log_e(1 \pm e^{(b+Kt_i)}) \right\}^2 \tag{4.43}$$

Partial differentiation of the right-hand side of (4.43) with respect to each parameter in turn, and supressing the summation limits, gives:

$$\frac{\partial \mathscr{S}^2}{\partial a} = -2.\sum \left\{ l_i - a + \frac{1}{n}.\ \log_e(1 \pm e^{(b+Kt_i)}) \right\} \tag{4.44}$$

$$\frac{\partial \mathscr{S}^2}{\partial b} = \pm\frac{2}{n}.\sum \left\{ l_i - a + \frac{1}{n}.\ \log_e(1 \pm e^{(b+Kt_i)}) \right\} \left\{ \frac{e^{(b+Kt_i)})}{1 \pm e^{(b+Kt_i)}} \right\} \tag{4.45}$$

$$\frac{\partial \mathscr{S}^2}{\partial K} = \pm\frac{2}{n}.\sum \left\{ l_i - a + \frac{1}{n}.\ \log_e(1 \pm e^{(b+Kt_i)}) \right\} \left\{ \frac{e^{(b+Kt_i)}t_i}{1 \pm e^{(b+Kt_i)}} \right\} \tag{4.46}$$

$$\frac{\partial \mathscr{S}^2}{\partial n} = -\frac{2}{n^2}.\sum \left\{ l_i - a + \frac{1}{n}.\ \log_e(1 \pm e^{(b+Kt_i)}) \right\} \left\{ \log_e(1 \pm e^{(b+Kt_i)}) \right\} \tag{4.47}$$

On equating these to zero, it is quite evident that equations (4.44) to (4.47) are non-linear and do not have explicit solutions (see page 70).

The Newton–Raphson method

Consider a function of a single variable, $f(x)$, in which the root or roots cannot be explicitly determined. If a guess of the value of the root is made, x_0, then it can be shown (e.g. Causton, 1977, page 118) that an improved estimate of the root, x_1, is given by

$$x_1 = x_0 - \frac{f(x_0)}{f'(x_0)} \tag{4.48}$$

The initial guess, x_0, is usually called the starting value of x, and x_1 the adjusted value. Equation (4.48) can be used as often as desired, and each application of it is called an iteration. For the ith iteration, the starting value will be the $(i-1)$th iteration's adjusted value; a succession of iterations may (but not always) show a convergence onto the required root. The scheme of iterations based on (4.48) is called the Newton-Raphson method.

The method can also be employed in the present situation for minimizing (4.43). Here, we have a function of 4 variables (the parameters) and we need to find the roots of the 4 first order partial derivatives. In other words, we require values for a, b, K, and n which make the right-hand sides of (4.44) to (4.47) equal to zero. Let $\mathbf{a}_0$ be a column vector of length 4 of starting values, and let $\mathbf{a}_1$ be a similar column vector of adjusted values. Then

$$\mathbf{a}_1 = \mathbf{a}_0 - \mathbf{M}^{-1}\mathbf{v} \tag{4.49}$$

where $\mathbf{v}$ is a column vector of length 4 of the first order partial derivatives (i.e. equations (4.44) to (4.47), and $\mathbf{M}$ is a symmetrical matrix of order 4 containing all the second order partial derivatives; all the elements of $\mathbf{v}$ and $\mathbf{M}$ are evaluated using the elements of $\mathbf{a}_0$. The second order partial derivatives are as follows:

$$\frac{\partial^2 \mathscr{S}^2}{\partial a^2} = 2h \tag{4.50}$$

$$\frac{\partial^2 \mathscr{S}^2}{\partial a \partial b} = \mp\frac{2}{n}.\Sigma\left\{\frac{e^{(b+Kt_i)}}{1 \pm e^{(b+Kt_i)}}\right\} \tag{4.51}$$

$$\frac{\partial^2 \mathscr{S}^2}{\partial a \partial K} = \mp\frac{2}{n}.\Sigma\left\{\frac{e^{(b+Kt_i)}t_i}{1 \pm e^{(b+Kt_i)}}\right\} \tag{4.52}$$

$$\frac{\partial^2 \mathscr{S}^2}{\partial a \partial n} = \frac{2}{n^2}.\Sigma\left\{\log_e(1 \pm e^{(b+Kt_i)})\right\} \tag{4.53}$$

$$\frac{\partial^2 \mathscr{S}^2}{\partial b^2} = \pm\frac{2}{n}.\Sigma\left[1 - a + \frac{1}{n}\left\{\log_e(1 \pm e^{(b+Kt_i)}) \pm e^{(b+Kt_i)}\right\}\right] \times \left[\frac{e^{(b+Kt_i)}}{(1 \pm e^{(b+Kt_i)})^2}\right] \tag{4.54}$$

$$\frac{\partial^2 \mathscr{S}^2}{\partial b \partial k} = \pm\frac{2}{n}.\Sigma\left[1 - a + \frac{1}{n}\left\{\log_e(1 \pm e^{(b+Kt_i)}) \pm e^{(b+Kt_i)}\right\}\right] \times \left[\frac{e^{(b+Kt_i)}t_i}{(1 \pm e^{(b+Kt_i)})^2}\right] \tag{4.55}$$

$$\frac{\partial^2 \mathscr{S}^2}{\partial b \partial n} = \mp\frac{2}{n^2}.\Sigma\left\{1 - a + \frac{2}{n}.\log_e(1 \pm e^{(b+Kt_i)})\right\}\left\{\frac{e^{(b+Kt_i)}}{1 \pm e^{(b+Kt_i)}}\right\} \tag{4.56}$$

$$\frac{\partial^2 \mathscr{S}^2}{\partial K^2} = \pm\frac{2}{n}.\Sigma\left[1 - a + \frac{1}{n}\left\{\log_e(1 \pm e^{(b+Kt_i)}) \pm e^{(b+Kt_i)}\right\}\right] \times \left[\frac{e^{(b+Kt_i)}t_i^2}{(1 \pm e^{(b+Kt_i)})^2}\right] \tag{4.57}$$

$$\frac{\partial^2 \mathscr{S}^2}{\partial K \partial n} = \mp\frac{2}{n^2}.\Sigma\left\{1 - a + \frac{2}{n}.\log_e(1 \pm e^{(b+Kt_i)})\right\}\left\{\frac{e^{(b+Kt_i)}t_i}{1 \pm e^{(b+Kt_i)}}\right\} \tag{4.58}$$

$$\frac{\partial^2 \mathscr{S}^2}{\partial n^2} = \frac{4}{n^3}.\Sigma\left\{1 - a + \frac{3}{2n}.\log_e(1 \pm e^{(b+Kt_i)})\right\}\left\{\log_e(1 \pm e^{(b+Kt_i)})\right\} \tag{4.59}$$

The disadvantage of this method is that the initial estimates provided must be close to the least squares solution or the method may converge slowly or not at all, or may, indeed, converge on incorrect values. The provision of good starting values is, therefore, very important.

Nelder (1961) obtained starting values by the use of the empirical method given by Richards (1959), however, with irregular or inextensive data this method gives very poor results. Another disadvantage in using Richards' method is that in the subjective parts of the procedure one tends automatically to give equal weighting to each point (each point being an L- or $\bar{L}$-value rather than an l- or $\bar{l}$-value), indeed, it would be very difficult and subjective to do anything else; but this could very markedly affect the starting values obtained, particularly that of n.

Causton (1969) proposed an alternative procedure, which was an adaptation of Hartley's (1948) method, using the generating law of the Richards function (equation (4.10)). This was an improvement, and designed to be an automatic algorithm to precede the least squares Newton-Raphson algorithm. However, Davies & Ku (1977) showed that starting values based on Causton's method were not always adequate in that when applied to the least squares algorithm, convergence might either be to some local minimum or complete instability might result with some sets of data. The least squares hypersurface has been little investigated for the Richards function, but it is undoubtedly not simple in form. In particular, Davies & Ku (1977) showed that the contour 'ellipses' (they are not always true ellipses) in the b–K plane are very elongated if b is high, producing a long trough rather than a sharply defined minimum.

Three alternative approaches can be adopted to try and overcome this instability. Firstly, the iterative method can be changed; secondly, the model can be re-parameterized (Gillis & Ratkovsky, 1978) or, thirdly, a further improvement of the starting values can be effected. The second option is the least attractive in that the new parameters would be less biologically meaningful, and the first approach will be discussed briefly later. page 105.

Hadley's method of obtaining starting values

A glance at equation (4.42) shows that for given values of b and K, values of a, $\frac{1}{n}$, and $\mathscr{S}^2$ may be estimated by ordinary linear regression. Hadley's (1978) method is simply a direct search involving a particular strategy for defining values of b and K, followed by linear regression; this pair of operations (cycle) may be repeated as many times as desired.

In more detail, the procedure is as follows. An initial arbitrary value, b_0, is assumed (usually $b_0 = 1$), and from this an initial K-value, K_{00}, is obtained from

$$K_{00} = -\frac{b_0}{\bar{t}} \tag{4.60}$$

where $\bar{t}$ is the mean of the t_i of the data. The reason for the double subscript for K is that its value is adjusted in two stages for a given b. From this pair of values, b_0 and K_{00}, $\mathscr{S}^2$ is calculated. Then, still maintaining b_0, K_{00} is increased by a small amount, ΔK, where

$$\Delta K = \frac{K_{00}}{D} \tag{4.61}$$

The value of D is usually taken as 20, but may be adjusted as required. With the new K-value, a new residual sum of squares, $\mathscr{S}^2$, is calculated and if this is less than the previous residual, then K_{00} is increased progressively by ΔK until the residual (calculated at each stage) reaches a minimum. On the other hand, if the minimum is more than before, on first adjustment of K_{00}, then K_{00} is progressively decreased by ΔK until a minimum is reached. At this point the K-value can be called K_{10}, and ΔK is decreased to a new value, defined by

$$\Delta K = \frac{K_{10}}{10D} \tag{4.62}$$

The K-value is then adjusted up or down as before, using the value of ΔK defined in (4.62), until the residual reaches a new minimum, K_1; K_1 is considered to be the optimal K-value for the value of b_0, with a minimum residual, $\mathscr{S}_1^2$.

Following this, the value of $b_i (i = 2, \ldots)$ is progressively increased or decreased (according to the change in $\mathscr{S}^2$ consequent upon the initial change of b_i) and, using the same procedure as above, an optimal K-value, $K_i (i = 2, \ldots)$, is obtained for each b_i. In practice, b_i is changed in steps of unity or less; for each b_i the initial K-value is K_{i-1}, and this value is adjusted using

$$\Delta K = \frac{K_{i-1}}{10D} \tag{4.63}$$

to obtain the optimal K_i.

For each parameter pair (b_i, K_i) the residual sum of squares is examined. If $\mathscr{S}_i^2 < \mathscr{S}_{i-1}^2$, b_i is changed to b_{i+1}. However, if $\mathscr{S}_i^2 > \mathscr{S}_{i-1}^2$, b_i is decreased in smaller steps (usually 0.1 of the previous b_i increments) until the residual again reaches a minimum. To increase the precision in this final stage, ΔK may be decreased further by increasing the denominator on the right-hand side of equation (4.62).

The method is thus one of simple search; the strategy is simple in concept but may be time consuming, even in terms of electronic computing time.

However, starting values of any desired level of accuracy can be acquired. It may be asked why any least squares iterations need to be done in these circumstances: the answers are that a greater time would be required by search for the ultimate precision obtainable by a few iterations, and that at least one iteration is required in order to obtain variances and covariances of the parameter estimates.

Other iterative methods

As previously mentioned, the Newton-Raphson technique is efficient only when accurate starting values are supplied and, furthermore, convergence towards the minimum is ensured only if **M** is a positive definite matrix (Box, Davies & Swann, 1969). These authors suggest some improvements to the technique which may overcome this latter problem, but it can be suggested that unless good starting values are available an alternative approximation method for minimizing (4.43) may be more profitable. Reviews of the different methods of non-linear optimization and their application to curve fitting are given in Box, Davies & Swann (1969), Dixon (1972) and Chambers (1973), and Mead & Pike (1975) discuss the methods in a biological context. The direct search techniques are most usually employed by statisticians in function minimization.

Namkoong & Matzinger (1975) employed a method attributed to Marquardt (1963) when using Richards functions to describe the growth of *Nicotiana tabacum* in selection experiments. This fitting method overcomes many problems inherent in previous methods, and combines the best features of both the gradient (credited to Cauchy, 1847, cited by Dixon, 1972) and Taylor series methods (which include the Newton-Raphson). In gradient methods starting values are supplied, and revised values are found by assessing the slope of $\Sigma(l_i - \hat{l})^2$, i.e. $\mathscr{S}^2$. Whereas Taylor series methods converge rapidly, providing that the starting values are not too far removed from the minimum, the gradient methods are able to converge on true parameters even when starting values are far removed; convergence, however, even under good conditions, is slow.

Convergence problems centre on the configuration of $\mathscr{S}^2 = f(\beta_k)$ $k = 1, \ldots, p$. When the parameters, β_k, are linear, the residual sum of squares as a function of the p parameters hypersurface can be represented by a series of ellipsoid contours, and $\mathscr{S}^2$ will be a minimum at their centre. However, when the parameters are non-linear the contours are distorted, although they may approach the ellipsoid form in the vicinity of the minimum. This means that the hypersurface can be elongated in some directions and attenuated in others so that the minimum lies at the bottom of a trough. From the work of Davies & Ku (1977) this can be inferred to occur in the Richards function when b (and usually also K and n) is large.

During the initial stages of Marquardt's (1963) method the gradient approach predominates, but as the minimum is neared the procedure approximates to a Taylor series method. To illustrate Marquardt's method, first rearrange (4.49) to

$$\mathbf{a}_0 - \mathbf{a}_1 = \mathbf{M}^{-1}\mathbf{v}$$

then let $\Delta\,\mathbf{a} = \mathbf{a}_0 - \mathbf{a}_1$, and premultiply both sides by $\mathbf{M}$ giving

$$\mathbf{M}.\Delta\,\mathbf{a} = \mathbf{v} \tag{4.64}$$

Marquardt modifies equation (4.64) to

$$(\mathbf{M} + \lambda\mathbf{I})\,\Delta\,\mathbf{a} = \mathbf{v} \tag{4.65}$$

where $\mathbf{I}$ is the unit matrix, and λ is a quantity which changes with each iteration. When λ is small, equation (4.65) approaches (4.64) – the Newton-Raphson equation – and when λ is large, equation (4.65) approaches $\lambda\mathbf{I}.\Delta\mathbf{a} = \mathbf{v}$. As $\lambda \to \infty$, the method approaches that of steepest descent. The strategy adopted to change λ is given in Marquardt's paper and the method is described, in relation to biological data, by Conway, Glass and Wilcox (1970).

The two main problem areas in non-linear regression are: (a) the non-applicability of the method when $\mathbf{M}$ is not a positive definite matrix, and (b) the existence of extended troughs in the least squares hypersurface; the existence of either or both of these problems for a particular set of data can be inferred by examining the eigenvalues of matrix $\mathbf{M}$. If all the eigenvalues of $\mathbf{M}$ are positive, then $\mathbf{M}$ is positive definite; moreover, the more nearly equal the eigenvalues the more nearly circular are the ellipsoids of the least squares hypersurface, and directions of elongated troughs may be exposed.

Concluding assessments of the estimated function

Our current procedure is to use the Newton-Raphson method for giving least squares adjustments to starting values provided by Hadley's method. The two parts of the process have been combined into one computer program, and problems arise only with poor data (i.e. where there is an incomplete spread of data along the major part of the curve, or highly variable data). When a satisfactory fit has been obtained, there remain a number of other computations for completion of the analysis, and these are detailed in this section.

Variances and covariances of the parameter estimates

One advantage of using the Newton-Raphson method is that matrix $\mathbf{M}$ leads directly to the information matrix, and hence to the variance-covariance matrix of the parameter estimates. Analogous to the linear regression situation, we have

$$\mathbf{I} = -\frac{1}{2\sigma^2}\cdot\mathbf{M} \tag{4.66}$$

and $-\mathbf{I}^{-1}$ is the required variance-convariance matrix. An estimate of σ^2 is given by

$$s^2 = \frac{\mathscr{S}^2}{h-4} \tag{4.67}$$

However, it should be pointed out that in non-linear regression the variances and convariances of the parameter estimates, obtained in this way, will not be exact.

The variance of $\hat{l}$

Again in the non-linear regression context of the Richards function, only an approximate result can be obtained for the variance of an estimate of l from the fitted function, $\mathscr{V}(\hat{l})$. The approximation formula is that given by Kendall & Stuart (1977), and in the Appendix to this book, equation A.2. For the Richards function, $k = 4$, and $\theta_1 = \alpha$, $\theta_2 = \beta$, $\theta_3 = K$, and $\theta_4 = v$, are the corresponding expected values of $x_1 = a$, $x_2 = b$, $x_3 = K$, and $x_4 = n$; $f(\theta_1, \ldots, \theta_k)$ is, of course, equal to $\hat{l}$. The partial derivatives are evaluated at their expected values and so, in practice, using the estimated values of the parameters; the derivatives are:

$$\frac{\partial \hat{l}}{\partial a} = 1 \tag{4.68}$$

$$\frac{\partial \hat{l}}{\partial b} = \mp\frac{e^{(b+Kt)}}{n(1 \pm e^{(b+Kt)})} \tag{4.69}$$

$$\frac{\partial \hat{l}}{\partial K} = \mp\frac{e^{(b+Kt)}t}{n(1 \pm e^{(b+Kt)})} \tag{4.70}$$

$$\frac{\partial \hat{l}}{\partial n} = \frac{1}{n^2}\cdot\log_e(1 \pm e^{(b+Kt)}) \tag{4.71}$$

Thus (A.2) may be evaluated at any desired value of t.

Venus & Causton (1979c) have given an example of the use of the above method in relation to leaf growth data, and assessed its statistical validity in a simulation study. From the latter, they concluded that (A.2) provided a very reasonable estimate of $\mathscr{V}(\hat{l})$ under a wide range of replicate and harvest numbers; the standard deviation of the simulated l-values was, however, kept constant at 0.2.

The estimated value of relative growth rate and its variance

Mathematically, we have

$$\mathrm{R} = \mp\frac{Ke^{b}}{n(\mathrm{e}^{-Kt} \pm \mathrm{e}^{b})} \tag{4.72}$$

but $\hat{\mathrm{R}} \neq \mathrm{R}$ because the parameter estimates, b, K, and n have not been obtained for this function, but for the $l(t)$ form of the Richards function (4.41). In the case of polynomials used to describe growth data, which are linear in their parameters, the relative growth rate function is also a polynomial and so $\hat{\mathrm{R}}$ does equal R here (page 52); but for the Richards function, which in non-linear in its parameters, $\hat{\mathrm{R}}$, as well as $\mathscr{V}(\hat{\mathrm{R}})$, has to be obtained by an approximation formula similar to (A.2), and this is also given in the Appendix, equation A.1.

For the R(t) form of the Richards function, $k = 3$, and $\theta_1 = \beta$, $\theta_2 = K$, and $\theta_3 = v$, are the corresponding expected values of $x_1 = b$, $x_2 = K$, and $x_3 = n$; $f(\theta_1, \ldots, \theta_k)$ is given by (4.72). The first order partial derivatives are:

$$\frac{\partial \mathrm{R}}{\partial b} = \mp\frac{K\mathrm{e}^{(b-Kt)}}{n(\mathrm{e}^{-Kt} \pm \mathrm{e}^{b})^2} \tag{4.73}$$

$$\frac{\partial \mathrm{R}}{\partial K} = \mp\frac{\mathrm{e}^{b}\{\mathrm{e}^{-Kt}(1+Kt) \pm \mathrm{e}^{b}\}}{n(\mathrm{e}^{-Kt} \pm \mathrm{e}^{b})^2} \tag{4.74}$$

$$\frac{\partial \mathrm{R}}{\partial n} = \mp\frac{K\mathrm{e}^{b}}{n^2(\mathrm{e}^{-Kt} \pm \mathrm{e}^{b})} \tag{4.75}$$

and the second order partial derivatives are:

$$\frac{\partial^2 \mathrm{R}}{\partial b^2} = \frac{K\mathrm{e}^{(b-Kt)}(\mathrm{e}^{b} \mp \mathrm{e}^{-Kt})}{n(\mathrm{e}^{-Kt} \pm \mathrm{e}^{b})^3} \tag{4.76}$$

$$\frac{\partial^2 \mathrm{R}}{\partial b.\partial K} = \mp\frac{\mathrm{e}^{(b-Kt)}\{\mathrm{e}^{-Kt} \pm \mathrm{e}^{b} + Kt(\mathrm{e}^{-Kt} \mp \mathrm{e}^{b})\}}{n(\mathrm{e}^{-Kt} \pm \mathrm{e}^{b})^3} \tag{4.77}$$

$$\frac{\partial^2 R}{\partial b.\partial n} = \frac{K\mathrm{e}^{(b-Kt)}}{n^2(\mathrm{e}^{-Kt} \pm \mathrm{e}^{b})^2} \tag{4.78}$$

$$\frac{\partial^2 \mathrm{R}}{\partial K^2} = \mp\frac{\mathrm{e}^{(b-Kt)}\{2(\mathrm{e}^{-Kt} \rightarrow \mathrm{e}^{b}) + Kt(\mathrm{e}^{-Kt} \mp \mathrm{e}^{b})\}\,t}{n(\mathrm{e}^{-Kt} \pm \mathrm{e}^{b})^3} \tag{4.79}$$

$$\frac{\partial^2 \mathrm{R}}{\partial K.\partial n} = \pm\frac{\mathrm{e}^{b}\{\mathrm{e}^{-Kt}(1+Kt) \pm \mathrm{e}^{b}\}}{n^2(\mathrm{e}^{-Kt} \pm \mathrm{e}^{b})^2} \tag{4.80}$$

$$\frac{\partial^2 R}{\partial n^2} = \mp \frac{2Ke^b}{n^3(e^{-Kt} \pm e^b)} \tag{4.81}$$

Thus $\hat{R}$ and $\mathscr{V}(\hat{R})$ may be evaluated at any desired value of t.

Testing the appropriateness of the Richards function

Where there are several replicate l-values at each t, the adequacy of the Richards function in describing the data can be assessed directly in a manner analogous to the test of linearity, described in Chapter 3 (Table 3.1). The only changes to be made in Table 3.1 occur in the 'Regression' and 'Deviations' rows (where we now write l instead of y) which become:

Deviations sum of squares $= N \, . \sum_{i=1}^{h} (\bar{l}_i - \hat{l}_i)^2$

Regression sum of squares $= N\left\{\sum_{i=1}^{h} (\bar{l}_i - \bar{l})^2 - \sum_{i=1}^{h} (\bar{l}_i - \hat{l}_i)^2\right\}$

Deviations degrees of freedom $= h - 4$

Regression degrees of freedom $= 3$

However, the test is not exact here, since Draper & Smith (1966, page 282) imply that the mean square ratio M_R/M_D is only approximately an F-distribution in non-linear regression, presumably because the mean squares themselves are not distributed as χ^2. Thus it would seem that the test is not as rigorous as for a function with linear parameters, but we provide further evidence on this question below.

The distributions of the estimates of the parameters and their variances: a simulation study

In a linear regression situation, the parameter estimates are normally distributed; in a non-linear situation, the parameter estimates are not, in general, normally distributed, but may be approximately so. Non-normality creates difficulties when testing hypotheses and setting confidence intervals on the parameters, and so it is important to gain some insight into the distributions of the parameter estimates and their variances. Accordingly, a simulation study was undertaken to assess: (a) whether the parameter estimates of a Richards function are approximately normally distributed; and (b) whether the variances of the parameter estimates approximately follow the χ^2-distribution. Other interesting items of information have also been obtained from the results.

Two quite different underlying Richards functions were used: one was the function fitted to sunflower leaf pair 3 (dry weight), in which the value of n lies between 0 and 1; and the other function was that fitted to wheat leaf 5 (area), in which the value of n, and hence those of b and K because of high correlation of the estimates, are numerically relatively high (Table 4.1). Three different experimental situations were simulated: (i) using the sunflower underlying relationship with 'harvests' on alternate days, between days 8 and 34 inclusive; (ii) using the wheat relationship with harvests on alternate days, between days 6 and 24 inclusive; and (iii) the same as (ii) but with daily harvests. For each of the three experiments, 200 sets of simulated data were generated, which were normally distributed about the underlying relationship with a standard deviation of 0.2; 10 replicates were taken at each harvest.

Results are shown in Tables 4.1 and 4.2. In Table 4.1 we see that very good parameter estimates were obtained for the sunflower and wheat (daily harvests) simulations, but were less good in the case of wheat (alternate-day harvests). Variances of the parameter estimates were very small in the case of sunflower; whereas they were larger, but still acceptable, for wheat with daily harvests (except for $\mathcal{V}(a)$, which was smaller than for sunflower). In the case of wheat with alternate-day harvests, the variances of b, K and n were ridiculously large, whereas that for a was still small.

Table 4.1 Parameter estimates, their variances, and the variances of the estimated Richards functions in a simulation study, using homoscedastic data generated from two different underlying functions. Each figure is a mean value obtained from 200 sets of generated data.

Data type	Sunflower		Wheat		
	ACTUAL	Alternate day harvests	ACTUAL	Daily harvests	Alternate day harvests
Parameter Estimates					
a	−1.114	−1.114	2.767	2.769	2.775
b	6.430	6.407	45.49	48.11	52.93
K	−0.3898	−0.3890	−2.675	−2.828	−3.112
n	0.4395	0.4377	3.206	3.392	3.725
Variances of estimates					
a		0.000438		0.000311	0.000616
b		0.1650		260.3	5.57×10^7
K		0.000377		0.8963	1.7×10^5
n		0.001136		1.313	1.6×10^5
Variance of curve		0.03932		0.03995	0.03950

The results in Table 4.2 were calculated as follows. The set of 200 of each of the four parameter estimates in each of the three experiments were sorted into 26 size classes, and the observed relative frequency in each class was then compared with the probability of the corresponding size class for the normal distribution with mean of the 200 estimated values of the parameter

Table 4.2 Properties of the distributions of the parameter estimates and their variances for the fitted Richards functions in a simulation study, using homoscedastic data generated from two different underlying functions. Each figure is based on 200 sets of generated data.

Date type	Sunflower			Wheat: daily harvests			Wheat: alternate day harvests		
	$\chi^2_{[18]}$	Skewness	Kurtosis	$\chi^2_{[18]}$	Skewness	Kurtosis	$\chi^2_{[16]}$	Skewness	Kurtosis
a	23.1	0.39	0.13	15.6	0.02	−0.41	24.1	−0.14	−0.56
b	23.2	0.08	−0.25	34.3*	2.81	12.94	294.6***	3.17	9.94
K	18.8	−0.03	−0.19	38.2**	−2.80	17.97	294.3***	−3.17	9.92
n	18.1	0.12	−0.28	35.7**	2.81	13.06	278.9***	3.17	9.92
$\mathscr{V}(a)$	12.4	0.31	−0.06	30.2*	−0.32	−0.08	17.3	0.11	−0.09
$\mathscr{V}(b)$	17.0	0.43	0.12	55.7***	9.89	108.30	1146***	7.63	64.26
$\mathscr{V}(K)$	23.0	0.43	0.03	57.1***	9.90	108.70	970.5***	8.19	72.57
$\mathscr{V}(n)$	22.0	0.45	0.05	55.0***	9.92	109.10	1104***	5.96	35.72
Variance of curve	16.6	−0.02	−0.38	15.9	−0.17	0.51	14.45	0.09	−0.36
Deviations sum of squares	16.8	0.77	0.39	39.3**	0.62	0.69	21.3	0.57	−0.36
Within groups sum of squares	14.1	0.54	−0.54	15.6	−0.06	0.58	15.3	0.06	−0.43
Number of significant F-ratios		6.8 %			4.9 %			5.4 %	

and variance of the 200 estimated values about their mean. A χ^2-goodness-of-fit test of the form

$$\chi^2_{[c-3]} = \sum_{i=1}^{c} \frac{(\text{observed} - \text{expected})^2}{\text{expected}}$$

where c is the number of terms in the summation, was used to assess the comparison; size classes with expected frequencies less than 5 were amalgamated with adjacent classes in the usual way. For the variances of the parameter estimates, a similar procedure was followed in principle. Because the degrees of freedom associated with each parameter estimate was, respectively, 136, 96, and 186 for sunflower, wheat (alternate-day harvests), and wheat (daily harvests), the normal approximation of χ^2 was used. Each of the 200 variances of a parameter estimate in one experiment was transformed by doubling its value and taking the square root; the transformed values were again sorted into 26 size classes, and the observed relative frequency in each class was compared with the probability of the corresponding size class for the normal distribution with mean of the 200 transformed values and variance of the 200 transformed values about their mean. The same goodness-of-fit test, as above, was used to assess the comparison. The same procedure was used for the variance of curve, deviations means square, and residual mean square. Skewness and kurtosis values were also calculated in every case. Finally, the number of 'significant' F-ratios at $P < 0.05$ for M_D/M_W were counted.

From Table 4.2, it is quite evident that the parameter estimates are approximately normally distributed, and that the variances of the estimates, the variances of the curves, and the deviations and residual mean squares, approximately follow χ^2-distributions in the case of the sunflower simulated data. For the wheat data, the situation is quite different: a and its variance are approximately normally and χ^2-distributed, respectively, but the remaining three parameters and their variances are very far removed from these theoretical distributions, χ^2 being highly significant in every case; the distributions are positively skew (negative in the case of K, because K-values are always negative) and extremely leptokurtic. An examination of the distributions of the values giving very χ^2 -values reveals that these are really pathological distributions, with a peak of probability of occurrence around the mean together with additional peaks at the extremes of the range. Overall, however, the variances of the curves, the deviations and residual mean squares do approximately follow χ^2-distributions. The percentage of significant F-ratios seems unduly high for sunflower where, from the above results, one might have expected a figure very close to 5 per cent; for wheat, however, the numbers of significant F-ratios are quite close to 5 per cent.

The results of this simulation study are gratifyingly clear cut for the two situations investigated. In the wheat-derived data, where the underlying values of b, K, and n are high, the distributions of the estimates of these parameters and their variances are very peculiar, and this is regardless of whether the harvests were taken daily or on alternate days. On the other hand, in the sunflower-derived data, where the underlying values of b, K, and n are low, there is no evidence that the distributions of the parameter estimates and their variances are different from those that theoretically occur in a linear regression situation.

The actual parameter estimates and their variances for wheat-derived data are interesting (Table 4.1). With alternate-day harvests, the parameter estimates are rather poor, and their variances are unrealistically large; with daily harvests, however, the situation is greatly improved. Evidently, in 'difficult' cases the intensity of information along the course of the curve is vital. These results extend those of the small simulation experiment carried out by Causton *et al.* (1978), in which they showed the importance of an adequate spread of data along the curve for good parameter estimation, especially at the lower end of the curve, but the effect of harvest interval was not investigated.

It should, however, be remembered that general statistical theory shows that efficient estimation is achieved by concentrating information at a few points rather than spreading it over a large number. The best known instance is in linear regression, where the most precise estimate of the gradient of a straight line (i.e. having the smallest variance of estimate) is obtained by dividing the total number of potential observations into two groups and by placing these as far apart as possible with regard to the x-axis; by this means, $\Sigma(x-\bar{x})^2$ is maximized and so $\mathscr{V}(b)$ is minimized (equation (3.21)). However, in the absence of information regarding the most efficient placing of points along the time-axis for the Richards function, which would doubtless differ between the four parameters, a good spread of points is at present recommended.

The results of this simulation study serve to reinforce previous experience with fitting the Richards function. We have rarely had problems of estimation where $0.05 < n < 0.6$ approximately, the more usual range of n for individual leaf growth data; convergence is quick with good starting values, and the parameter variances are small compared with the values of the estimates. In the more unusual cases, where $n > 1$, the variances of the parameter estimates are usually much larger in comparison with the values of the parameter estimates themselves, and sometimes there are convergence problems in the fitting process. Fortunately, most plant growth data, particularly those of individual leaves, have parameter values in the range where there are few difficulties (see Appendix tables), but the problems which arise when the parameters b, K, and n have high numerical values have yet to be solved.

Testing the estimation model assumptions

The assumptions of the model used in the estimation of the Richards function are given on page 100. We must now enquire whether the structure of typical individual leaf growth data at least approximate to the model assumptions since, if there is unconformity in this respect, inappropriate and possibly misleading parameter estimates will result.

We have already noted that assumption (d) is almost certainly adhered to, owing to the destructive harvesting programme of sampling. On the other hand, whether the growth of individual leaves accords precisely with assumption (b) is less certain. Although a requirement of the statistical model, the assessment of the occurrence of assumption (b) for actual growth data is a biological problem. It is most unlikely that the growth of a *single* individual leaf would follow the Richards or any other regular function, but it may be that the mean of a whole popuation of a particular leaf on plants growing under identical and uniform conditions would do so, and so the deviations in the size of individual leaves from the trend set by a Richards function at any time could be considered as random fluctuations. This leads on to assumption (a): growth is a multiplicative rather than an additive process, so for this reason alone growth attributes would tend to be log-normally rather than normally distributed. Whether systematic deviations of the growth of *individual* single leaves from a Richards function trend might markedly affect the overall deviations of l_{ij} about their means, $\bar{l}_i$, from those of normal deviates is impossible to judge, but it should be borne in mind. Assumption (c) is related to (a), in that if growth data are log-normally distributed and the coefficient of variation remains constant, then assumption (c) would hold.

Although assumption (b) cannot be statistically examined, assumptions (a) and (c) can, and we propose to do this in the present section, using examples from data which will be fully described in the next chapter.

Testing for normality of the logarithmically transformed data

The samples involved in the present experiments are small, with a maximum of 10 replicates (Table 2.7), and this means that testing for normality of the underlying population will not give very reliable results, in that only extreme departures from normality in a sample of observations will be exposed. Because the assumption is made that growth data are log-normally distributed, the logarithmically transformed data will be tested for normality at a given harvest time, i.e. testing for normality of l_{ij} for a given i.

An initial graphical examination of two representative sets of logarithmically transformed growth data – leaf pair 3 of sunflower, and leaf 6 of birch (Fig. 4.5) – gives little evidence for acceptance or rejection of the hypothesis

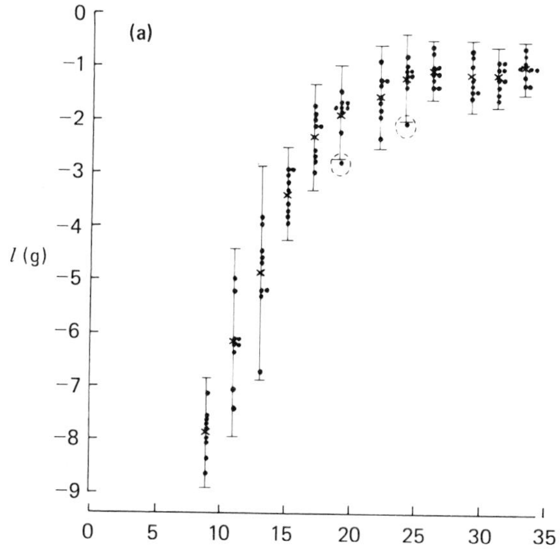

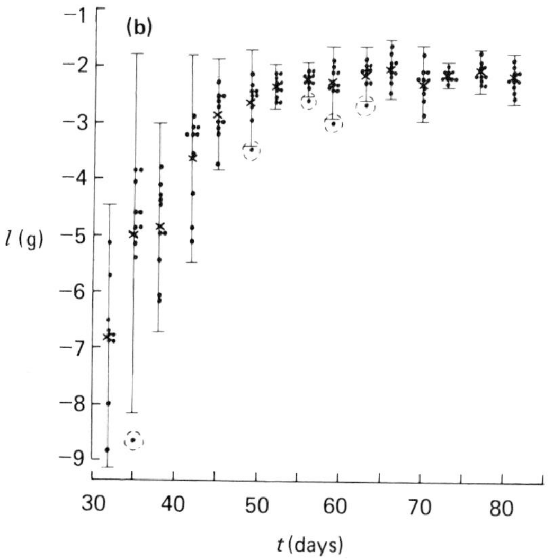

Fig. 4.5 Primary logarithmically transformed data with intervals of ± students-t × standard deviation of sample, observations lying outside these intervals are ringed. Data for (**a**) sunflower weight leaf pair 3 and (**b**) birch weight leaf 6.

of normality. Data, such as those for birch leaf 6 at day 32 and sunflower leaf pair 3 at day 11, are often spread around the mean in a manner indicative of underlying normality; i.e. a denser concentration of points around the mean and then trailing off above and below the mean. Other data sets, including sunflower leaf pair 3 at day 15 and birch leaf 6 at day 35, appear to have skew or perhaps even bimodal distributions. Included on the graphs of Fig. 4.5 are bars above and below the sample mean values, representing plus and minus $t_{[N-1]} \cdot s$, where s is the sample standard deviation (note that it is the standard deviation rather than the standard error that is involved here). It would be expected that 95% of all sample values would lie within these bands; but the present examples show a higher percentage inclusion (platykurtic distributions?), as for birch leaf 6 only 4 points out of 150 are excluded (3%) and for sunflower leaf pair 3 only 2 out of 108 observations are excluded (2%). However, the overall numbers of observations involved are relatively small, and values of 97% and 98% of points included may not be unreasonable in relation to the hypothesis of normality; clearly, however, a more objective assessment is necessary.

The Kolmogorov-Smirnov test

The Kolmogorov-Smirnov test for rankable scores is a non-parametric method which compares the observed sample distribution of replicated points around their mean with that expected if the distribution was normal, the null hypothesis being that there is no difference between the two distributions. The procedure is described, *inter alia*, by Meddis (1975).

For a given harvest, i, firstly the mean and standard deviation of the sample observations are calculated. The observations themselves are then transformed to standard normal deviates in the usual way:

$$z_{ij} = \frac{l_{ij} - \bar{l}_i}{s_i} \tag{4.82}$$

The z-values are ranked in ascending order, and two further quantities – the observed cumulative proportion (P_j), and the expected cumulative proportion (E_j) – are obtained for each ranked z-value. Each observed cumulative proportion is given simply by

$$P_j = j/N \tag{4.83}$$

where j is the ranked replicate number, and N is the total number of observations in the sample, and the expected cumulative proportion corresponding to each z_{ij} is obtained from a standard table of the normal deviate. Finally, compute

$$d_j = P_j - E_j \tag{4.84}$$

for all j, and the largest absolute value of d_j, D, is assessed by special tables (e.g. Meddis, 1975, page 63); a value of D smaller than the tabulated value does not indicate the occurrence of a rare event on the basis of the null hypothesis.

A large number (> 100) of single leaf growth samples were tested for normality using this technique; however, not one reached significance, and therefore only the following four examples are included here: sunflower leaf pair 2 for day 3 (Table 4.3); sunflower leaf pair 3 for day 15 (Table 4.4), already discussed above; birch leaf 3 for day 21 (Table 4.5); and birch leaf 6 for day 35 (Table 4.6), again discussed above. Although the results of the Kolmogorov-Smirnov test are all non-significant, a little reflection as to how the common deviations from normality (skewness and kurtosis) would affect the d-values enables us to visualize trends in the sample values. Thus, the two sunflower data sets show a tendency to positive skewness, while birch leaf 3 at day 21 shows a lesser tendency to negative skewness, and birch leaf 6 at day 35 data set is slightly platykurtic. However, there is no evidence to suggest that assumption (a) in the Richards function estimation model is untenable.

The failure of any of this large number of test statistic values to reach significance is worrying, however, since it suggests that the test may not be working properly; one would expect about 5% of the results to be significant purely by chance. The test assumptions and procedures given by Meddis (1975) were completely adhered to, so we are unable to offer an explanation for this anomaly.

Assessing the assumption of constant variance

The assumption has been made that, although untransformed growth data are heteroscedastic in that the variance of the populations increases with the size of the attribute measured, logarithmic transformation of the data will change the variability to a homoscedastic situation. Graphical examination of a typical set of leaf growth data (sunflower leaf pair 4, Fig. 4.6), however, indicates that this may not be generally true; the assumption regarding untransformed data appears to be valid (Fig. 4.6a,b), but not that of the logarithmically transformed data (Fig. 4.6c,d). Because of this indication that the logarithmically transformed leaf growth data may not be homoscedastic, a more rigorous test was conducted on a wider range of data.

Bartlett's test of homogeneity of variance

Bartlett's (1937) test assesses whether a given set of samples could conceivably have been drawn from underlying populations having the same variance. A set of data in our present context comprises the h samples (one at each harvest), each of N replicates, of growth data for a particular leaf;

Table 4.3 Results of the Kolmogorov-Smirnov test for normality of growth data (logarithmically transformed). Sunflower leaf pair no. 2 at day 3, dry weight.

$\bar{x} = -9.6819$ $s = 0.4679$ $N = 9$

Original values	z	Ranked z	P	E	d
−10.1519	−1.0045	−3.1955	0.1111	0.0010	0.1011
−10.0778	−0.8461	−1.0045	0.2222	0.1587	0.0635
−9.0364	1.3796	−0.8461	0.3333	0.1977	0.1356
−9.4088	0.5837	−0.4314	0.4444	0.3336	0.1108
−9.6566	0.0541	0.0541	0.5555	0.5199	0.0356
−9.8837	−0.4313	0.5837	0.6666	0.7190	−0.0524
−10.3815	−3.1955	0.7102	0.7777	0.7610	0.0167
−9.3496	0.7102	1.0502	0.8888	0.8531	0.0357
−9.1908	1.0502	1.3796	1.0000	0.9146	0.0854

$D = 0.1356$ N.S.

Table 4.4 Results of the Kolmogorov-Smirnov test for normality of growth data (logarithmically transformed). Sunflower leaf pair no. 3 at day 15, dry weight.

$\bar{x} = -3.1910$ $s = 0.3777$ $N = 9$

Original values	z	Ranked z	P	E	d
−3.0345	0.9439	−1.4194	0.1111	0.778	0.0333
−3.7339	−0.9077	−1.0326	0.2222	0.1515	0.0707
−3.5972	−0.5459	−0.9077	0.3333	0.1814	0.1519
−2.9662	1.1247	−0.5459	0.4444	0.2946	0.1498
−3.9271	−1.4194	0.1324	0.5555	0.5517	0.0039
−3.1797	0.5594	0.5594	0.6666	0.7122	−0.0456
−3.3410	0.1324	0.9439	0.7777	0.8364	−0.0587
−3.7810	−1.0236	1.1247	0.8888	0.8686	0.0202
−2.9584	1.1427	1.1427	1.0000	0.8728	0.0272

$D = 0.1519$ N.S.

Fig. 4.6c illustrates one such set. The test statistic is

$$\chi^2_{[h-1]} = (Nh - 1)(h \, . \log_e \overline{s^2} - \sum_{i=1}^{h} \log_e s_i^2)$$

where $\overline{s^2} = (1/h) \, . \sum_{i=1}^{h} s_i^2$, and s_i^2 is the variance of the sample taken at the ith

Table 4.5 Results of the Kolmogorov-Smirnov test for normality of growth data (logarithmically transformed). Birch leaf no. 3 at day 21, dry weight.

$\bar{x} = -5.9548$ $s = 0.8076$ $N = 10$

Original values	z	Ranked z	P	E	d
−6.9280	−1.1903	−1.6394	0.1	0.0505	0.0495
−5.8091	0.1804	−1.1903	0.2	0.1170	0.0830
−5.5940	0.4468	−0.3653	0.3	0.3557	−0.0557
−4.7307	1.5157	0.1804	0.4	0.5714	−0.1714
−5.0768	1.0872	0.2369	0.5	0.5910	−0.0910
−5.5090	0.5520	0.4468	0.6	0.6740	−0.0740
−6.2503	−0.3658	0.5520	0.7	0.6985	−0.0015
−7.2788	−1.6394	1.0872	0.8	0.8621	−0.0621
−5.7635	0.2369	1.0964	0.9	0.8670	0.0330
−6.6077	1.0964	1.5157	1.0	0.9400	0.0600

$D = 0.1714$ N.S.

Table 4.6 Results of the Kolmogorov-Smirnov test for normality of growth data (logarithmically transformed). Birch leaf no. 6 at day 35, dry weight.

$\bar{x} = -4.9957$ $s = 1.3810$ $N = 10$

Original values	z	Ranked z	P	E	d
−4.6387	0.2585	−2.2663	0.1	0.0116	0.0884
−4.9009	0.0686	−0.3051	0.2	0.3783	−0.1783
−5.4171	−0.3051	−0.1079	0.3	0.4563	−0.1562
−4.8783	0.1174	0.0686	0.4	0.5279	−0.1279
−4.0570	0.6797	0.1174	0.5	0.5438	−0.0438
−3.8490	0.8303	0.2585	0.6	0.6026	−0.0026
−4.5953	0.2900	0.2900	0.7	0.6141	0.0859
−8.6226	−2.2663	0.6797	0.8	0.7517	0.0483
−3.8538	0.8269	0.8269	0.9	0.7939	0.1061
−5.1447	−0.1079	0.8303	1.0	0.7939	0.2061

$D = 0.2061$ N.S.

harvest. The null hypothesis is that $\sigma_i^2 = \sigma_j^2, i \neq j$, giving a small value of χ^2, and so the test is one-tailed.

Results for sunflower and wheat are given in Table 4.7. Most of the χ^2-values are very highly significant, providing substantial evidence that: (a) for the data sets examined, the assumption of homoscedasticity must be discarded; and (b) that the heteroscedastic structure of single leaf growth

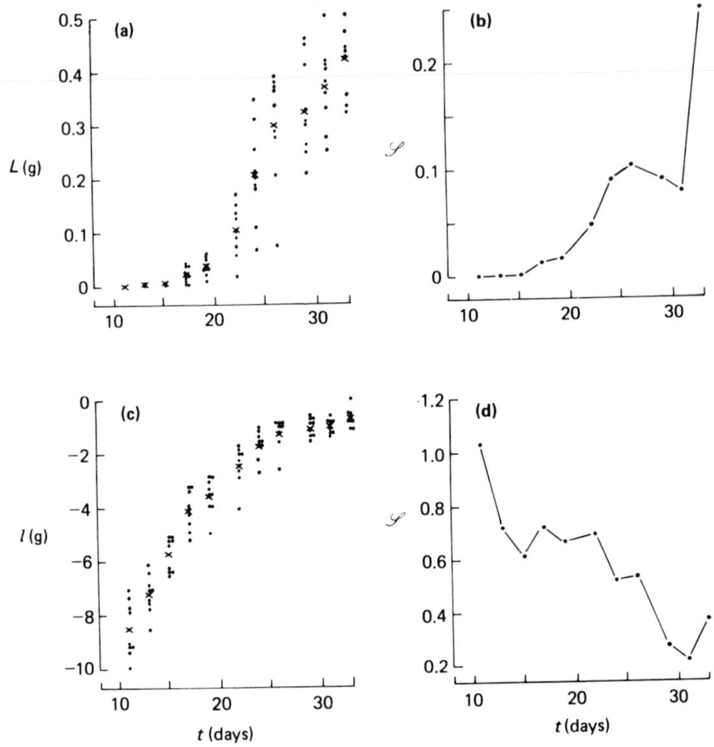

Fig. 4.6 Primary data for the weight of sunflower leaf pair 4 against time: (**a**) untransformed data points with mean values, x, (**b**) standard deviations of untransformed data, (**c**) $\log_e$ (transformed data) with mean values, x, and (**d**) standard deviations of transformed data.

Table 4.7 Results of Bartlett's test of variance homogeneity applied to sunflower and wheat leaves (dry weight).

Leaf pair number	Sunflower χ^2	Degrees of freedom	Leaf number	Wheat χ^2	Degrees of freedom
1	39.8***	15	1	27.2*	16
2	41.0***	14	2	60.3***	16
3	33.4***	11	3	76.7***	14
4	23.8**	9	4	87.8***	11
5	24.4**	9	5	20.4**	7
6	6.7	8	6	34.4**	5
7	7.7	5			
8	4.2	5			

data is widespread, if not universal. Thus, the estimation method for the Richards function already described is not strictly applicable. In the few cases where the χ^2-value is not significant only a small part of the growth curve is covered by the data.

Before leaving this section, it should be pointed out that Bartlett's test is very susceptible to non-normality of the data. However, the fact that the Kolmogorov-Smirnov test nowhere detected departure from normality among our data, together with the predominantly significant χ^2-values obtained in the Bartlett's test, gives considerable confidence in our assertion of a heterogeneous variance structure.

A supplementary experiment

In an attempt to categorize the heteroscedastic nature of individual leaf growth data more precisely, a small supplementary experiment was undertaken using the same batch of plants throughout, thus avoiding the complications inherent in destructive sampling at each harvest time. This necessitated the use of leaf area as growth attribute. A different species, *Hordeum vulgare* cv Zephyr (a uniculm barley variety) was chosen: firstly, because Kemp (1960) has proposed a formula for the estimation of the area of a grass leaf from length and width measurements only; secondly, a large grass species, such as a cereal, is decidedly advantageous; thirdly, a uniculm (i.e. non-tillering) plant having a simple shoot morphology simplifies operations involving repeated handling, and seeds of this uniculm barley cultivar were to hand.

Twenty plants were grown in a controlled environment cabinet for a month, and every one or two days the length and width of each leaf on every plant was measured. These measurements were converted into area estimates, using Kemp's (1960) formula:

$$L_{\mathrm{A}kij} = 0.905fb \tag{4.85}$$

where $L_{\mathrm{A}kij}$ is the estimated area of the kth leaf on the jth replicate plant at the ith harvest, and f and b are the length and width, respectively, of that particular leaf. Results are shown in Fig. 4.7, from which it can be inferred that the change of standard deviation with time may follow a rectangular hyperbola, and that the change of standard deviation with $l_{\mathrm{A}ki}$ may be linear. Possible uses of these results will be discussed later (page 124).

The Richards function with heteroscedastic data

The weighted maximum likelihood estimation of the function

Reviewing the four assumptions for the previous method of fitting the Richards function to data (page 100), we see that all the assumptions hold

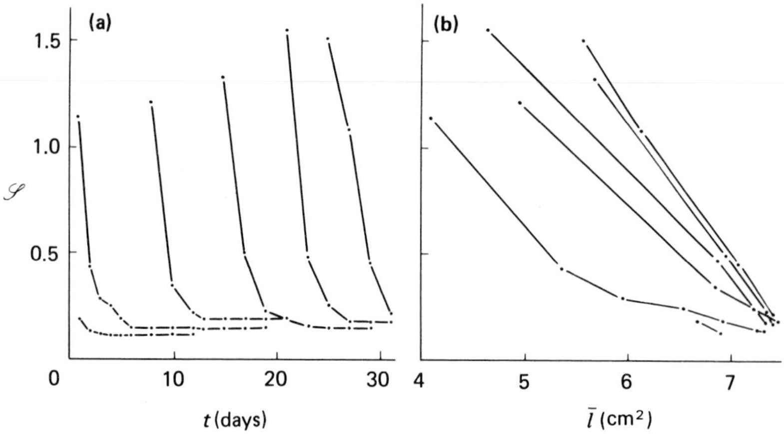

Fig. 4.7 Standard deviations of leaves from a continuous sampling barley experiment (var. uniculm Zephyr). (**a**) Standard deviation against time and (**b**) against $\bar{l}$ (see text for details).

for individual leaf growth data, at least approximately, except (c). Fortunately, in principle, it is easy to adjust the estimation method to allow the fitting of a function to heteroscedastic data, and we now show how this is done for the Richards function using the method of maximum likelihood. Assumptions (a), (b), and (d) are assumed to be unchanged, and the new assumption reads:

(c) the population variance of l-values at the ith harvest is σ_i^2, and $\sigma_i^2 \neq \sigma_j^2$, $i \neq j$.

Again, we shall consider the situation where there is only one observation of l at each of h harvests.

The likelihood of the h observations is

$$L = \frac{1}{(2\pi)^{h/2} \cdot \prod_{i=1}^{h} \sigma_i^2} \cdot \prod_{i=1}^{h} \exp\left[-\frac{1}{2} \left\{ \frac{l_i - \alpha + (1/v) \cdot \log_e(1 \pm e^{(\beta + Kt_i)})}{\sigma_i} \right\}^2 \right]$$

and so (4.86)

$$\log_e L = -\frac{h}{2} \cdot \log_e(2\pi) - \frac{1}{2} \cdot \sum_{i=1}^{h} \log_e \sigma_i^2 - \frac{1}{2} \cdot \sum_{i=1}^{h} \frac{\{l_i - \alpha + (1/v) \cdot \log_e(1 \pm e^{(\beta + Kt_i)})\}^2}{\sigma_i^2} \quad (4.87)$$

Partial differentiation of (4.87) with respect to α, β, K, v, and each of the σ_i^2

in turn, and suppressing the summation limits gives:

$$\frac{\partial(\log_e L)}{\partial\alpha} = \sum\left\{l_i - \alpha + \frac{1}{v}.\log_e(1 \pm e^{(\beta+Kt_i)})\right\}\frac{1}{\sigma_i^2} \tag{4.88}$$

$$\frac{\partial(\log_e L)}{\partial\beta} = \mp\frac{1}{v}.\sum\left\{l_i - \alpha + \frac{1}{v}.\log_e(1 \pm e^{(\beta+Kt_i)})\right\}\left\{\frac{e^{(\beta+Kt_i)}}{(1 \pm e^{(\beta+Kt_i)})\sigma_i^2}\right\} \tag{4.89}$$

$$\frac{\partial(\log_e L)}{\partial K} = \mp\frac{1}{v}.\sum\left\{l_i - \alpha + \frac{1}{v}.\log_e(1 \pm e^{(\beta+Kt_i)})\right\}\left\{\frac{e^{(\beta+Kt_i)}}{(1 \pm e^{(\beta+Kt_i)})\sigma_i^2}\right\} \tag{4.90}$$

$$\frac{\partial(\log_e L)}{\partial v} = \frac{1}{v^2}.\sum\left\{l_i - \alpha + \frac{1}{v}.\log_e(1 \pm e^{(\beta+Kt_i)})\right\}\left\{\frac{\log_e(1 \pm e^{(\beta+Kt_i)})}{\sigma_i^2}\right\} \tag{4.91}$$

$$\frac{\partial(\log_e L)}{\partial\sigma_i^2} = -\frac{1}{2\sigma_i^2} + \frac{\left\{l_i - \alpha + (1/v).\log_e(1 \pm e^{(\beta+Kt_i)})\right\}^2}{2\sigma_i^4} \quad i = 1, \ldots, h \tag{4.92}$$

Apart from the usual difference between least squares partial derivatives (containing the factor -2) and maximum likelihood partial derivatives (containing the factor $1/\sigma^2$), the only difference between the set of equations (4.44) to (4.47) on the one hand and the set (4.88) to (4.91) on the other is that the latter equations all contain the factor $1/\sigma_i^2$, whereas the former do not. This means that the factor $1/\sigma_i^2$ does not disappear when the partial derivatives (4.88) to (4.91) are equated to zero; whereas if $\sigma_i^2 = \sigma_j^2$ for all i, j, then σ_i^2 in the above partial derivatives *is* a constant, σ^2, and would disappear when the right-hand sides of equations (4.88) to (4.91) are equated to zero.

In terms of the method of least squares, we are minimizing

$$\mathscr{S}^2 = \sum\left\{l_i - a + \frac{1}{n}.\log_e(1 \pm e^{(b+Kt_i)})\right\}^2.\frac{1}{\sigma_i^2} \tag{4.93}$$

In other words, each l_i is being 'weighted' inversely to its variance. This is sensible, because if an observation comes from a population with a small variance the observation is likely to be near the population mean, whereas an observation from a population with a large variance is less likely to be near to the population mean; therefore, more 'notice' should be taken of the former observation than the latter, and this is done by giving the former observation a higher weighting than the latter.

There is, however, one important difference between (4.93) and (4.43) in that the former contains variance terms whereas the latter does not. In order to make the magnitudes of the two expressions comparable, the right-hand side of (4.93) is multiplied through by $\overline{\sigma^2}$ ($= h^{-1} . \Sigma\sigma_i^2$). Indeed, we found in practice that without such a scaling factor (in the form of $\overline{s^2}$) the least squares iterations algorithm failed to converge.

On equating (4.92) to zero, we find that

$$\hat{\sigma}_i^2 = \{l_i - \hat{\alpha} + (1/\hat{\nu}) . \log_e(1 \pm e^{(\hat{\beta} + \hat{\kappa} t_i)}\}^2 \quad (4.94)$$

which is simply the square of the deviation of l_i from the fitted curve, $\hat{l}_i$. This is almost certainly a biased estimateof σ_i^2 and, in any case, would be highly unreliable as it is based on only one observation. On the other hand, with replicated data at each harvest, we have

$$\hat{\sigma}_i^2 = \frac{1}{N} . \sum_{j=1}^{N} \left\{ \overline{l}_i - \hat{\alpha} + (1/\hat{\nu}) . \log_e(1 \pm e^{\hat{\beta} + \hat{\kappa} t_i}) \right\}^2 \quad (4.95)$$

which gives an estimate (although biased) of σ_i^2; division by $N-4$ instead of N would give an unbiased estimate. Obtaining the variances and covariances of the parameter estimates is the same as before: each second order partial derivative contains the factor $\overline{\sigma^2}/\sigma_i^2$, and so there is no common factor of $\overline{\sigma^2}/\sigma^2$ as a scalar multiplier of the information matrix.

The practical application of the above minimization requires estimates of σ_i^2. An obvious course of action, when there are several l-values at each harvest, is to replace σ_i^2 by s_i^2 which is calculated from the data of the ith harvest. This is an attractive method since the data at each harvest provide their own precision, but a possible disadvantage is that sample variances of l_{ij} fluctuate from harvest to harvest about some underlying 'true' trend because of sampling variation; thus the latter will be superimposed on what otherwise may be a steady change of population variance as the leaf grows.

The alternative is to assume a plausible relationship between the variance of the dependent (or response) variate and the independent (or regressor) variable. A possible relationship was suggested on page 121 from experimental evidence. If we modify that suggestion slightly, and assume that the change of variance of l with time follows a rectangular hyperbola, then the simplest adequate relationship would be

$$s^2 = \frac{1}{t} + c \quad (4.96)$$

where c is the minimum variance of l at the asymptotic end of the $l(t)$ curve. Hence, we have

$$\frac{1}{s_i^2} = \frac{t_i}{1 + ct_i} \quad (4.97)$$

and this function of time could replace $1/\sigma_i^2$ in the estimation equations. An even simpler alternative would be to have $s^2 = 1/t$ and so $1/s_i^2 = t_i$; thus, t_i would replace $1/\sigma_i^2$ in the estimation process. It would also be possible to assume that the variance of l is a function of l itself; the suggestion was that standard deviation is a declining linear function of l, say $s = a - bl$, where a and b are simply constants of this linear relationship and are nothing to do with parameter estimates of associated Richards functions; then

$$\frac{1}{s_i^2} = \frac{1}{(a - bl_i)^2} \tag{4.98}$$

but simpler procedures would be to assume that $1/s_i^2 = l_i^2$ or even $1/s_i^2 = l_i$, but this last does imply a change in the basic assumption of a linear relationship between s_i and l_i. A complication of weighting using a relationship involving the response variate arises when there are several replicate l-values at each harvest. Here, should l_{ij} or $\overline{l}_i$ be substituted for l_i in a relationship such as (4.98)? There is no clear answer to this question, although there is nothing to suggest that the largest replicate at any one time is more reliable than the other replicates and weighting with l_{ij} does imply this.

Because of the various difficulties and approximations involved with any form of systematic weighting, it is probably best to use $1/s_i^2$ as the weighting factor, since in practice we always have several replicate l-values at each harvest with which to estimate population variances at those times. Sampling fluctuations in s_i^2 along the curve will be smoothed out by the curve fitting process from the point of view of quantities such as the variances of the $\hat{l}_i$, whose magnitudes partly depend on the population variance of the l_{ij}. Thus, we shall solely use this weighting method in the remainder of the book.

Finally, there is the assessment of the appropriateness of the Richards function fitted by weighting to heteroscedastic data where there are N replicate l-values at each harvest. The scheme is shown in Table 4.8, and it is essentially the same as that shown in Table 3.1 for the linear function, but with certain of the sums of squares expressed in different ways. For the deviations sum of squares, we have utilized the fact that $\mathscr{S}^2$, which is given by

$$\mathscr{S}^2 = \overline{\sigma^2} \, . \sum_{i=1}^{h} \sum_{j=1}^{N} (l_{ij} - \hat{l}_i)^2 \, . \frac{1}{\sigma_i^2} \tag{4.99}$$

where $\hat{l}_i = \hat{\alpha} - (1/\hat{\nu}) \, . \log_e(1 \pm e^{(\hat{\beta} + \hat{\kappa} t_i)})$, is the sum of the deviations and the within groups sums of squares (page 80); and then the regression sum of squares is the between groups minus the deviations sums of squares. The remainder of the table and the assessments of the F-ratios are then as before, but bearing in mind that the ratios of the mean squares may be only approximately distributed as F (but see page 109).

Table 4.8 Scheme for assessing the appropriatness of the Richards function fitted by weighting to heteroscedastic data. Compare with Table 3.1.

	Sum of squares	Degrees of freedom	Mean Square	F
Between groups	$\overline{s^2} N \cdot \sum_{i=1}^{h} (\overline{l}_i - \overline{l})^2 \cdot \frac{1}{s_i^2}$	$h-3$	M_B	$\frac{M_B}{M_W}$
Regression	$\overline{s^2} \left\{ \sum_{i=1}^{h} \sum_{j=1}^{N} (l_{ij} - \overline{l})^2 \cdot \frac{1}{s_i^2} - \sum_{i=1}^{h} \sum_{j=1}^{N} (l_{ij} - \hat{l}_i)^2 \cdot \frac{1}{s_i^2} \right\}$	3	M_R	$\frac{M_R}{M_W}$
Deviations	$\overline{s^2} \left\{ \sum_{i=1}^{h} \sum_{j=1}^{N} (l_{ij} - \hat{l}_i)^2 \cdot \frac{1}{s_i^2} - \sum_{i=1}^{h} \sum_{j=1}^{N} (l_{ij} - \overline{l})^2 + N \cdot \sum_{i=1}^{h} (\overline{l}_i - \overline{l})^2 \cdot \frac{1}{s_i^2} \right\}$	$h-4$	M_D	$\frac{M_D}{M_W}$
Within groups	$\overline{s^2} \left\{ \sum_{i=1}^{h} \sum_{j=1}^{N} (l_{ij} - \overline{l})^2 \cdot \frac{1}{s_i^2} - N \cdot \sum_{i=1}^{h} (\overline{l}_i - \overline{l})^2 \cdot \frac{1}{s_i^2} \right\}$	$h(N-1)$	M_W	—
TOTAL	$\overline{s^2} \cdot \sum_{i=1}^{h} \sum_{j=1}^{N} (l_{ij} - \overline{l})^2 \cdot \frac{1}{s_i^2}$	$Nh-1$	–	—

A simulation study to assess the performance of weighted and non-weighted estimations with heteroscedastic data

This study was similar to the one carried out on homoscedastic data (pages 109–113); the differences being that only one underlying relationship was used here (sunflower leaf pair 3, dry weight), weighted and non-weighted estimation procedures were compared, and the standard deviation of the generated data was made to vary along the length of the curve. To achieve this last condition, the standard deviation adjustments to the normal deviates at the ith harvest were made by using the equation

$$s_i = 0.083 - 0.105\lambda_i \tag{4.100}$$

This relationship was arrived at by assuming that the standard deviation equalled unity at the first harvest, and declined linearly to 0.2 at the final harvest; this appeared to be realistic when experimental data were examined (Figs. 4.6 and 4.7). One example of a set of generated data is shown in Fig. 4.8, together with their mean and underlying Richards function values

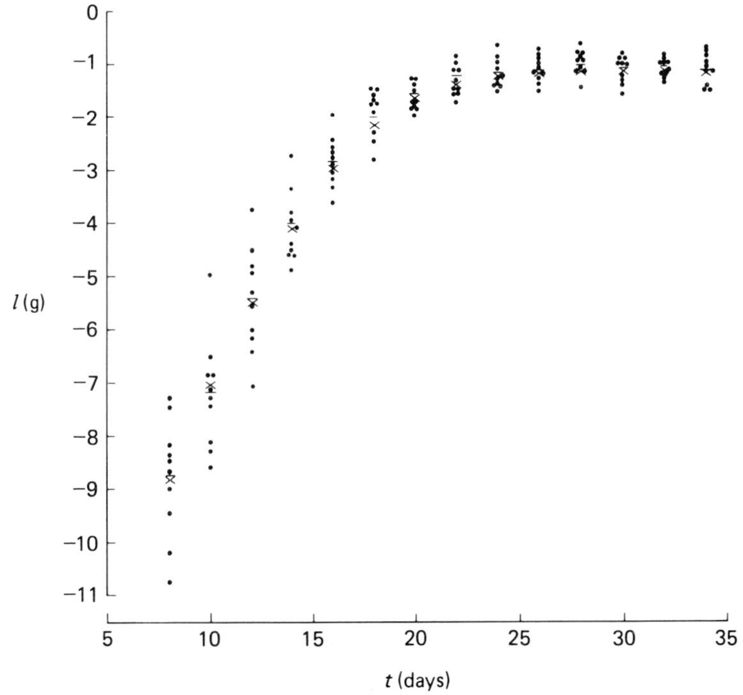

Fig. 4.8 An example of simulated data for the investigation of the weighted least squares Richards function fitting method: individual points (•), means of the simulated data (—) and the underlying relationship values (×).

at each harvest; and the same data set together with details of the two Richards functions, using weighted and non-weighted estimation methods, are given in Table 4.13 (page 139).

Table 4.9 Parameter estimates, their variances, and the variances of the estimated Richards functions in a simulation study, using heteroscedastic data generated from one underlying function (sunflower, c.f. Table 4.1). Each figure is a mean value obtained from 200 sets of generated data.

	ACTUAL	Non-weighted estimation procedure	Weighted estimation procedure
Parameter estimates			
a	−1.114	−1.100	−1.117
b	6.430	6.894	6.533
K	−0.3898	−0.391	−0.396
n	0.4395	0.4375	0.4491
Variances of estimates			
a		0.002369	0.000470
b		0.9194	0.5852
K		0.002094	0.001137
n		0.006318	0.004988
Variance of curve		0.2098	0.0653

Results are shown in Tables 4.9 and 4.10, and in Fig. 4.9. Table 4.9 shows that the parameter estimates given by the weighted method are not quite as close to the original as those given by the unweighted method, except for parameter a. On the other hand, we see in Fig. 4.9 that both sets of estimates are in reality very close to the underlying values, and it is evident that the difference between the two estimates of a particular parameter is quite insignificant when compared with their confidence intervals. This conclusion is strengthened by the fact that the first 108 sets of simulated data gave the following results when estimated by the weighting procedure: $a = -1.116$, $b = 6.442$, $K = -0.3916$, and $n = 0.4417$; these were the best set of estimates obtained, taken as a whole. By contrast, the remaining 92 sets of simulated data gave the worst estimates of the underlying parameters, and this shows that the results appearing in Table 4.9 are due to the vagaries of sampling variability. The variances of the estimated parameters and of the estimated function are all consistently smaller by the weighting method (Table 4.9).

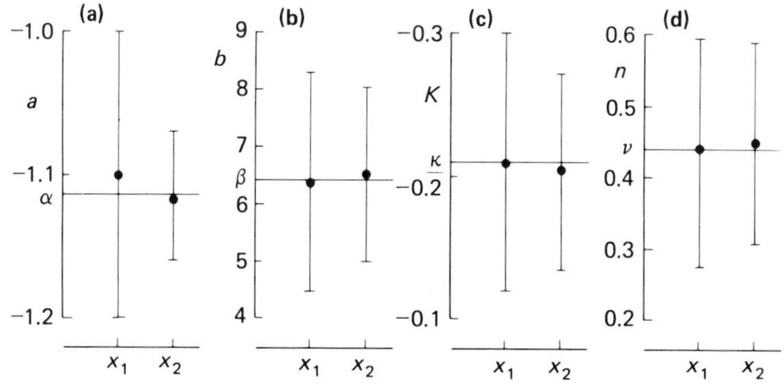

Fig. 4.9 Comparison of Richards function parameter estimates by non-weighted (x_1) and weighted (x_2) methods with underlying values indicated by a horizontal line.

In Table 4.10, we see that there are no significant χ^2-values associated with the parameter estimates and their variances, except for a which seems extraordinarily abnormal. Using the weighting procedure, we find that the variance of the curve and the within groups sum of squares both have distributions very markedly different from a χ^2-distribution, resulting from pathological distributions of values, see p. 112; this being so, it is scarcely surprising that the number of significant F-ratios is as high as 10%. The F-statistic must be totally useless in this situation.

How do the results of this simulation study relate to the practical situation of estimating the Richards function with real data? We have already established that, in general, leaf growth data are heteroscedastic (Table 4.7), and so the weighted estimation method is the appropriate one to use. On the other hand, the parameter estimates appear to be not significantly different between the two methods, but the variances of their estimates are consistently lower by the weighted procedure. Further, the distribution of a seems to be markedly anormal by the non-weighted method, but against this is the fact that the distributions of the variance of the curve and the within groups sum of squares are singular by the weighted method, thus rendering the F-statistic unusable as a test of appropriateness of the Richards function for a given set of data. Apart from the fact that the weighted method is statistically the correct method to use for leaf growth data, it would seem that there is little to choose between either approach. If the unweighted method is used, the confidence interval of a must be regarded with suspicion, whereas no testing of the function's appropriateness can be undertaken in connection with the weighting method. It is, of course, necessary to reiterate that this is just one simulation study, and using a relationship in which ν is low. While the conclusions may tentatively be

Table 4.10 Properties of the distributions of the parameter estimates and their variances for the fitted Richards functions in a simulation study, using heteroscedastic data generated from one underlying function (sunflower, c.f. Table 4.2.). Each figure is based on 200 sets of generated data.

	Non-weighted estimation procedure			Weighted estimation procedure		
	$\chi^2_{[18]}$	Skewness	Kurtosis	$\chi^2_{[18]}$	Skewness	Kurtosis
a	630.4***	13.09	173.80	10.99	0.056	0.067
b	15.2	0.11	−0.13	28.62	0.513	0.515
K	24.3	−1.35	6.53	19.57	−0.613	0.568
n	18.7	0.12	−0.28	26.00	0.632	0.542
$\mathscr{V}(a)$	27.2	1.00	1.68	11.70	0.687	1.473
$\mathscr{V}(b)$	16.6	0.87	0.85	18.00	0.735	1.920
$\mathscr{V}(K)$	22.4	0.94	1.14	23.27	1.135	1.915
$\mathscr{V}(n)$	21.1	0.72	0.50	22.46	0.944	1.482
Variance of curve †	17.4	0.50	0.66	309.2***	9.639	93.81
Deviations sum of squares	26.5	0.20	6.57	10.86	0.559	−0.029
Within groups sum of squares	22.7	0.48	0.41	336.1***	9.542	90.91
Number of significant F-ratios		4.8 %			10 %	

† The variances of curve as estimated by the two fitting methods are not strictly comparable; that for the non-weighted method being estimated as $\frac{1}{Nh-4}\left\{\sum_{i=1}^{h}\sum_{j=1}^{N}(l_{ij}-\hat{l}_i)^2\right\}$, whereas for the weighted fitting the quantity $\frac{\bar{s}^2}{Nh-4}\left\{\sum_{i=1}^{h}\sum_{j=1}^{N}(l_{ij}-\hat{l}_i)^2\frac{1}{s_i^2}\right\}$ has been used.

regarded as correct for Richards functions with a low value of v, they may well be untrue for functions with high values of this parameter. In the previous simulation study with homoscedastic data (Tables 4.1 and 4.2), quite different results were obtained according to whether the underlying function had a low or a high value of v; thus, it would be most unwise to attempt an extrapolation of our present results to a situation involving a Richards function with a high value of v.

Comparison of the weighted and non-weighted methods with leaf growth data

To provide a little more information about the applicability of the two approaches to Richards function estimation of leaf growth data, we give two instances where a comparison has been made. These examples are sunflower leaf pair no. 3, dry weight, and wheat leaf no. 5, area, both of whose non-weighted estimated parameters have already been used as actual parameter values in the previous simulation studies. The forms of the growth curve differ quite markedly between the two data sets, in that the sunflower example gives rise to low values of n and that of wheat high values of n (Tables 4.11a and 4.12a).

When experimental data are analysed, the underlying relationship is not known, and so the interpretation of comparisons of fitting methods is more difficult. However, in common with the simulation study, the variances of the parameter estimates, the variance of the fitted function, and the value of a are all lower in the weighted estimation procedure than in the non-

Table 4.11 Sunflower leaf pair no. 3, dry weight. Comparison of Richards function estimations, using non-weighted and weighted estimation procedures.

(a) Parameter estimates and their variances

Procedure	a	b	K	n	$\mathcal{V}(a)$	$\mathcal{V}(b)$	$\mathcal{V}(K)$	$\mathcal{V}(n)$	Variance of curve
Non-weighted	−1.114	6.430	−0.389	0.439	0.0068	2.935	0.0065	0.0213	0.2089
Weighted	−1.072	5.482	−0.348	0.358	0.0025	2.084	0.0039	0.0149	0.1108

(b) Data and curve details

t(days)	s_i^2	$\bar{l}_i$	$\hat{l}_i$ unweighted	$\hat{l}_i$ weighted	$\mathcal{V}(\hat{l}_i)$ unweighted	$\mathcal{V}(\hat{l}_i)$ weighted
9	0.1982	−7.87	−7.88	−7.89	0.0195	0.0203
10			−7.05	−7.01	0.0101	0.0131
11	0.5908	−6.19	−6.24	−6.18	0.0084	0.0137
12			−5.46	−5.39	0.0093	0.0156
13	0.7554	−4.88	−4.73	−4.65	0.0099	0.0152
14			−4.06	−3.99	0.0090	0.0124
15	0.1427	−3.39	−3.45	−3.40	0.0077	0.0089
16			−2.92	−2.89	0.0075	0.0066
17	0.1904	−2.34	−2.48	−2.47	0.0084	0.0060
18			−2.12	−2.12	0.0095	0.0063
19	0.1516	−1.88	−1.84	−1.85	0.0097	0.0066
20			−1.63	−1.64	0.0086	0.0062
21			−1.48	−1.49	0.0068	0.0053
22	0.1742	−1·57	−1.37	−1.37	0.0052	0.0041
23			−1.29	−1.29	0.0041	0.0030
24	0.1269	−1.21	−1.23	−1.23	0.0036	0.0022
25			−1.19	−1.18	0.0036	0.0016
26	0.0570	−1.07	−1.17	−1.15	0.0039	0.0014
27			−1.15	−1.13	0..043	0.0014
28			−1.14	−1.11	0.0048	0.0014
29	0.0891	−1.18	−1.13	−1.10	0.0052	0.0016
30			−1.13	−1.09	0.0056	0.0017
31	0.0612	−1.17	−1.12	−1.09	0.0053	0.0019
32			−1.12	−1.08	0.0061	0.0020
33	0.0460	−1.01	−1.12	−1.08	0.0063	0.0021

weighted method. Conversely, the estimates of β, K (absolute value), and $\bar{\nu}$ are lower by the weighted method for both sets of leaf data, but the reverse is true for the simulated data; the differences are much smaller in the simulated data, being based on 200 data sets, than in the leaf data, which comprise only two individual data sets. In Table 4.11b, we see that $\mathscr{V}(\hat{l})$ is greater for low $\hat{l}$-values and smaller for high $\hat{l}$-values in the weighted estimation than in the non-weighted method; this is the expected result. In wheat (Table 4.12), however, only two $\mathscr{V}(\hat{l})$-values in the weighted estimation are higher than their counterparts in the non-weighted case; but the wheat data are less extensive, and the range of sample variances is greater than in the sunflower data.

In terminating these comparisons of non-weighted and weighted estimation methods to heteroscedastic data, we may conclude that the latter procedure should give a better estimate of α since the high weighting is at the asymptotic end of the $l(t)$ function. One cannot be as sanguine about the

Table 4.12 Wheat leaf no. 5, area. Comparison of Richards function estimates, using non-weighted and weighted estimation procedures.
(a) Parameter estimates and their variances

Procedure	a	b	K	n	$\mathscr{V}(a)$	$\mathscr{V}(b)$	$\mathscr{V}(K)$	$\mathscr{V}(n)$	Variance of curve
Non-weighted	2.767	45.49	−2.675	3.206	0.0083	227.4	7.704	12.23	0.1945
Weighted	2.807	28.49	−1.697	1.961	0.0008	222.7	0.725	1.326	0.0317

(b) Data and curve details

t(days)	s_i^2	$\bar{l}_i$	$\hat{l}_i$ unweighted	$\hat{l}_i$ weighted	$\mathscr{V}(\hat{l}_i)$ unweighted	$\mathscr{V}(\hat{l}_i)$ weighted
12	0.1021	−1.33	−1.41	−1.34	0.0296	0.0168
13			−0.58	−0.48	0.0139	0.0121
14	1.078	0.00	0.26	0.38	0.0116	0.0205
15	0.2107	1.27	1.09	1.23	0.0212	0.0296
16			1.90	2.00	0.0283	0.0170
17	0.0443	2.55	2.55	2.53	0.0320	0.0082
18			2.75	2.74	0.0091	0.0038
19	0.0783	2.65	2.77	2.79	0.0079	0.0010
20			2.77	2.80	0.0082	0.0008
21	0.0109	2.84	2.77	2.81	0.0083	0.0008
22			2.77	2.81	0.0083	0.0008
23	0.0211	2.76	2.77	2.81	0.0083	0.0008
24			2.77	2.81	0.0083	0.0008
25			2.77	2.81	0.0083	0.0008
26			2.77	2.81	0.0083	0.0008
27	0.0162	2.83	2.77	2.81	0.0083	0.0008

other parameters since, although the weighted method should *a priori* give better estimates, good estimation is achieved only by having plenty of information (i.e. data) at the lower end of the curve (Causton *et al.*, 1978). In having low weighting at the lower end of the curve, one is in effect lessening the amount of information at the lower end, and on this criterion the estimates of β, K, and ν might be expected to be worse than if weighting is absent. Possibly a balance is achieved in the long run, which is indicated by these estimates in the simulation study being scarcely different from one another and, within the variability of the data and the size of the study, either estimation procedure appeared to give acceptable results. Apart from the fact that the weighted estimation method is the correct one to use with individual leaf growth data, which are mostly heteroscedastic, on balance it seems to be the preferable method in practice because of the lower variances of the estimates obtained, more realistic confidence intervals for the $\hat{l}$ and, possibly, a better sampling distribution of a. We shall use the weighted method for estimating Richards functions for individual leaf growth data in the examples of application in the remainder of this book.

The nature of the heteroscedasticity of individual leaf growth data

Structural and developmental variability

Consideration of experimental individual leaf growth data will reveal the conditions producing heterogeneity of variances. A batch of seed put to germinate under as constant conditions as possible will never all germinate simultaneously; germination may be spread over several days and therefore, during subsequent growth, plants will be at somewhat different developmental stages at any one time. This developmental variation has been reduced in the experiments described in this book by selecting plants, for inclusion in the experiments, that had germinated within a restricted period of time; but the developmental variation still existed, even if only to a limited extent.

When examining a Level 1 or 1a component or entity, such as foliage or whole plant, these developmental differences have little noticeable effect. This is because growth of these parts is indeterminate, and often tends to a linear form with time (exponential) for extended periods. Problems, however, arise with components whose growth is determinate, because size increase is sooner or later curvilinear and ultimately asymptotic; relative growth rates can thus change enormously in magnitude over short periods during the growth of such a component.

Consider the hypothetical situation where the growth of all replicates of a certain single leaf follow the same growth curve. If the developmental stages of all the plants were identical at any chosen time, then the variability of the single leaf at any point during its development would be zero. If, however, the plants are not all at the same developmental stage then the variability of the single leaf growth data will still approach zero as all the replicate individuals approach the asymptotic size. Prior to this stage the variability could be quite considerable, depending on the magnitude of the developmental variability, as the sample will comprise both developmentally young plants where the leaf in question will be small, together with more advanced plants with relatively large leaves, and a range of intermediates. Variability will gradually decrease with time as more replicate individual leaves approach the asymptote. Figure 4.10 shows the expected trends for two magnitudes of developmental variability.

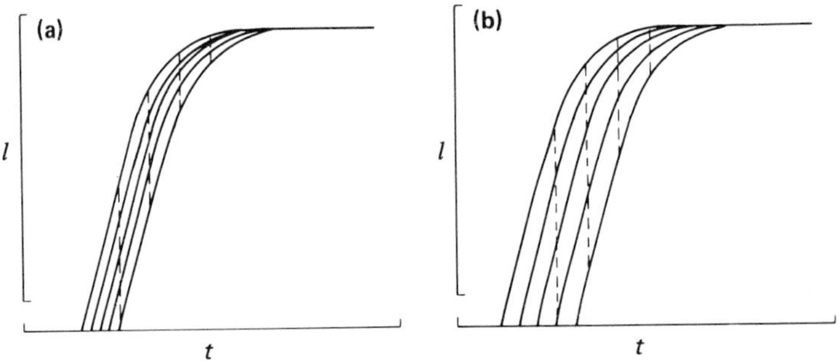

Fig. 4.10 Diagramatic representation of developmental variation when there is no structural variability: (**a**) small developmental variation, and (**b**) a greater amount of developmental variation.

In the hypothetical example just described, we can say that the *structural variability* of the leaf in question was zero and that variability, where present, could be described as *developmental variability*. With real growth data, both structural and developmental variability exist, but the assumption has usually been made that the overall variability is constant. The idea of a constant structural variability is not unreasonable, and in its simplest form can be thought of solely as a difference in α-values between replicate leaf growth Richards function curves. It is more probable, however, that none of these growth curves have identical shapes and that structural variability involves changes in all four parameters, but in such a way that structural variability stays constant.

In practice, therefore, there are four possible ways in which structural and developmental variability may jointly occur.

1. *Structural and developmental variability both constant*

Differences in the variability of the replicated data for the different harvests are produced only because the plants were of different developmental ages at the beginning of the experiment. The developmental ages of the plants do not change in relation to each other, but, as shown in Fig. 4.10 for the special case of constant zero structural variability, as the growth of the replicates of the single leaf progresses the developmental differences between the plants has a progressively reduced effect until, when all replicates have reached maximum size, there are no developmental differences between them. The variability at this stage is thus wholly structural. The change of variability with time under these conditions would resemble that shown in Fig. 4.7a. The simplest situation giving rise to this would be that the Richards function curve of each replicate would be of the same shape (specifically, v and K constant between replicates), but with structural variability occasioned by different α-values, and developmental variability by differing β-values between replicates.

2. Structural variability constant; developmental variability changing

This situation would be similar to the last, except that the developmental ages of the plants would change in relation to one another because the replicate leaves are growing at different rates. The pattern of change of variability with time would still resemble that shown in Fig. 4.7a, but the gradient of the descending portion would differ from the constant variability situation. If the leaves of the plants that were developmentally young at some given initial time grew faster than those which were developmentally older at the same time, then the gradient of the descending portion of the variance-time relationship would be steeper than if all replicates grew at the same rate, and *vice versa.* In terms of the Richards function parameters, changing developmental variability could be accommodated by differing K-values between replicates, leaving only v constant.

3 & 4. Structural variability changing, developmental variability constant or changing

This implies that if all developmental variation was removed, the data would still be heteroscedastic. The variance-time relationship would probably assume a less simple form than implied under 1 and 2 above, and there would seem to be little incentive in attempting to analyse variability changes under these conditions at this stage.

In what follows, the assumption will be made that heteroscedasticity may be attributed primarily to developmental variability, and that structural variability plays a minor role, if any.

The possibility of removing developmental variation

Given a set of single leaf growth data in which heteroscedasticity exists, and that the assumption has been made that heteroscedasticity is due entirely to developmental variability, can the two components of variability be separated? Apart from the intrinsic biological interest in the magnitudes of the two components, such a separation could assist with Richards function estimation, in that if developmental variability could be removed, the data would be homoscedastic and the non-weighted estimation procedure would be the correct method to use.

The separation of structural and developmental variability involves establishing a strategy of operations which aims firstly to define the magnitude of structural variation in the vicinity of the asymptote, and then to move individual data points parallel to the time-axis to eliminate developmental variation. For example, if the leaf in question of a plant at a particular harvest is unduly large in relation to the sample mean at that harvest and with regard to the magnitude of the structural variance, then that plant is considered to be developmentally more advanced than the others in that sample, and so the point on the graph representing the plant's leaf is advanced along the time-axis by an amount determined by the strategy laid down. We now describe a strategy which we have devised for this purpose, present some results of its use, and then discuss the performance and biological adequacy of the strategy.

Method

In this strategy, all operations concern the leaf in question, k, and no reference to any other leaf is made. The strategy can be described by the following steps.

1. The variance of the leaf sample at each harvest is calculated.

2. The structural variance for the leaf is defined as the mean variance of harvests in the asymptotic region. This is found by comparing the average of the variances for all h harvests with the average of the variances of all harvests except the first, i.e. $h-1$ harvests, with that for all harvests except the first two, and so on. If the average of $h-i$ harvest variances is less than that for $h-(i-1)$ and also less than that for $h-(i+1)$ harvests, then the average variance of the $h-i$ harvests is designated as the structural variance, V_s.

3. At each harvest, an interval is defined within which 95% of the data points would be expected to lie if structural variability alone were present. The length of the interval is $2t_{[N-1]}\sqrt{V_s}$ centred on the sample mean at any harvest (Fig. 4.11). Any individual point lying outside this

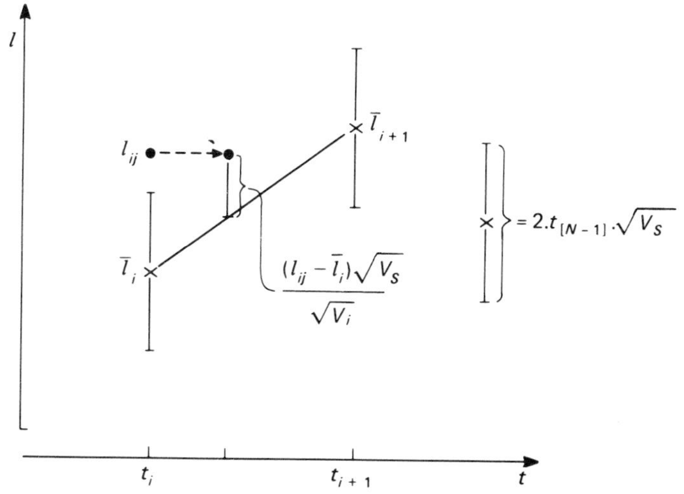

Fig. 4.11 Representation of point-moving technique (see text for details).

interval at a particular harvest is moved to a more 'appropriate' position, using the following scheme.

4. If
$$l_{ij} \geqslant \bar{l}_i + t_{[N-1]} \sqrt{V_s}$$

then the straight line joining $\bar{l}_i$ with $\bar{l}_{i+1}$ is used in the moving procedure. The gradient of the line is given by

$$b = (\bar{l}_{i+1} - \bar{l}_i)/(t_{i+1} - t_i) \tag{4.101}$$

and the assumption is made that the ratio between the standard deviation at harvest i and the structural standard deviation equals the ratio between the actual deviation of l_{ij} from $\bar{l}_i$ and the deviation of l_{ij} from the line joining $\bar{l}_i$ and $\bar{l}_{i+1}$, l_{mij}, after the replicate leaf has been moved, i.e. that

$$\frac{s_i}{\sqrt{V_s}} = \frac{l_{ij} - \bar{l}_i}{l_{ij} - l_{mij}}$$

(Fig. 4.11), and so

$$l_{mij} = l_{ij} - \frac{(l_{ij} - \bar{l}_i)\sqrt{V_s}}{s_i} \tag{4.102}$$

Now since $$b=(l_{mij}-\bar{l}_i)/(t_{mij}-t_i)$$

then $$t_{mij}=\frac{l_{mij}-\bar{l}_i}{b}+t_i \qquad (4.103)$$

Substituting for l_{mij} (4.102) and for b (4.101) in equation (4.103) finally yields

$$t_{mij}=\frac{l_{ij}-\bar{l}_i-\{(l_{ij}-\bar{l}_i)\sqrt{V_s}\}/s_i}{(\bar{l}_{i+1}-\bar{l}_i)/(t_{i+1}-t_i)} \qquad (4.104)$$

5. If $$l_{ij}\leqslant \bar{l}_i-t_{[N-1]}\sqrt{V_s}$$
then the straight line joining $\bar{l}_{i-1}$ and $\bar{l}_i$ is used in a similar moving procedure. In the rare case where $l_{ij}>\bar{l}_h+t_{[N-1]}\sqrt{V_s}$, where $\bar{l}_h$ is the mean leaf size at the last harvest, then that replicate measurement is eliminated from the analysis.

6. Richards function estimation is then carried out by the normal unweighted method.

Results

The point moving method has been applied to three sets of data: (a) one of the generated sets used in the simulation study, illustrated in Fig. 4.8 and further detailed in Table 4.13b; (b) wheat leaf 1, dry weight, illustrated in Fig. 4.12a, b; (c) maize leaf 3, dry weight, illustrated in Fig. 4.12c, d.

For data set (a), results are compared with those obtained by the ordinary non-weighted and weighted methods. The point-moving technique produces inferior results for this data set, compared with the weighted estimation procedure, as may be seen by contrasting the results given with the underlying parameter values and the λ-values at each t. Except for parameter a, the confidence intervals produced by the point-moving method do not even encompass the underlying parameter values. The variance of the curve should be an estimate of the structural variance, 0.04 for these data; the estimate for this example is rather on the low side, despite the fact that the structural variance calculated from the data by method 2 above is 0.0507 (Table 4.13b).

For the two leaf data examples presented, the sample variances at each harvest, together with the calculated structural variances, are shown in Table 4.14. The results of fitting by the point-moving method, as compared with the direct unweighted procedure, are shown in Fig. 4.12 and Table 4.15. For each of these examples, the variance of curve approximates quite closely to the structural variance calculated in Table 4.14. A curious feature is the very high variance of n given by the point-moving technique for wheat leaf

Table 4.13 Comparison of three Richards function estimation methods (non-weighted, weighted, and point-moved) on one set of heteroscedastic data.

(a) Parameter estimates and their variances

Procedure	a	b	K	n	$\mathscr{V}(a)$	$\mathscr{V}(b)$	$\mathscr{V}(K)$	$\mathscr{V}(n)$	Variance of curve
ACTUAL	−1.114	6.430	−0.390	0.440					
Non-weighted	−1.113	7.528	−0.445	0.526	0.0061	3.342	0.0083	0.0227	0.3025
Weighted	−1.113	6.800	−0.411	0.465	0.0009	1.499	0.0029	0.0126	0.0678
Point-moved	−1.144	8.008	−0.454	0.588	0.0008	0.3797	0.0009	0.0026	0.0306

(b) Data and curve details

t_i	λ_i	$\bar{l}_i$	s_i^2	s_{20-34}^2	$\hat{l}_i$ unweighted	$\hat{l}_i$ weighted	$\hat{l}_i$ point-moved	$\mathscr{V}(\hat{l}_i)$ unweighted	$\mathscr{V}(\hat{l}_i)$ weighted	$\mathscr{V}(\hat{l}_i)$ point-moved
8	−8.73	−8.65	1.2090		−8.70	−8.74	−8.98	0.0237	0.0833	0.0023
10	−7.05	−7.16	1.0890		−7.06	−7.05	−7.46	0.0095	0.0309	0.0010
12	−5.46	−5.44	0.9821		−5.49	−5.45	−5.97	0.0112	0.0190	0.0010
14	−4.06	−4.08	0·4401		−4.05	−4.02	−4.57	0.0111	0.0154	0.0011
16	−2.92	−2.82	0.2119		−2.87	−2.87	−3.33	0.0094	0.0079	0.0009
18	−2.12	−1.98	0.1741		−2.03	−2.06	−2.37	0.0111	0.0036	0.0011
20	−1.63	−1.63	0.0481		−1.55	−1.58	−1.75	0.0089	0.0024	0.0010
22	−1.37	−1.35	0.0653		−1.30	−1.33	−1.41	0.0050	0.0014	0.0006
24	−1.23	−1.19	0.0719		−1.19	−1.21	−1.26	0.0038	0.0007	0.0004
26	−1.17	−1.13	0.0411		−1.15	−1.16	−1.19	0.0042	0.0006	0.0005
28	−1.14	−1.00	0.0489	0.0507	−1.13	−1.13	−1.16	0.0050	0.0006	0.0006
30	−1.13	−1.21	0.0297		−1.12	−1.12	−1.15	0.0055	0.0007	0.0007
32	−1.12	−1.12	0.0217		−1.12	−1.12	−1.15	0.0058	0.0008	0.0007
34	−1.12	−1.10	0.0791		−1.11	−1.11	−1.14	0.0060	0.0008	0.0008

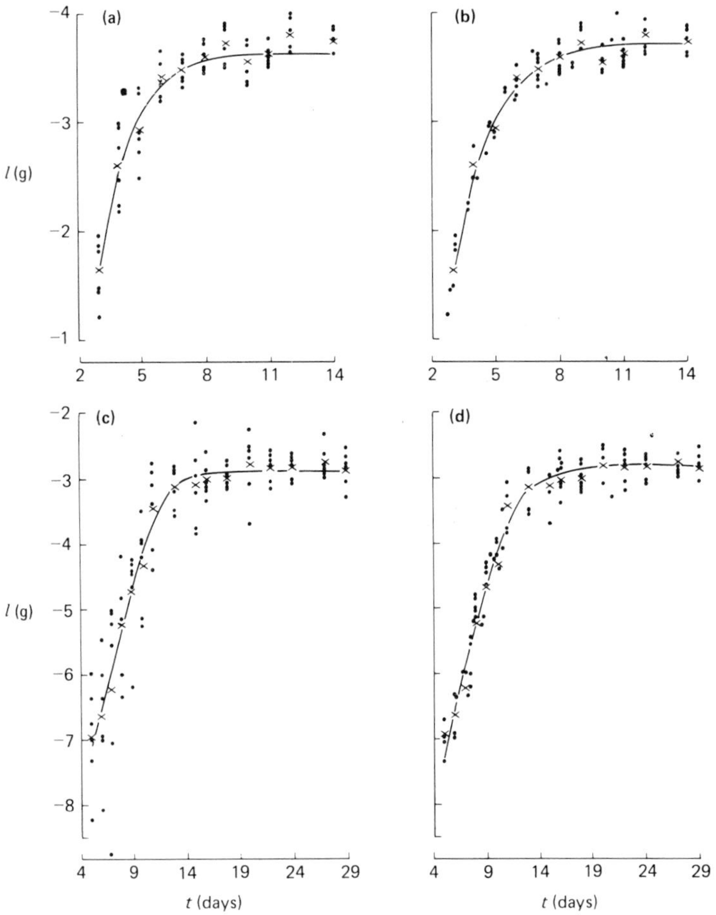

Fig. 4.12 Richards function fittings by the ordinary least squares method (**a**) and (**c**), and the point-moved method (**b**) and (**d**); for wheat leaf 1 (**a**) and (**b**), and maize leaf 3 (**c**) and (**d**).

1; this is a reversal of the trend found so far by using this point-moving strategy, namely, the reduction in parameter estimate variances occasioned by this technique. On the other hand for maize, where n is high as estimated by the unweighted technique, the parameter estimates produced by the weighted method are nearer to those given by the point-moving technique than to the results yielded by the non-weighted procedure.

Discussion

At the outset, we must emphasize that the point-moving technique is not, at this stage, being offered as an alternative estimation strategy to the

Table 4.14 Variances of the growth data at each harvest for the leaves used to illustrate the point-moved estimation of the Richards function, together with calculated structural variance.

t_i	Wheat leaf 1		Maize leaf 13	
1	0.085			
2	0.012			
3	0.103			
4	0.031		0.614	
5	0.015	0.0174 (harvests 5–12)	0.846	
6	0.016		3.115	
7	0.029		0.645	
8	0.023		0.549	
9	0.006		0.488	
10	0.023		0.447	
12	0.010		0.100	
14			0.242	
15			0.069	0.0799 (harvests 15–29)
17			0.031	
19			0.239	
21			0.055	
23			0.035	
27			0.055	
29			0.075	

weighted maximum likelihood method already described and tested. Apart from a paucity of evidence from application, the strategy adopted is probably inefficient due to the following considerations.

The idea of structural and developmental variability is a biological one and so, to take full advantage of this concept, any point-moving strategy should also be based on biological criteria. The strategy suggested above relies entirely on statistical and geometric considerations, and the only advantage of this is that the strategy can be applied to the basic leaf growth data as they stand, without any more detailed knowledge about the plants comprising the sample. From a purely statistical point of view, weighting the observations is the correct method of dealing with heteroscedastic data, and so if only non-biological criteria are involved in a point-moving strategy, there is really no incentive to use it.

In the above strategy, a leaf datum point was moved if the leaf was considered to be exceptionally too advanced or too retarded, from a developmental viewpoint, than the average of the replicates of the leaf in question in that particular harvest. Whatever may be the merits or disadvantages in such data adjustment in general, the undesirable aspect of this strategy is that the developmental age of a leaf has been based on

Table 4.15 Comparison of three Richards function estimation methods (non-weighted, weighted, and point-moved) on two sets of leaf data.

(a) Wheat leaf 1

	Non-weighted estimation procedure	Weighted estimation procedure	Point-moved estimation procedure
PARAMETER ESTIMATES			
a	−4.381	−4.349	−4.284
b	0.914	0.399	−0.910
K	−0.768	−0.706	−0.587
v	0.389	0.274	0.702
VARIANCES OF ESTIMATES			
a	0.00009	0.0004	0.0008
b	6.4757	9.2730	0.0120
K	0.0606	0.0378	0.0013
n	0.4046	0.3705	38.6600
Variance of curve	0.0554	0.0267	0.0185

(b) Maize leaf 3

PARAMETER ESTIMATES			
a	−2.904	−2.867	−2.834
b	7.513	3.321	2.808
K	−0.948	−0.501	−0.463
n	1.578	0.679	0.546
VARIANCES OF ESTIMATES			
a	0.0087	0.0024	0.0022
b	61.2280	5.8213	0.0110
K	0.8295	0.0464	0.0004
n	2.8921	0.2599	0.0006
Variance of curve	0.2880	0.0939	0.0832

geometric rather than biological criteria. An obvious refinement to the strategy would thus be to use the plastochron (Erickson and Michelini, 1957; Lamoreaux, Chaney and Brown, 1978), or some other developmental index, instead of the geometric argument used above.

Whether a point-moving method of fitting the Richards function to single leaf growth data is a desirable approach in itself can only be decided after the performance of the method has been assessed in simulation studies. If a strategy could be evolved which yielded estimated functions as good as or better than those obtained by weighting the observations, then the growth analyst would have the advantage of an estimate of the structural variability of the plants sampled, which in turn would lead to an assessment of developmental variability at different points along the growth curve. Whether this state of affairs can be achieved, and whether any useful biological information would emerge from such results is quite uncertain at the present time; the concept and methods of this section are purely speculative, and should be treated as such.

5

Single leaf growth and the Richards function: applications

The growth of both dry weight and area of single leaves can be investigated using Richards functions fitted to the transformed primary data by the weighted least squares method described in the previous chapter (page 121). In general, fitting caused no problems; but when the data do not adequately represent all growth stages of a leaf, fitting is not possible. This situation is found for the final leaves to be initiated during an experimental period, i.e. those that had not reached the asymptotic size, and applies to, for example, leaf pairs six, seven, eight and nine of sunflower; also, for the early initiated leaves when the first harvest was not sufficiently early to cover the initial development of these leaves. Causton, Elias & Hadley (1978) found similar fitting problems which they then investigated using simulated data. They showed that:

(*a*) the precision of the *a*-estimates decreased when data were removed from the upper end of the curve;
(*b*) the precision of the *n*-estimates and of the growth rate estimates were decreased when data were removed from the lower end of the curve;
(*c*) for the data type investigated, removal of data from the lower end of the curve had a more profound effect than a corresponding number of harvests removed from the upper part of the curve (this type of data had a reasonable asymptotic phase);
(*d*) the general magnitudes of the standard errors of the parameters reflected the size of the variability of the underlying data.

These conclusions of Causton *et al.* (1978) should be borne in mind throughout the discussions in the present chapter, as they explain any poor fits that are obtained, and also the large standard errors found for some parameters.

When a Richards function fitting was not possible the appropriate degree of polynomial was used, selected by the method described in Venus & Causton (1979a) (in this case for foliage, root or stem growth). These polynomial curves will not be described in detail in the present chapter, and were fitted purely to complete the 'runs' of leaf fittings required in the model of Chapter 7. The polynomials are summarized in the Appendix (page 274). Richards function curves were also fitted to the total tiller leaf lamina of wheat and total branch leaves of birch. It was not possible to fit curves to individual branch leaves because the branching pattern and branch development were so variable as to make individual fitting not feasible.

The Richards function single leaf fittings will be considered both in terms of their parameters and also the curves generated. Time scales for this work were taken as the first experimental harvest occurring at day 1.

Biologically relevant parameters

The Richards function fitting yields estimates of the parameters α, β, κ and v; of which only α and v can be considered to be biologically meaningful. Parameter α gives the asymptotic maximum size of the leaf, and v describes the shape of the curve (page 93, Fig. 4.1). The constant β has no biological significance, it is concerned purely with the positioning of the curve in relation to the time-axis. Finally, κ is a rate constant but its interpretation depends upon the value of v. Richards (1959) proposed three different combinations of parameters which have biological significance. The first of these, a weighted mean relative growth rate over the whole growth period, was derived on page 92, as $\kappa/(v+1)$. By a parallel argument, again using a weighting proportional to the absolute growth rate, a weighted mean absolute growth rate can be obtained – $A\kappa/\{2(v+2)\}$. The third useful parameter combination is $\{2(v+2)\}/\kappa$, which represents the time required for the major portion of growth to occur and is usually described as the duration of growth – although this can only be approximate as it is derived from an asymptotic function.

The last two derived parameters have recently been usefully employed by Dennett, Auld & Elston (1978) and Dennett, Elston & Milford (1979) in the description of leaf growth in *Vicia faba* (broad bean). We consider that the weighted mean relative growth rate can also yield useful information and we will use all three derived quantities.

Correlations between parameter estimates

It is obviously important to examine trends in parameters α and v, but β and κ are less informative. However, it has already been mentioned on page 88, that all the parameters of the Richards function are highly correlated, and thus an examination of α and v does to some extent reflect trends in β and κ. The correlations between parameter estimates for four different fittings are summarized in Table 5.1. It can be seen that the parameters b, k and n are all highly correlated (greater than 0.96). The correlations of a with the other parameters are somewhat smaller and are indeed non-significant for wheat leaf 5 area. The correlations are predominately positive but a is negatively correlated with b and n. It might be possible to reduce the high correlations between parameters by using an alternative parameterization of the model,

Table 5.1 Correlation matrices between parameters for Richards function fittings to single leaves.

(a) Birch leaf 8 – weight

	a	b	k
b	−0.66		
k	0.71	0.99	
n	−0.60	0.99	0.96

(b) Sunflower leaf pair 3 – weight

	a	b	k
b	−0.58		
k	0.62	0.99	
n	−0.57	0.99	0.98

(c) Wheat leaf 5 – area

	a	b	k
b	−0.20		
k	0.21	1.00	
n	−0.19	1.00	0.99

(d) Maize leaf 3 – area

	a	b	k
b	−0.53		
k	0.58	0.99	
n	−0.52	0.99	0.97

but this is not biologically appealing as none of the parameters might then be meaningful.

The high correlations of b, k and n, together with lack of biological meaning of the first two, lead us to present detailed results only of the trends in parameters a and n.

Parameter combinations: 'secondary' parameters

The parameter combinations $\kappa/(\nu+1)$, $A\kappa/\{2(\nu+2)\}$, and $\{2(\nu+2)\}/\kappa$ need to be estimated for each single leaf fitting, and this involves the use of approximation formulae (Kendall & Stuart, 1977), as given in the Appendix, page 261, to give the estimates and variances of the estimate for each parameter.

Mean relative growth rate

For this parameter, $\overline{\mathrm{R}}$, mathematically we have that

$$\overline{\mathrm{R}} = \frac{k}{(n+1)} \tag{5.1}$$

and the estimate of $\overline{\mathrm{R}}$, $\hat{\overline{\mathrm{R}}}$, is given by

$$\hat{\overline{\mathrm{R}}} \simeq \frac{k}{(n+1)} + \frac{1}{2}\cdot\frac{\partial^2\overline{\mathrm{R}}}{\partial k^2}\cdot\mathscr{V}(k) + \frac{1}{2}\cdot\frac{\partial^2\overline{\mathrm{R}}}{\partial n^2}\cdot\mathscr{V}(n) + \frac{\partial^2\overline{\mathrm{R}}}{\partial k\,.\,\partial n}\cdot\mathscr{C}(k,n) \tag{5.2}$$

and an estimate of the variance is

$$\mathscr{V}(\overline{\mathrm{R}}) \simeq \left(\frac{\partial\overline{\mathrm{R}}}{\partial k}\right)^2.\mathscr{V}(k) + \left(\frac{\partial\overline{\mathrm{R}}}{\partial n}\right)^2.\mathscr{V}(n) + 2.\left(\frac{\partial\overline{\mathrm{R}}}{\partial k}\cdot\frac{\partial\overline{\mathrm{R}}}{\partial n}\right).\mathscr{C}(k,n) \tag{5.3}$$

where the first and second order partial derivatives are:

$$\frac{\partial \overline{R}}{\partial k} = \frac{1}{(n+1)} \tag{5.4}$$

$$\frac{\partial \overline{R}}{\partial n} = -\frac{k}{(n+1)^2} \tag{5.5}$$

$$\frac{\partial^2 \overline{R}}{\partial k^2} = 0 \tag{5.6}$$

$$\frac{\partial^2 \overline{R}}{\partial n^2} = \frac{2k}{(n+1)^3} \tag{5.7}$$

$$\frac{\partial^2 \overline{R}}{\partial k . \partial n} = -\frac{1}{(n+1)^2} \tag{5.8}$$

Mean absolute growth rate

For this parameter, $\overline{A}$, we have mathematically that

$$\overline{A} = \frac{Ak}{2(n+2)} \tag{5.9}$$

or

$$\overline{A} = \frac{e^a . k}{2(n+2)} \tag{5.10}$$

However, we know that $\hat{\overline{A}} \neq \overline{A}$, and equations similar to (5.2) and (5.3) can be proposed for this parameter. The first and second order partial derivatives required to evaluate $\hat{\overline{A}}$ and $\mathscr{V}(\hat{\overline{A}})$ are:

$$\frac{\partial \overline{A}}{\partial a} = \frac{e^a . k}{\{2(n+2)\}} \tag{5.11}$$

$$\frac{\partial \overline{A}}{\partial k} = \frac{e^a}{\{2(n+2)\}} \tag{5.12}$$

$$\frac{\partial \overline{A}}{\partial n} = \frac{e^a . k}{\{2(n+2)^2\}} \tag{5.13}$$

$$\frac{\partial^2 \overline{A}}{\partial a^2} = \frac{e^a . k}{2(n+2)} \tag{5.14}$$

$$\frac{\partial^2 \overline{A}}{\partial k^2} = 0 \tag{5.15}$$

$$\frac{\partial^2 \overline{A}}{\partial n^2} = \frac{2e^a . k}{2(n+2)^3} \tag{5.16}$$

$$\frac{\partial^2 \bar{\mathrm{A}}}{\partial a.\partial k} = \frac{\mathrm{e}^a}{2(n+2)} \tag{5.17}$$

$$\frac{\partial^2 \bar{\mathrm{A}}}{\partial a.\partial n} = -\frac{\mathrm{e}^a.k}{2(n+2)^2} \tag{5.18}$$

$$\frac{\partial^2 \bar{\mathrm{A}}}{\partial k.\partial n} = -\frac{\mathrm{e}^a}{2(n+2)^2} \tag{5.19}$$

Duration of growth

Mathematically, this parameter, D, is given by

$$\mathrm{D} = \frac{2(n+2)}{k} \tag{5.20}$$

This can be estimated as $\hat{\mathrm{D}}$, using approximate equations similar to (5.2) and (5.3) to obtain $\hat{\mathrm{D}}$ and $\mathscr{V}(\hat{\mathrm{D}})$. The required first and second order partial derivatives are:

$$\frac{\partial \mathrm{D}}{\partial k} = -\frac{2(n+2)}{k^2} \tag{5.21}$$

$$\frac{\partial \mathrm{D}}{\partial n} = \frac{2}{k} \tag{5.22}$$

$$\frac{\partial^2 \mathrm{D}}{\partial k^2} = \frac{4(n+2)}{k^3} \tag{5.23}$$

$$\frac{\partial^2 \mathrm{D}}{\partial n^2} = 0 \tag{5.24}$$

$$\frac{\partial^2 \mathrm{D}}{\partial k \partial n} = -\frac{2}{k^2} \tag{5.25}$$

We can thus obtain estimates of all three derived parameters and an approximate variance for each; and we are now in a position to examine in detail Richards function descriptions of single leaf growth in terms of five parameters: a, n, $k/(n+1)$, $Ak/\{2(n+2)\}$ and $\{2(n+2)\}/k$. To date, the reality of the varance estimates for the last three parameters has not been investigated and it is possible that the combination of the use of the approximation formula together with the fact that n and k are highly correlated may make the variances very approximate indeed.

Parameter comparisons within and between species

The single leaf growth of four species, maize, wheat, sunflower and birch, will be described in the present section. Details of the experiments have already been given (page 43), but it is useful to remember that the durations of the experiments vary considerably; from 25 days for maize to 81 days for birch. This, together with differing rates of leaf production, means that the number of leaves described by Richards functions are different for each species. In sunflower, for example, only 10 leaves completed sufficient growth to be fittable; however, in this species leaves are produced in pairs and it was impossible to find a criterion for the separation of the two leaves of a pair, thus only five fittings were carried out (A therefore represents twice the asymptotic leaf size).

In order to facilitate species comparisons, trends in a and n for the weight and area data will be discussed first, and then we will move on to the derived parameters. All parameter estimates and their variances are summarized in the Appendix, pages 266–273.

Primary parameters

Dry weights

Parameter a initially increases with leaf number for all four species (Fig 5.1 for maize and wheat and Fig. 5.2 for sunflower and birch). The sunflower a-values increase for all leaf pairs examined as do those for wheat, although for leaves 5 and 6 of wheat and leaf pairs 3, 4 and 5 of sunflower the increase is not significant because of the greater size of the confidence intervals; indeed $\mathscr{V}(a)$ for sunflower leaf pair 5 is very large. In maize the a-values rise until leaf 6 and then decline, although not significantly; a similar situation is found for the birch experiment with a maximum a for leaf 11 and the values attained by subsequent leaves gradually decrease (Borrill, 1959; Dormer, 1972; Williams, 1975; Causton *et al.*, 1978).

There are two alternative explanations as to why the first leaves produced by plants are small: either there is an insufficient supply of assimilate in the young plant to support larger leaves, or the final size of the leaf is determined in the primordium and thus the small leaf size is a function of the small size of the first primordium to be produced (Cutter, 1971, page 50). The later decline in leaf size found in maize, and presumably present in the other short-lived species, can be explained in terms of competition for assimilate. As time progresses the majority of assimilate is channelled into stem extension and flower development, and not into the production of large amounts of leaf tissue. This is not to say that because less leaf material is produced the photosynthetic capacity of the plant falls; indeed there is ample evidence to the contrary. The flag leaf of many grass species

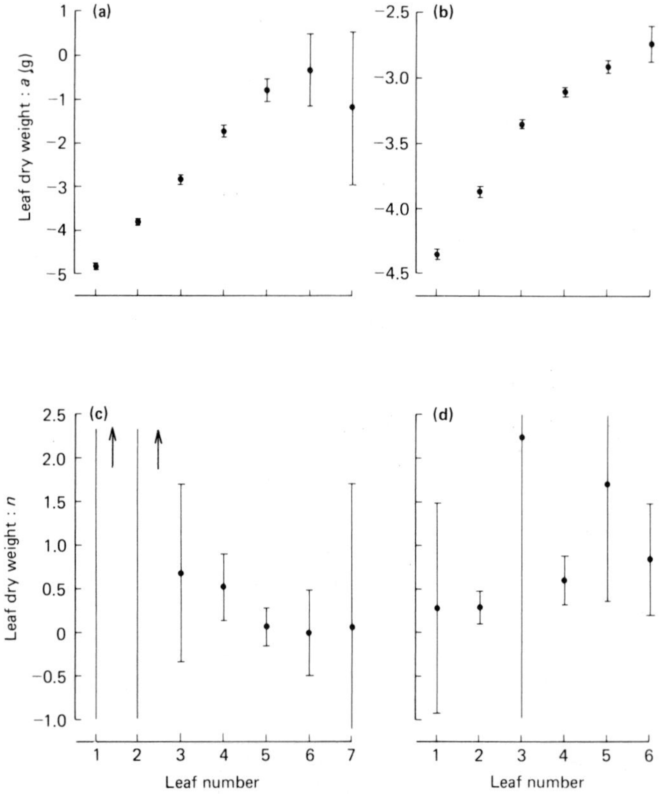

Fig. 5.1 Richards function parameter estimates with 95% confidence intervals for dry weight: *a*, (**a**) and (**b**), and *n*, (**c**) and (**d**); of maize, (**a**) and (**c**), and wheat, (**b**) and (**d**). (Arrows indicate values out of range.)

maintains a high photosynthetic rate for prolonged periods (King, Wardlaw & Evans, 1967; Evans & Rawson, 1970). The demand for assimilates from the flag leaf by the ear maintains the rate of production by a source – sink control mechanism, possibly mediated through an auxin system (Sweet & Wareing, 1966a) or by changes in sugar concentration.

The birch plants did not flower and it is, therefore, more difficult to explain the decrease in *a*-values from leaf 12 onwards, except in terms of an overall strategy of seasonal growth with leaf size gradually declining towards the end of the growing season.

An examination of the magnitude of the *a*-values for the different species reveals that while the first leaf of maize reaches a slightly smaller asymptotic size than that of wheat, the largest leaf of maize is considerably larger than

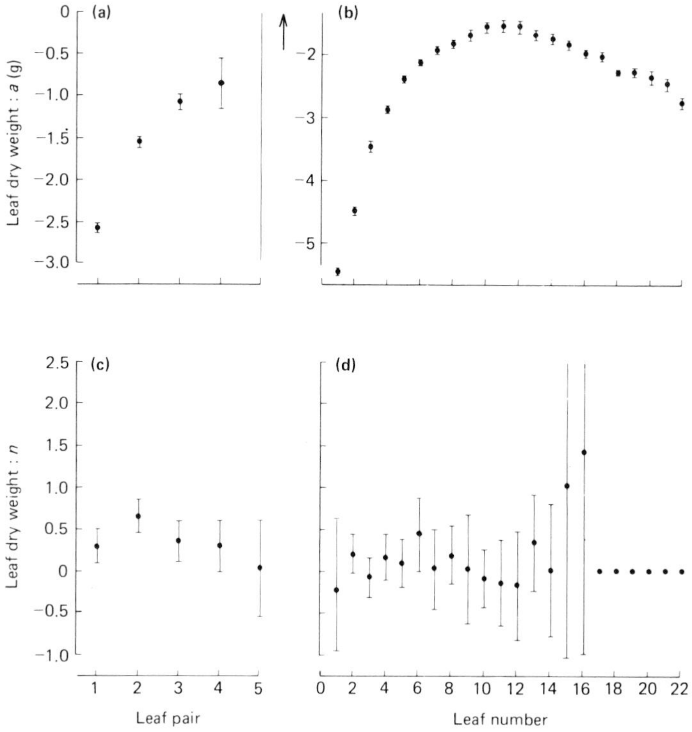

Fig. 5.2 Richards function parameter estimates with 95% confidence intervals for dry weight: *a*, (**a**) and (**b**), and *n*, (**c**) and (**d**); of sunflower, (**a**) and (**c**), and birch (**b**) and (**d**). (Arrow indicates values out of range.)

that of wheat when mature. Maize, in fact, produces the largest leaves (in terms of dry weight) of all the species examined, followed by sunflower, birch and finally wheat. The seed of maize is large in comparison to the other species but this is not reflected in leaf growth as the first leaf is quite small. The small size of the first leaf of birch, on the other hand, could possibly be attributed to the very small seed.

Trends with leaf number in the parameter *n* are difficult to discern for any of the species (Fig. 5.1c,d and Fig. 5.2c,d); however, in maize *n* appears to decline with leaf number although the differences are not significant. The parameter *n* defines the shape of the curve and it is interesting to find that in birch none of the *n*-values are significantly different from zero, and only one is significantly different for maize. In birch the final 6 leaf fittings give *n*-values indistinguishable from zero and this produces problems in the estimation of the relative growth rate curve, which could only be overcome by using a mathematical evaluation of R rather than estimating $\hat{R}$, see page 108. In these extreme situations it would perhaps have been advantageous to use the Gompertz function (page 98) to describe growth. This would then have meant attempting to compare results from two types of fitting, which is not altogether satisfactory. An alternative approach would be to fit all the leaves by the Gompertz function; however, if this function is fitted when v, of the underlying relationship, is not very close to zero then the resulting estimates of a, b and k may be unrealistic (Venus, unpublished). Thus problems arise from this approach. Other workers have also found that the parameter n is very near to zero for other species: Hackett & Rawson (1974) for *Nicotiana tabacum* and Causton *et al.* (1978) for *Impatiens parviflora*. The former workers then adopted the Gompertz function for all further work as it is easier to handle statistically.

The sunflower and wheat fittings give n-values that are slightly higher than the two species discussed above, values are between 0.27 and 1.7 for wheat. A higher n-value implies a longer period of initial, nearly constant, relative growth rate than a lower value, followed by a brief period of fast-changing growth rate and then the asymptotic phase. Causton *et al.* (1978) attempt to relate the relative magnitudes of the parameter n to the aggregated effect of the interrelationships between cell division and extension. Thus, if two leaves commence growth at the same time and reach the same maximum size, but in one leaf cell multiplication is prolonged and extension reduced, then this leaf would have a higher value of n. Causton *et al.* discuss this relationship in respect of different environmental regimes acting on the same leaf number but the idea may well extend to differences between species. Steer (1971) found in *Capsicum frutescens* (pepper) that in later-initiated leaves cell division played a more important role than in earlier initiated leaves, but his work did not involve Richards function descriptions of growth. It would be interesting to investigate the interaction of these two hypotheses; indeed, a very detailed investigation of growth, development, chemical composition and various physiological activities occurring within a single leaf could well reveal a closer correspondence between the Richards function description and the growth processes.

When we examine the magnitude of the confidence intervals for the parameters a and n it can be seen that the later leaves (maize 6 and 7, wheat 6 and sunflower 4 and 5) give less reliable estimates of a, presumably because of the lack of data in the asymptotic region. The large variances of n given by some leaf fittings can usually be traced back to a lack of data in the ascending phase of growth, maize 1, 2, 3 and wheat 1, 3, 5 etc. These high variances will be reflected in the secondary parameters and thus only a limited number of leaves will give reliable estimates for all parameters, for maize only leaves 4, 5 and 6. This emphasizes the need for extensive data collection over the entire development of each leaf in order to draw sound conclusions from the results.

Areas

The estimates of parameter a for the area data tell a similar story to those for the weight data (Fig. 5.3 for maize and wheat and Fig. 5.4 for sunflower and birch). It was impossible to obtain realistic fittings for the area of leaves 1 and 2 of maize and thus results start at leaf 3.

In contrast to leaf dry weight, where the maximum size was greatest for maize, in terms of area the largest size was given by the leaf pairs of sunflower. The pattern of a-values increasing with leaf number up to a certain point and then declining has been found in a wide variety of species for leaf area: for example, *Gossypium hirsutum* (Portsmouth, 1937), *Lycopersicum esculentum* (Cooper, 1959), *Glycine max* (Shibles, Anderson & Gibson, 1975; Woodward, 1976) and *Vicia faba* (Dennett, Elston & Milford, 1979); presumably, the reasons given in relation to leaf dry weight to explain the distribution of a-values with respect to leaf number apply equally to leaf area, page 149.

For the area fittings, more of the n-values are significantly different from zero including four leaves of birch. The n-values for the earlier leaves of birch tend to be higher for the area fittings than for the dry weight, but the nine final area fittings give n estimates that are indistinguishable from zero. The area of leaf pair 2 of sunflower gives a high value of n (others are higher but have very wide confidence bands), 1.37, and it does, indeed, appear that sunflower tends to have higher n-values than the other species examined.

Secondary parameters

Dry weights

The derived parameters for the weight fittings, $k/(n+1)$, $Ak/\{2(n+2)\}$, and $\{2(n+2)\}/k$, are presented in Figs. 5.5 (for maize and wheat) and 5.6 (for sunflower and birch).

The estimates of weighted mean relative growth rate are more variable for maize and wheat than for the dicotyledons – sunflower and birch. There is a tendency for $\hat{\bar{R}}$ to decline with leaf number in maize and sunflower (more pronounced in the latter); whereas in birch $\hat{\bar{R}}$ is high for the first two leaves then declines but gradually increases for the later leaves; however, diferences are rarely significant.

The most useful comparisons to be made from these results are of the range of values and mean values for the different species (Table 5.2). It can be seen that wheat gives the highest average mean relative growth rate, by a considerable margin, followed by maize, sunflower and finally birch. These average values have least meaning for the sunflower fittings as the $\hat{\bar{R}}$ values show real trends in this species, but for the other species the average comparison is realistic. It might be expected that birch would have lower single leaf mean relative growth rates because it is a perennial species; but

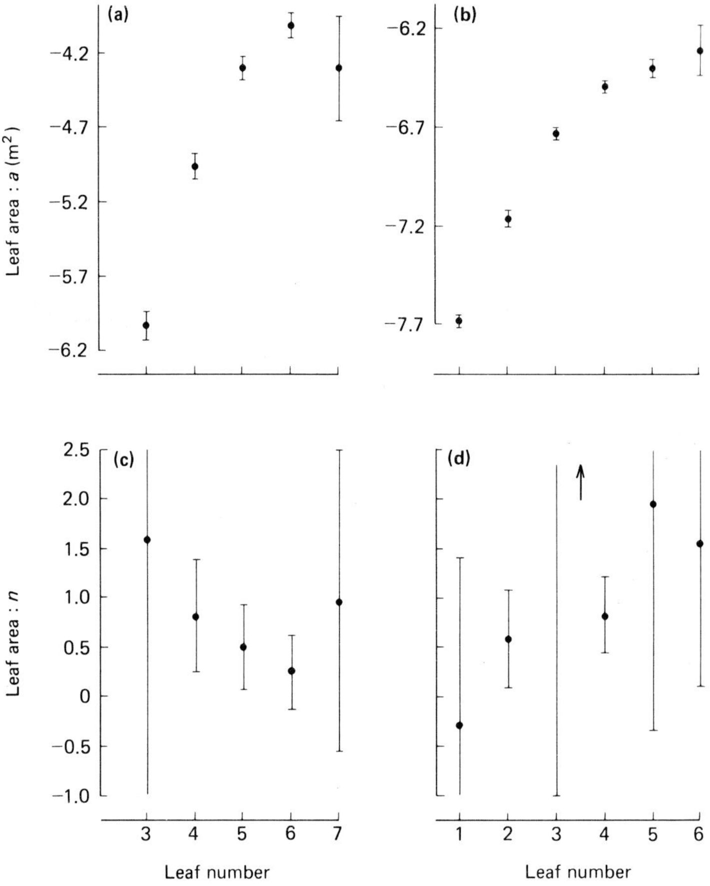

Fig. 5.3 Richards function parameter estimates with 95% confidence intervals for leaf area: a, (**a**) and (**b**), and n, (**c**) and (**d**); of maize, (**a**) and (**c**), and wheat, (**b**) and (**d**). (Arrow indicates values out of range.)

why the average for wheat should be so much higher than that for maize, a C_4 plant, and sunflower, known to be a highly productive C_3 plant, is difficult to explain.

It is difficult to make comparisons of weighted mean absolute growth rates, within or between species, (Figs. 5.5c,d and 5.6c,d) because the magnitude of this parameter depends to a large extent upon the size of a. There is, however, a tendency for $\hat{\bar{A}}$ to rise for the first leaves and then decline for later leaves. These differences are, in many instances, significant. It is interesting to note that these trends are not directly related to trends in a; for example, for maize the largest value of a is for leaf 6 but $\bar{A}$ is largest for

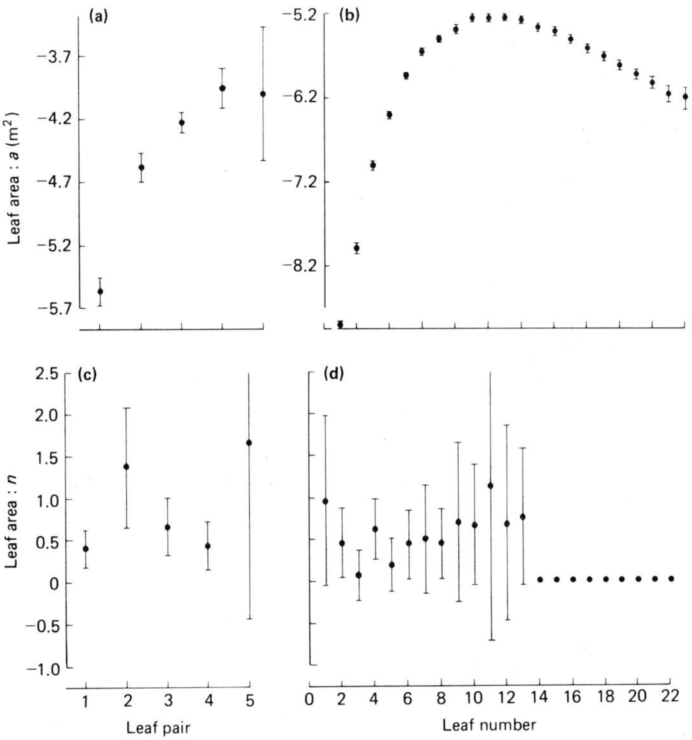

Fig. 5.4 Richards function parameter estimates with 95% confidence intervals for leaf area: *a*, (**a**) and (**b**), and *n*, (**c**) and (**d**); of sunflower, (**a**) and (**c**), and birch, (**b**) and (**d**).

Table 5.2 Summary of weighted mean relative growth rates for weight and area of maize, wheat, sunflower and birch. Maximum, minimum and average of estimates obtained for single leaves.

WEIGHT

	Minimum	Maximum	Average
Maize	0.1220	0.6540	0.2921
Wheat	0.3684	0.6108	0.5057
Sunflower	0.0556	0.2982	0.2160
Birch	0.1213	0.2847	0.1764

AREA

	Minimum	Maximum	Average
Maize	0.2752	0.4282	0.3555
Wheat	0.3995	2.1350	0.8199
Sunflower	0.2096	0.3495	0.2774
Birch	0.1528	0.3104	0.2281

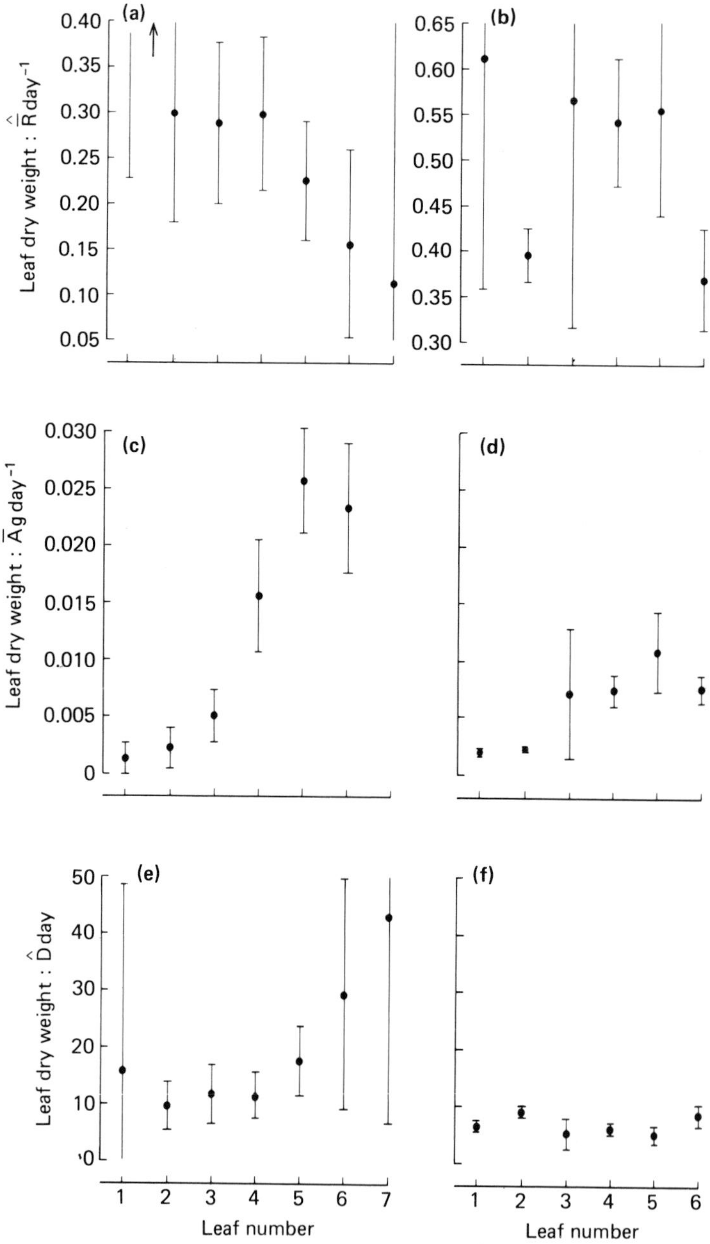

Fig. 5.5 Richards function parameter estimates with 95 %
confidence intervals for dry weight: $\hat{\bar{R}}$ (a) and (b), $\hat{\bar{A}}$ (c) and (d), and
$\hat{D}$, (e) and (f); of maize, (a), (c) and (e), and wheat, (b), (d) and (f). (Arrow
indicates values out of range.

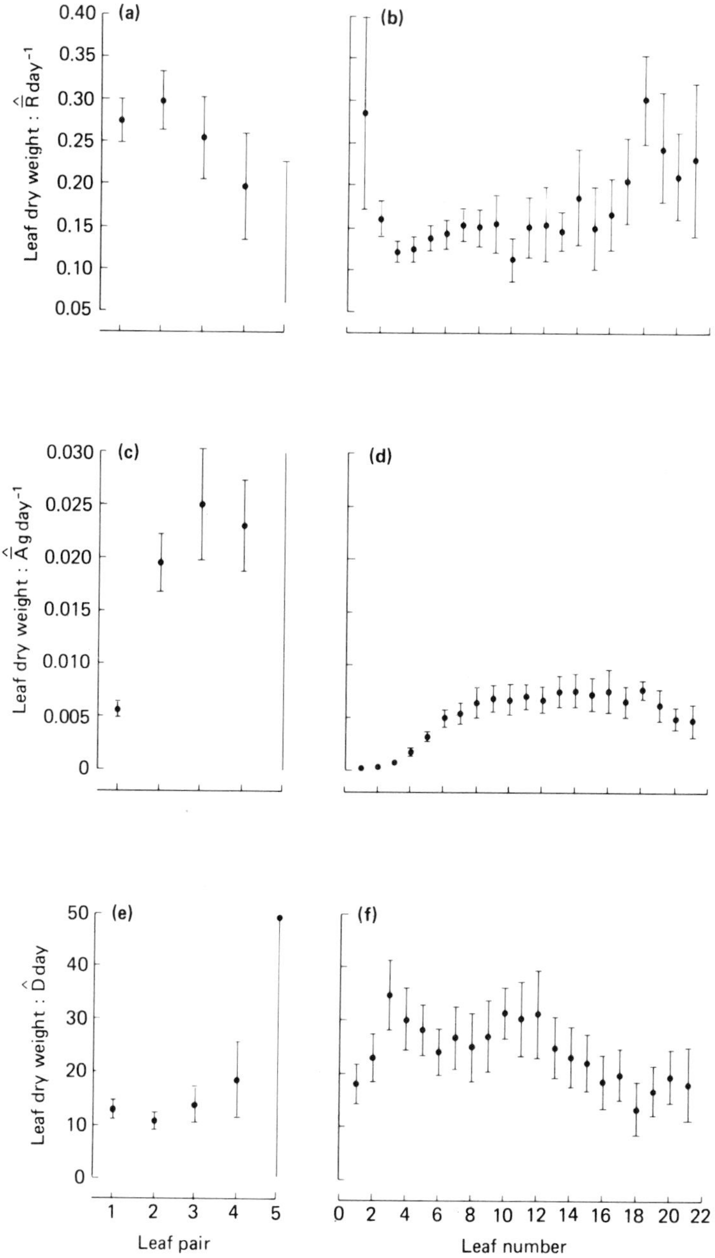

Fig. 5.6 Richards function parameter estimates with 95% confidence intervals for dry weight: $\hat{R}$ (**a**) and (**b**), $\hat{A}$, (**c**) and (**d**), and $\hat{D}$, (**e**) and (**f**); of sunflower (**a**), (**c**) and (**e**), and birch, (**b**), (**d**) and (**f**).

leaf 2. In wheat and sunflower a continues increasing with leaf number but mean absolute growth rate tends to decline for the final leaves fitted. The results for birch are interesting as $\bar{A}$ remains more or less constant for leaves 9 to 18 whilst a-values change considerably over these leaves.

Weighted mean absolute growth rates are lowest for the leaves of birch, although the maximum leaf size, as estimated by a, is smaller in wheat. It is, therefore, possible to obtain useful information from $\hat{\bar{A}}$-estimates especially when the relative magnitudes of the a-values are taken into consideration.

Duration of leaf growth (Figs. 5.5e, f and 5.6e, f) shows marked trends and species differences, many of which are significant. Duration is shortest for leaf growth in wheat, less than 10 days for all leaves, and there appear to be no trends with leaf number in this species. The values of $\hat{D}$ for maize and sunflower are of similar magnitude, as well as exhibiting similar trends: a decrease in duration of leaf growth from leaf 1 (leaf pair in the case if sunflower) to leaf 2 and then a gentle rise. However, the width of the confidence intervals, especially for maize, means that there are no significant differences between any leaves for either of the species.

The results of $\hat{D}$ for birch (Fig. 5.6 f) show that duration of leaf growth in this species is larger than in the other species under investigation, especially for leaves 3 to 12. There is also a change in $\hat{D}$ with leaf number: an increase from leaf 1 to 3, a more or less constant section (excepting leaf 6), then a decline from leaf 12 to leaf 16 and finally a levelling off One possible explanation of the trends exhibited by birch is given by Schoch (1974), who stated that the duration of growth may be determined by the production of younger competing leaves. Schoch found that removal of young leaves could increase the area and duration of growth of leaves that were still developing. It will be shown later that the rate of leaf production in birch increases with time, thus fitting this hypothesis.

Areas

The estimates of $\hat{\bar{R}}$, $\hat{\bar{A}}$ and $\hat{D}$ are given on Fig. 5.7, for maize and wheat, and Fig. 5.8, for sunflower and birch. The trends are very similar to those described above for the weight fittings and will be considered only briefly.

At first sight the trends in weighted mean relative growth rates are similar for the weight and area fittings, however, the $\hat{\bar{R}}$-values are larger in magnitude for the area data.

The absolute growth rate estimates cannot be compared, in magnitude, between leaf dry weight and area because of the involvement of the a-estimate in the formula; but trends can be compared, and we see that for maize the trends of $\hat{\bar{A}}$ are similar for weight and area. The wheat $\hat{\bar{A}}$-values are low for the first two leaf areas and then rise to a larger value for the remaining 4 leaves, this bears little relation to trends in the a-values;

similarly, for leaf pairs 2 to 5 of sunflower. Turning to the birch results we find that for the area fittings the $\bar{A}$-estimates do, to some extent, reflect trends in *a*, with a maximum for leaf 11 and then a gradual decline with leaf number, this contrasts with the results given for leaf dry weight. Dennett, Auld & Elston (1978) also found that $\bar{A}$ followed similar trends as *a* for the area of single leaves of *Vicia faba*, although the changes were most obvious for *a*.

Duration of growth of leaf area is shorter than that for leaf weight for all leaves in all four species. This is very usual and is discussed further in relation to single leaf specific leaf area, see page 168. These differences between leaf area and dry weight are least obvious for wheat, where the leaves grow very quickly. The downward trend of $\hat{D}$ with leaf number is less pronounced in the area fittings.

The pattern of change of duration of leaf growth with leaf number, in common with the other parameters examined, varies considerably for the different species under investigation, and other workers' results are also variable. Dennett, Elston & Milford (1979), when showing that for single leaves of *Vicia faba* the duration of growth was linearly related to temperature, found that $\bar{D}$ decreased with increase in leaf number (with a maximum duration of 40 days for leaf 2). Wolf (1947), on the other hand, found that in *Nicotiana tabacum* duration of growth did not depend on leaf positions, a result also obtained by Hackett & Rawson (1974), and again found in seedlings of *Populus* × *euramericana* (a popular hybrid) irrespective of the final size attained (Pieters, 1974).

Curves derived from fitted Richards functions

Richards function curve fitting to single leaf growth data yields realistic descriptions of the changes of leaf dry weight and area with time. The characteristics of these curves have been thoroughly examined in the preceeding sections, in terms of the parameters, and there is nothing to be gained from a presentation of the curves themselves. It is, however, possible to derive relative growth rate curves from the fitted relationships, page 108; and these are worthy of individual consideration.

Up to this point, the descriptions of leaf dry weight and area growth of a single leaf have been considered independently, but the two growth facets are, of course, intimately linked. The ratio of leaf area to leaf weight, a single leaf specific leaf area, exhibits interesting changes with time and also with leaf number.

The calculation of the specific leaf area ratio estimates, together with their variances, at consecutive time values is carried out in a manner analogous to that used for the classical growth analysis ratio calculation, where the ratio components are described by curves fitted to logarithmic data, page 54.

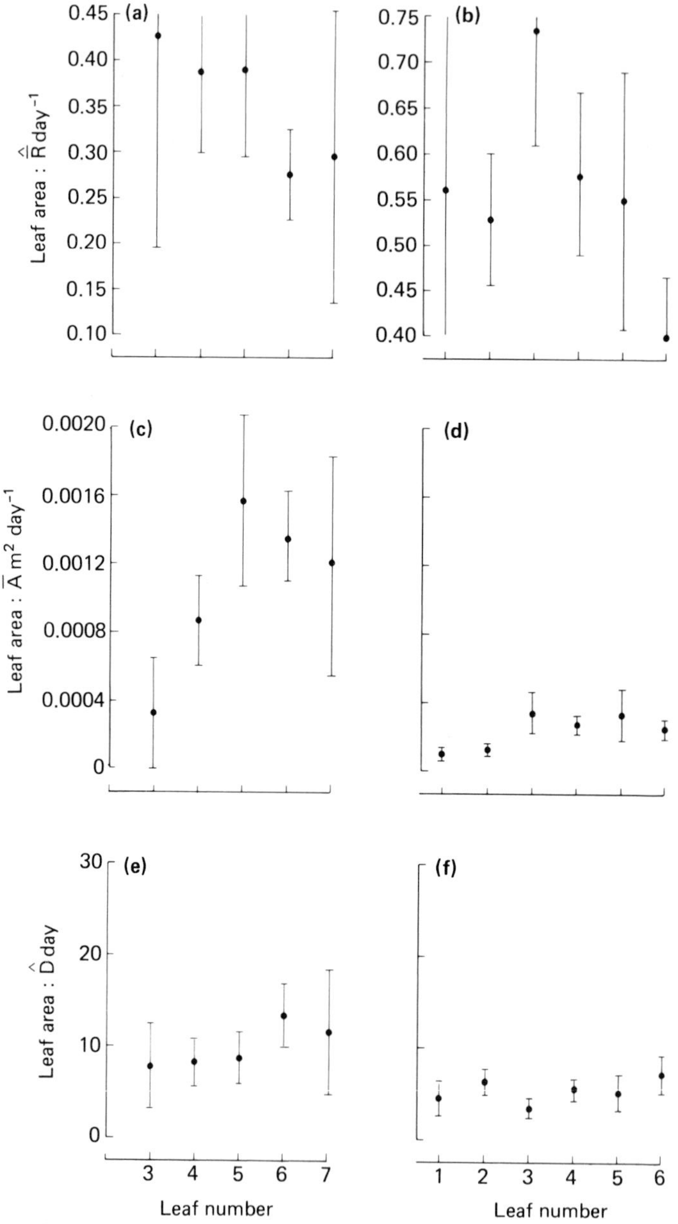

Fig. 5.7 Richards function parameter estimates with 95% confidence intervals for leaf area: $\hat{R}$, (**a**) and (**b**), $\hat{A}$, (**c**) and (**d**), and $\hat{D}$, (**e**) and (**f**); of maize (**a**), (**c**) and (**e**), and wheat, (**b**), (**d**) and (**f**).

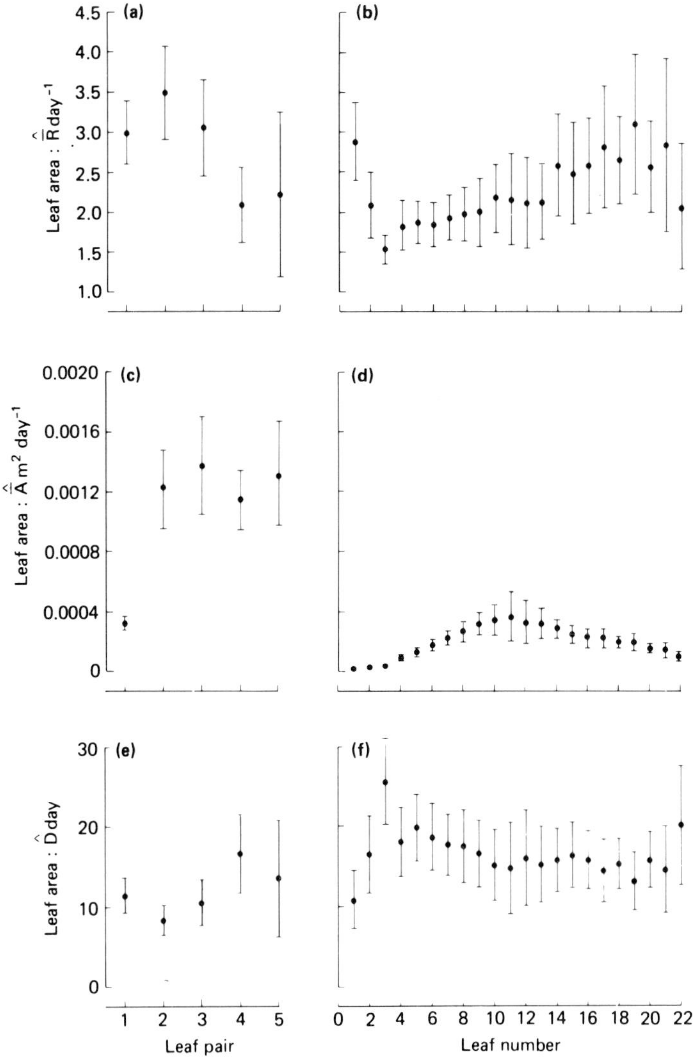

Fig. 5.8 Richards function parameter estimates with 95% confidence intervals for leaf area: $\hat{R}$, (**a**) and (**b**), $\hat{A}$, (**c**) and (**d**), and $\hat{D}$, (**e**) and (**f**); of sunflower, (**a**), (**c**) and (**e**), and birch, (**b**), (**d**) and (**f**).

Relative growth rate

Sunflower

Relative growth rate curves for the leaf pairs of sunflower are given in Fig. 5.9. The initial weight relative growth rates for leaf pairs 1, 3 and 4 are all

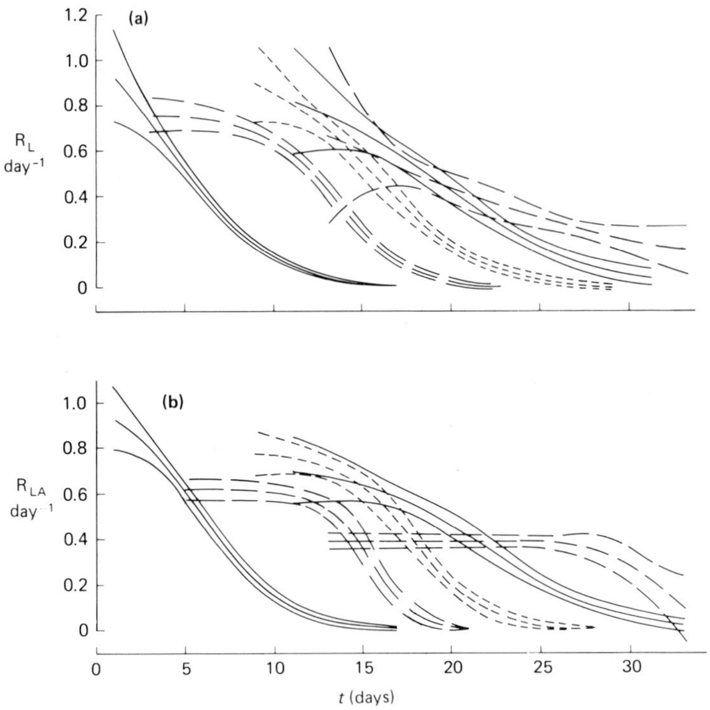

Fig. 5.9 Relative growth rate curves with 95% confidence bands for (**a**) leaf dry weight and (**b**) leaf area of leaf pairs 1 ——, 2 — —, 3 ----, 4 —— and 5 — — from sunflower.

very similar and those for leaf pairs 2 and 5 are not significantly lower. These latter two pairs of leaves appear to have a short period of 'constant' relative growth rate at the start of development. Leaf pair 2 has a relatively high n-value (0.65) which is associated with this type of growth; leaf pair 5, in contrast, has a very low n-estimate and thus the relative growth rate curve is surprising in form and perhaps the consequence of high parameter variances for this fitting.

The leaf area relative growth rate curves show similar trends to those described above, except that in this case the leaf pair 5 curve is very clearly defined with a significantly lower initial value than other leaves, and a long asymptotic initial section to the curve that is associated with a large n-value (1.67). The n-values for all leaf pairs are larger for the area data fittings and this is reflected in the slower initial rate of decline of area relative growth rate, when compared to dry weight.

As leaf pair number increases there appears to be a decrease in the rate of decline of relative growth rate over the greater part of each leaf pair's growth. This corresponds to the changes in weighted mean relative growth rate estimates with leaf number that have been described above.

Wheat

The relative growth rate curves for wheat (Fig. 5.10) have wider confidence bands than those for sunflower but the results are clear to see.

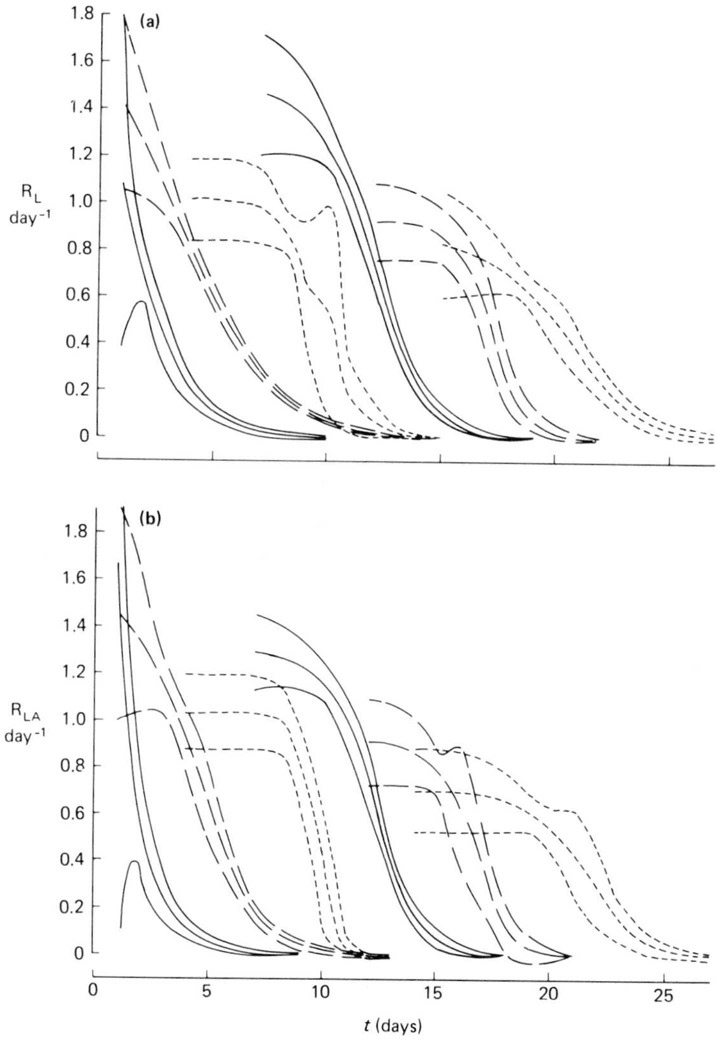

Fig. 5.10 Relative growth rate curves with 95% confidence bands for (**a**) leaf dry weight and (**b**) leaf area of leaves 1 ——, 2 — —, 3 ----, 4 ——, 5 — — and 6 ---- from wheat.

Following a short 'constant' phase, longest for leaves 3 and 5 where n equals 2.2 and 1.7 respectively, leaf dry weight relative growth rate declines very quickly; this reflects the small duration of growth, D, estimates given by this species, page 158. A change in initial relative growth rate with leaf number is difficult to define, although initial values for leaves 5 and 6 are certainly smaller than those for earlier leaves. An increased competition for assimilates when these later leaves develop, due to the presence of a large number of tiller leaves and an elongating stem, may be responsible for the lower initial relative growth rates. The high initial relative growth rate of leaf 4 is difficult to explain.

It is of interest to note that at any one time it was very rare for more than two main stem leaves to have relative growth rates of greater than 0.1 day^{-1}. This indicates that a new leaf did not commence development until the penultimately initiated leaf had reached maximum size, and this may be governed directly or indirectly by auxin mediation.

Leaf area relative growth rates show similar features as leaf weight with just a few differences. Leaf 1 has a very high initial leaf area relative growth rate, but also has a large confidence band. Secondly, the initial relative growth rate for leaf 4 is not significantly larger than that for leaf 3, and finally the decrease in initial relative growth rate from leaf 5 to leaf 6 is greater than for dry weight.

Maize

The relative growth rate curves, Fig. 5.11, for this species appear very muddled because at any one time at least three leaves have growth rates greater than 0.2 day^{-1}. The curves also have large confidence bands which add to the confusion. In respect of the dry weight curves it is possible to discern that the initial relative growth rate tends to increase with leaf number, this is in contrast to the wheat results.

The leaf area curves show no significant increase in initial rate with leaf number, indeed leaf 7 has the lowest initial value – but the fittings to leaf 7 may be unreliable as very little data were available in the asymptotic region. These results do not warrant further discussion because of the effects of high data variability on the fittings.

Birch

Relative growth rate curves are presented for alternate leaves only (1–19) in Fig. 5.12; this is purely to increase clarity, and the omitted leaves fit into the general trends that will be described below. The first striking feature of the curves is that, as time progressed, leaves were initiated at a faster rate, and thus more leaves were developing at any one time. None of the curves for birch have initial periods of 'constant' relative growth rate, and this reflects n-values near to zero for this species.

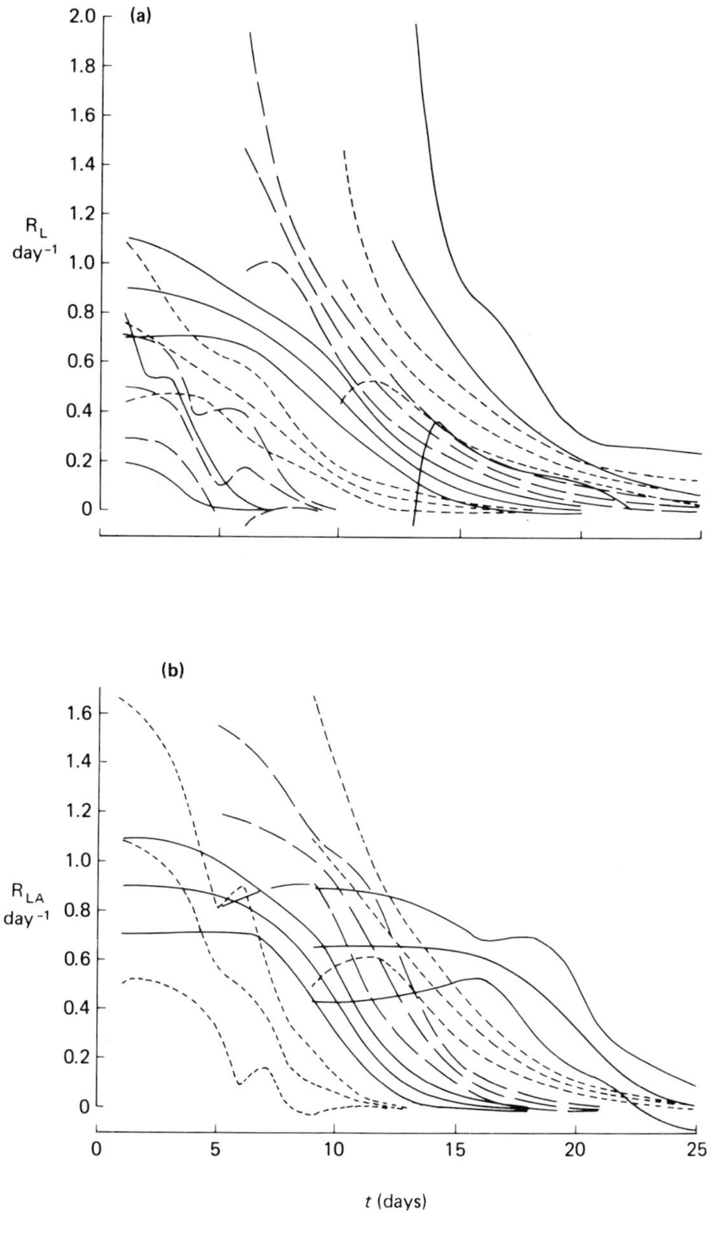

Fig. 5.11 Relative growth rate curves with 95% confidence bands for (**a**) leaf dry weight and (**b**) leaf area of leaves 1 ——, 2 — —, 3 ----, 4 ——, 5 — —, 6 ---- and 7 —— from maize (leaves 1 and 2 are not included for leaf area).

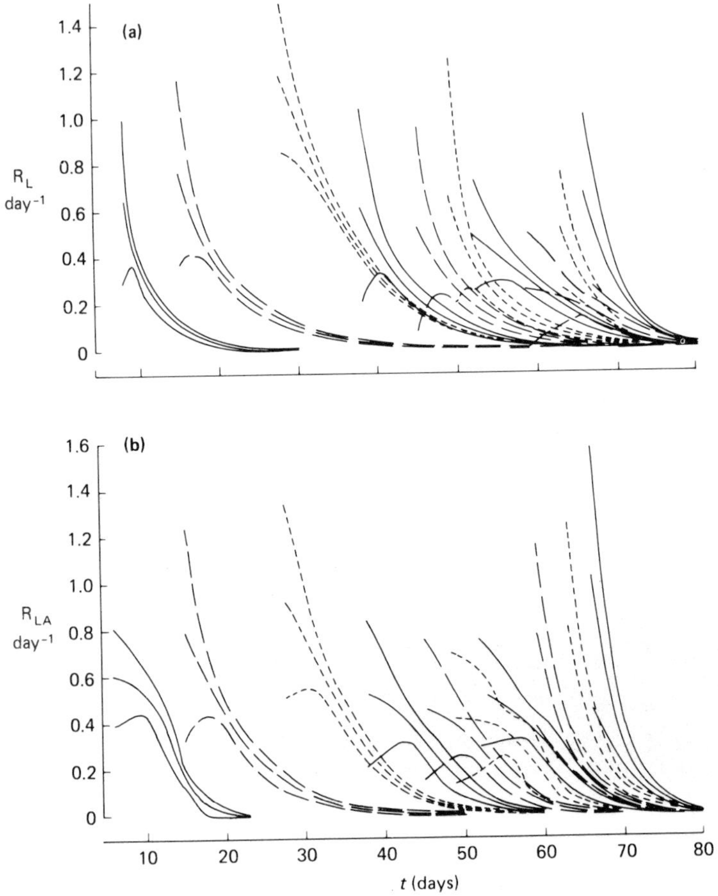

Fig. 5.12 Relative growth rate curves with 95% confidence bands for (**a**) leaf dry weight and (**b**) leaf area of leaves 1 ——, 3 — —, 5 ----, 7 ——, 9 — —, 11 ----, 13 ——, 15 — —, 17 ---- and 19 —— from birch.

Initial relative growth rates of birch leaves are lower than those for other species (similar results were given by weighted mean relative growth rates, page 153). It has been found that relative growth rates of woody tree species tend to be considerably lower than those for herbaceous species. Grime & Hunt (1975) found this, in a survey of maximum and average relative growth rates, for a large number of species; Jarvis & Jarvis (1964) obtained similar results when comparing the unit leaf rates of various tree species with that of sunflower. Although these other results apply to whole plant growth, this feature could well extend to single leaf growth.

Initial relative growth rates for birch leaves increase across the first three leaves (1, 3 and 5) and then are lower for subsequently initiated leaves. For leaf area growth the initial relative growth rates increase again for the later leaves. The initial relative growth rates of successive leaves are not significantly different for any of the species investigated, although various trends have been identified, for example a tendency to decrease with increasing leaf number in wheat. Work by Rogan & Smith (1975) using *Agropyron repens* (couch grass) revealed a progressive decline in growth rates of successively formed leaves, especially after leaf 6. This was related to the fact that up to the expansion of leaf 6 the rate of leaf formation exceeded the rate of leaf emergence, and primordia accumulated on the apex; this situation was reversed after leaf 6. The changes in the rates of formation and growth of leaves could be interpreted in terms of competition for assimilate between expanding leaves and developing primordia. There is no evidence to support these ideas for any of the species under current investigation, even for the grass species where corroboration with Rogan and Smith's working with *Agropyron* would be expected most. However, the number of leaves examined has been limited (except for birch) and it is possible that longer experiments might provide evidence of this type, although examination of shoot apices and primordia is obviously necessary to establish these kinds of growth interactions.

Specific leaf area

Sunflower

The specific leaf area ratio curves are of similar form for all leaf pairs (Fig. 5.13) and consist of a descending phase (not leaf 1), an ascent to a maximum

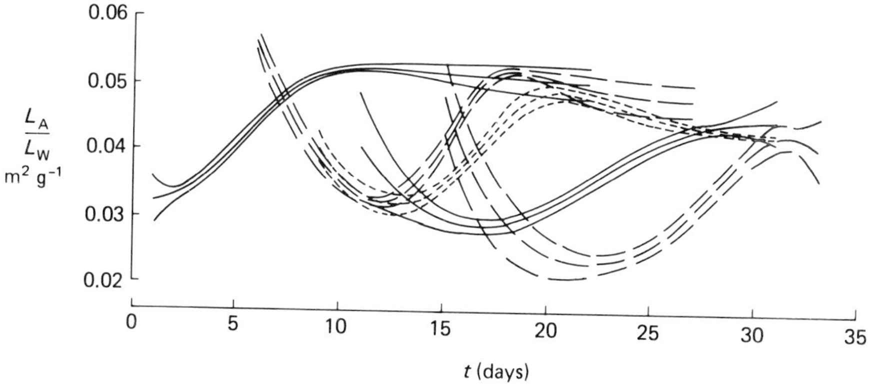

Fig. 5.13 Specific leaf area curves with 95% confidence bands for leaf pairs 1 ——, 2 — —, 3 ----, 4 —— and 5 — — of sunflower.

and then a slight fall to an asymptotic value. The traditional ideas of specific leaf area time trends do not include the initial declining phase, so apparent for this species, and can be explained in terms of leaf cellular development. During an initial period of growth, leaf area increases at a faster rate than leaf weight, and thus the ratio of area to weight rises; leaf area increase then slows down as a phase of predominately cell division is superceded by a phase of cell differentiation and deposition of strengthening materials such as lignin. Leaf weight in this period increases at a faster rate than leaf area and the specific leaf area falls, finally both weight and area cease to change and the ratio becomes constant. Thus physiological reasons can be found that interpret this form of curve. In the present examples the initial fall in specific leaf area is very pronounced and cannot just be dismissed as error due to handling of very small plants, although this may be a contributing factor. A biological reason seems to be more likely an answer as the trend is so widespread. Perhaps an initial part of the cell division phase involves a three-dimensional increase, when area does not change greatly – we are truly measuring projections – followed by two-dimensional cell division which would alter leaf area more than weight.

Growth of leaf pairs 1 to 3 has reached the period of asymptotic specific leaf area; leaf pairs 4 and 5 were still developing at the end of the experiment and have only reached the time of maximum specific leaf area, and even this is poorly defined.

Two trends of specific leaf area are apparent for this species. Firstly, both the maximum and asymptotic values appear to decline with increasing leaf pair number. Secondly, and associated with the first trend, there is an

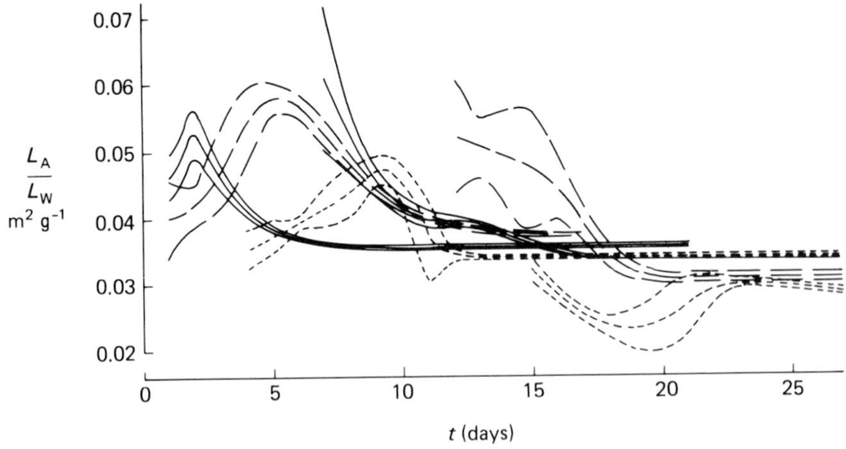

Fig. 5.14 Specific leaf area curves with 95% confidence bands for leaves 1 ——, 2 — —, 3 ----, 4 ——, 5 — — and 6 ---- of wheat.

increase in the difference between the maximum and asymptotic values. These differences are only significant between leaf pairs two and three.

Wheat

The specific leaf area trends (Fig. 5.14) for this species are very variable and somewhat difficult to explain. The final asymptotic specific leaf area decreases with increase in leaf number from leaf 2 to leaf 6; but that for leaf 1 is lower than that for leaf 2. Leaves 4 and 5 exhibit no clear maximum specific leaf area and if we assume that the initial value is near to the maximum specific leaf area there is no leaf to leaf trend in this facet of the curves. On the other hand the curves for leaves 4 and 5 may be spurious for some reason and in this case we can see that the maximum specific leaf area is inversely related to leaf number for leaves 2, 3 and 6. Also, the difference between maximum and asymptotic values decreases with leaf number for these three leaves; this is the reverse of the situation found in sunflower.

Maize

For this species, curves are only presented for the four leaves – 3, 4, 5 and 6 – this is because Richards function fitting to the leaf area of leaves 1 and 2 was not possible and the poorness of the fittings to dry weight and area of leaf 7 yielded a specific leaf area curve with enormous confidence bands (Fig. 5.15).

Maximum points on the specific leaf area curves for the four leaves are not significantly different, but that for leaf 4 appears to be lower than those for other leaves. The final asymptotic values give the appearance of declining with increasing leaf number, as in the other species examined.

Birch

Curves for alternate leaves only (1–19) are presented here (Fig. 5.16), as for the relative growth curves. Other leaves fit into the trends described below.

The final asymptotic specific leaf area tends to decrease with increase in leaf number. This is not a smooth trend as there is a fall from leaf 1 to 3, final values then only decline slightly until leaf 15 when the decrease accelerates. The asymptotic value of leaf 19 is larger than that for leaves 15 and 17 but this may just be that the final asymptotic size has not been reached.

The maximum specific leaf areas oscillate, although it could possibly be said that there is an overall decline with increasing leaf number. The curve for leaf 15 is rather different in shape to that for the other leaves and at first sight may have been a spurious result; however, the curve for leaf 16 is of a similar form and must be due to a different type of growth in these two leaves. An examination of the parameter estimates for these leaves (see Appendix A.8 and A.9, pages 270 and 273) reveals that whereas the area

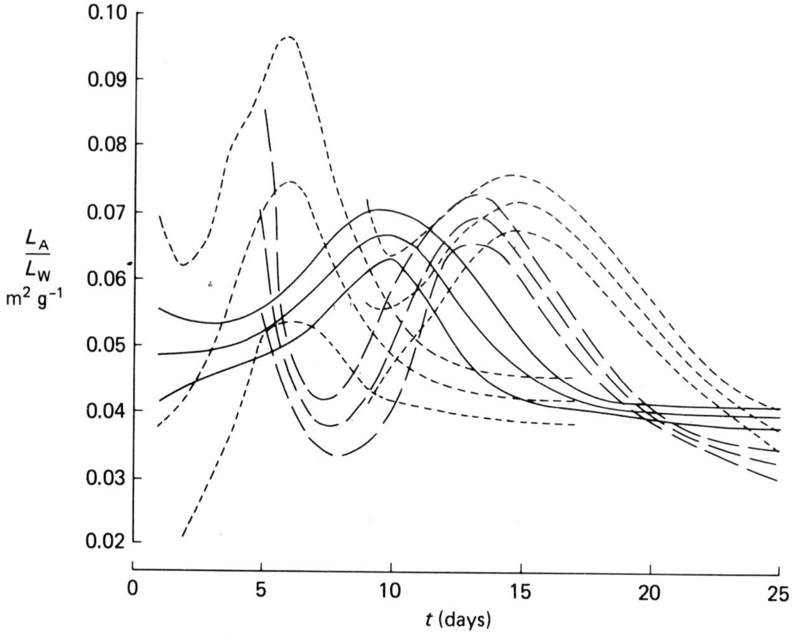

Fig. 5.15 Specific leaf area curves with 95% confidence bands for leaves 3 ----, 4 ——, 5 — — and 6 ---- of maize.

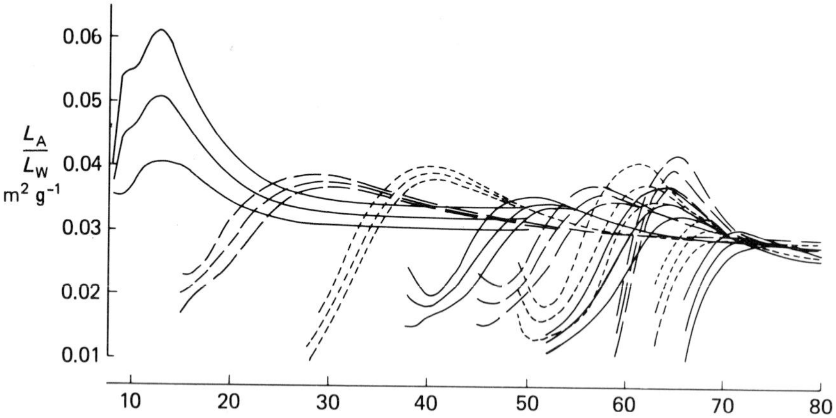

Fig. 5.16 Specific leaf area curves with 95% confidence bands for leaves 1 ——, 3 — —, 5 ----, 7 ——, 9 — —, 11 ----, 13 ——, 15 — —, 17 ---- and 19 —— of birch.

fittings give *n*-estimates that are indistinguishable from zero, the *n*-values for the weight fittings are greater than unity with large variances. This is in contrast to other leaves in which the *n*-values for the weight and area fittings are similar to one another, and this difference may well be responsible for the different specific leaf area curves.

The species compared

The maximum values attained for specific leaf area vary from species to species. Sunflower and wheat leaves give maximum specific leaf area values between 4.0 m^2g^{-1} and 6.0 m^2g^{-1}, maize has considerably higher values (6.6 m^2g^{-1} to 7.5 m^2g^{-1}), whereas the values for birch are lower (3.0 m^2g^{-1} to 4.0 m^2g^{-1} excluding leaf 1). The final asymptotic values vary less between species but they can be ranked from highest to lowest, although the differences are not significant: sunflower, maize, wheat and birch.

In sunflower the fall from maximum to asymptotic specific leaf area is often as small as 0.3 m^2g^{-1}, in birch it was larger (0.5 m^2g^{-1}) and in wheat often greater than 1.5 m^2g^{-1}. The differences were considerably larger for maize than for the other species, primarily because the maximum values were so large. Thus, each species has its own strategy of growth in terms of specific leaf area. Maize has a different photosynthetic pathway (C_4) to the other species (C_3) and a different leaf anatomy, and this may well be responsible for some of the differences. The strategy for sunflower is very different from that of the other species in that the distance between maximum and asymptotic values increases with leaf number.

In work with *Capsicum frutescens*, Steer (1971) found that the difference between minimum and maximum specific leaf area declined with increasing leaf number. This relationship is found, in the present work, for sunflower (leaf pairs 1 to 4) and wheat (leaves 2, 3 and 6). Steer found that the large changes in specific leaf area during lamina development, given by early-initiated leaves, were associated with a high degree of palisade cell expansion; later leaves had slower rates of cell expansion, but cell division continued throughout the period of lamina expansion. These conclusions are in line with other work on leaf development, that both cell division and expansion contribute to development throughout lamina expansion (Sunderland, 1960).

Earlier-initiated leaves tend to have a larger average cell size (Ashby & Wangermann, 1950a), and cell division has a less important role in leaf expansion in the early leaves when compared to later-initiated leaves. It has been found that the photosynthetic capacity of the mature leaves is higher when the palisade cells are smaller, so that later-initiated leaves have higher photosynthetic fixation rates; this has been shown for *Capsicum frutescens* (Steer, 1971) and *Lolium perenne* (Wilson & Cooper, 1969). On the other hand, Milthorpe and Newton (1963) have found a correlation between the number of cells present at unfolding and the ultimate size of the leaf in *Cucumis sativus*, but a similar correlation could not be established for *Capsicum* (Steer, 1971). Steer (1972) also found that differences in specific leaf area at unfolding gave indications of differences in cell contents. Later leaves of *Capsicum*, with smaller changes in specific leaf area, had higher rates of CO_2 fixation when the leaf was small than older leaves when they were of a similar size. This higher photosynthetic rate was associated with a greater content and activity of ribulose-1, 6-diphosphate carboxylase at the unfolding of the later-initiated leaves.

The final values of specific leaf area for birch leaves are lower than for the other species examined (3.0 to 3.75m^2g^{-1}). The greater longevity of these leaves may be responsible for this, in that there is greater deposition of strengthening and protective tissue in these leaves.

The Richards function fittings to the single leaves of these four species have given us biologically interpretable parameters and growth trends through which we have been able to examine single leaf performance in a more rigorous way than has previously been possible. Comparisons have been made both within species, between different leaves, and between species. Further information could be gained if all species had been subjected to identical environmental conditions or controlled environments, and further work under these regimes is necessary in order to realize the full potential of the Richards function in this type of experiment.

6

Relationships between plant parts

In this chapter we shall be concerned with relationships between components at a specified level (Table 1.1, page 3), and between components and the entity at that level; the components of Level 1a will predominate.

Very many years ago, zoologists had found that linear measurements made on two different parts of an animal (x and y) were often found to be related, to a good degree of approximation, by the equation

$$\log_e y = \log_e \alpha + \beta . \log_e x \tag{6.1}$$

or equivalently

$$y = \alpha x^{\beta} \tag{6.2}$$

where α is a positive constant (Huxley, 1924). Later (Huxley, 1932), the relationship was shown to have a much wider biological validity; subsequently, Sinnott (1936) found the relationship to hold between lengths and breadths in cucurbit fruits, and Whaley & Whaley (1942) provided similar data for the leaves of two *Tropaeolum* species. Again, earlier, Pearsall (1927) showed the relationship to hold between the dry weights of two parts of growing plants and, recently, the same relationship has been shown to be valid at the biochemical level (Jolicoeur & Heusner, 1971; Chanter, 1977).

Relationships of the form (6.1) and (6.2) are known as allometric relationships. Sometimes, however, when data comprising $\log_e y$ are plotted against $\log_e x$, a curvilinear rather than a linear relationship would appear to underlie the data, and so it is necessary to distinguish between linear (or simple) and curvilinear allometry. On other occasions, two or more linear segments would appear to describe the data better than a single linear or curvilinear relationship. Note also that logarithms to any base may be used in (6.1), but we shall adhere to natural logarithms for uniformity and mathematical simplicity.

In the past, the concept of simple allometry has often been applied uncritically. Unless data are sufficiently extensive and/or of low variability, a log-log plot of them will scarcely show definite trends away from linearity. Secondly, as Richards (1969) has emphasized, equation (6.1) is an emperical relationship, even in cases where the log-log plot of points might indicate a very definite linear trend. Thirdly, fitting even a linear relationship to data of this kind is not straightforward: both axes of the graph now represent random variables, and this fact introduces difficulties which do not exist when one is fitting a relationship between a single random variable and a mathematical (i.e. non random) variable (e.g. time), which has been our concern hitherto in this book.

In this chapter, we shall deal with allometric relationships between the dry weights of plant parts. The mathematical nature of, and the statistical problems associated with, allometry will be considered; so too will the physiological implications in connection with dry matter partitioning within the plant. As previously, experimental results will also be presented, and the material will serve as an introduction to the use of allometric relationships in modelling plant growth to be considered in the final chapter.

Linear allometry: mathematical aspects

Although it would be desirable to consider the physiological aspects of linear allometry first, it is logically necessary to begin with the mathematics of the situation because, at the outset, it is necessary to establish criteria to decide when linear allometry is *physically* possible and when it is not. Only then, in cases where linear allometry is a physical possibility may one critically examine data to see whether linear allometry does actually occur, and then to deduce the physiological consequences.

In this part of the chapter, the underlying mathematical relationships of allometry will be dealt with, that is, the presence of random (plant to plant) variation will be disregarded. To fix ideas, one can consider a single growing plant in which the relationships between all the parts are fully deterministic (rather than stochastic). The notation employed will be W_i to emphasize the fact that we are dealing with dry weights. The general form of functional relationship to be considered is

$$W_i = f(W_j) \qquad i \neq j \tag{6.3}$$

where W_i and W_j are the dry weights of two parts (in a vague sense at this stage) of the plant at time t; although time no longer enters explicitly into the function, it is always present in an implicit fashion. For our present purpose, the one important form of (6.3) is given in the following

Definition

If, over a finite interval of time, W_i and W_j are related in the following manner:

$$\left.\begin{aligned} W_i &= \alpha W_j^{\beta} \\ \log_e W_i &= \log_e \alpha + \beta \,.\, \log_e W_j \end{aligned}\right\} \tag{6.4}$$

then a linear allometric relationship is said to exist between the two parts, or, they are said to exhibit linear allometry.

The use of the term 'parts' has been deliberately vague, but at this stage our original divisions of a plant into component-entity systems at various levels (Table 1.1) should be reintroduced. The relevant mathematics of linear allometry will now be summarized in the form of three theorems, and the practical consequences of these theorems then discussed. For the purpose of the theorems, the terms 'entity' and 'component' should be considered for any one level in Table 1.1, and the term 'sub-component', in Theorem 6.3, then refers to a component in the next level down.

Theorems on allometry

Theorem 6.1 *In an entity consisting of Q components, if linear allometric relatinships may be shown to exist between one particular component and every other component, then linear allometric relationships exist within each of the ${}^{Q}C_2$ pairs of components.*

Let W_Q be the 'particular component' referred to above; then

$$W_Q = \alpha_i W_i^{\beta_i} \qquad i = 1, \ldots, (Q-1) \tag{6.5}$$

Eliminating W_Q between each pair of equations in (6.5),

$$\alpha_i W_i^{\beta_i} = \alpha_j W_j^{\beta_j} \qquad i, j = 1, \ldots, (Q-1)\ i \neq j$$

i.e.

$$W_i = (\alpha_j/\alpha_i)^{1/\beta_i} \,.\, W_j^{\beta_j/\beta_i} \tag{6.6}$$

which is linearly allometric. Hence, the theorem is proved.

We shall need the following lemma for the proofs of Theorems 6.2 and its corollary, 6.3, and part of 7.3. We present the result in the form most useful for our purposes.

Lemma Suppose that

$$t = \sum_{i=1}^{Q} \lambda_i t^{g_i} \tag{6.7}$$

for all t, where $\lambda_i > 0$ ($i = 1, \ldots, Q$). Then $g_1 = g_2 = \ldots = g_Q = 1$.

By grouping together the terms on the right-hand side of (6.7) with the same exponent, and renumbering, we have a relationship of the form

$$t = \sum_{i=1}^{r} \mu_i t^{h_i} \tag{6.8}$$

where $1 \leqslant r \leqslant Q$ and $h_1 < h_2 \ldots < h_r$. Any one μ_i is the sum of the λ_is of the terms in (6.7) having the same exponent (or is a single λ_i if there is only one term in (6.7) with a particular value of the exponent); hence $\mu_i > 0$ ($i = 1, \ldots, r$). We consider the cases $h_r > 1$, $h_r < 1$, and $h_r = 1$ separately.

If $h_r > 1$, divide both sides of (6.8) by t^{h_r}:

$$t^{1-h_r} = \sum_{i=1}^{r-1} \mu_i t^{h_i - h_r} + \mu_r \tag{6.9}$$

(if $r = 1$, then the summation term is absent). Now let $t \to \infty$ in (6.9). Since $1 - h_r < 0$ and $h_i < h_r$ ($i = 1, \ldots, r-1$), the left-hand side of (6.9) tends to zero and the right-hand side to μ_r. Hence $\mu_r = 0$, a contradiction.

If $h_r < 1$, we divide (6.8) by t, giving

$$1 = \sum_{i=1}^{r-1} \mu_i t^{h_i - 1}$$

All the $h_i - 1$ are negative, so as $t \to \infty$, the right-hand side tends to zero; again a contradiction.

If $h_r = 1$, dividing (6.8) by t now gives

$$1 = \sum_{i=1}^{r-1} \mu_i t^{h_i - 1} + \mu_r$$

Letting $t \to \infty$, as before, now gives $\mu_r = 1$ and, in the limit,

$$\sum_{i=1}^{r-1} \mu_i t^{h_i - 1} = 0 \tag{6.10}$$

Now divide (6.10) by $t^{h_{r-1}-1}$ which yields

$$\sum_{i=1}^{r-2} \mu_i t^{h_i - h_{r-1}} + \mu_{r-1} = 0$$

Letting $t \to \infty$ again, we have $\mu_{r-1} = 0$ and, in the limit,

$$\sum_{i=1}^{r-2} \mu_i t^{h_i - h_{r-1}} = 0$$

Repeating this procedure as often as necessary finally gives $\mu_r = 1$, and

$$\mu_{r-1} = \mu_{r-2} = \ldots = \mu_1 = 0$$

Thus the right-hand side of (6.8) consists of one term only, with exponent of unity. Hence, all the g_i ($i = 1, \ldots, Q$) must be 1, as required.

Theorem 6.2 *In an entity consisting of Q components, each of which has a linear allometric relationship with every other component, the relationship between any particular component and the entity will not be linearly allometric unless the exponents of all* ${}^{Q}C_2$ *relationships are unity.*

Let W_Q be the 'particular component' referred to above; then

$$W_i = \alpha_i W_Q^{\beta_i} \qquad i = 1, \ldots, (Q-1)$$

Define W to be the dry weight of the entity, then

$$W = \sum_{i=1}^{Q-1} \alpha_i W_Q^{\beta_i} + W_Q \tag{6.11}$$

If W has a linear allometric relationship with W_Q, then we may write

$$W_Q = \alpha_Q W^{\beta_Q} \tag{6.12}$$

Substituting (6.12) for W_Q in (6.11), we have

$$W = \sum_{i=1}^{Q-1} \alpha_i \alpha_Q^{\beta_i} W^{\beta_Q \beta_i} + \alpha_Q W^{\beta_Q}$$

identically. By the lemma,

$$\beta_Q = \beta_Q \beta_i = \ldots = \beta_{Q-1} = 1$$

that is, all the exponents are unity, and the theorem is proved.

Corollary *In an entity consisting of Q components, no more than (Q – 1) of these may have linear allometric relationships with the entity, unless all exponents of such relationships are unity.*

Suppose, if possible, that all Q components have linear allometric relationships with the entity, i.e.

$$W_i = \alpha_i W^{\beta_i} \qquad i = 1, \ldots, Q$$

then

$$W = \sum_{i=1}^{Q} W_i = \sum_{i=1}^{Q} \alpha_i W^{\beta_i}$$

By the lemma, all β_i are unity, thus proving the second part of the corollary.

Now let $\quad W_i = \alpha_i W^{\beta_i} \qquad i = 1, \ldots, (Q-1)$

Then

$$W = \sum_{i=1}^{Q-1} \alpha_i W^{\beta_i} + W_Q$$

which shows that $(Q-1)$ components may have linear allometric relationships with the entity, and thus proving the first part of the corollary.

Theorem 6.3 *An entity consists of Q components. One of them, W_Q (say), consists of q sub-components. One of these sub-components, $W_{Q,1}$ (say), has linear allometric relationships with all the other q sub-components, and also with each of the other (Q – 1) components. Under these conditions, the relationships between W_Q and W_j, $j = 1, \ldots, (Q-1)$, are not generally linear allometric.*

By definition, we have relationships of the form

$$W_{Q,i} = \alpha_i W_{Q,1}^{\beta_i} \qquad i = 2, \ldots, q \tag{6.13}$$

$$W_{Q,1} = \theta_j W_j^{\gamma_j} \qquad j = 1, \ldots, (Q-1) \tag{6.14}$$

and
$$W_Q = \sum_{i=1}^{q} W_{Q,i}$$

From (6.13)
$$W_Q = W_{Q,1} + \sum_{i=2}^{q} \alpha_i W_{Q,1}^{\beta_i}$$

and from (6.14)

$$W_Q = \theta_j W_j^{\gamma_j} + \sum_{i=2}^{q} \alpha_i \theta_j^{\beta_i} W_j^{\beta_i \gamma_j} \tag{6.15}$$

If W_Q and W_j are also linearly allometrically related, then there is a relationship

$$W_j = \alpha W_Q^{\beta} \tag{6.16}$$

Substituting for W_j in (6.15), using (6.16), we have

$$W_Q = \theta_j \alpha^{\gamma_j} W_Q^{\beta\gamma_j} + \sum_{i=2}^{q} \alpha_i \theta_j^{\beta_i} \alpha^{\beta_i \gamma_j} W_Q^{\beta\beta_i\gamma_j}$$

By the lemma, $\beta\gamma_j = \beta\beta_i\gamma_j = 1 \quad (j = 1, \ldots, (Q-1); i = 2, \ldots, q)$.

So
$$\beta_2 = \beta_3 = \ldots = \beta_q = 1$$

This means that if and only if the exponents of the linear allometric relationships between the sub-components of a component are all unity, then a linear allometric relationship *does* exist between W_Q and the other W_j. To see this, put $\beta_i = 1$ in (6.15), then

$$W_Q = \theta_j W_j^{\gamma_j} + \sum_{i=2}^{q} \alpha_i \theta_j W_j^{\gamma_j}$$

i.e.
$$W_Q = \theta_j \left(1 + \sum_{i=2}^{q} \alpha_i\right) W_j^{\gamma_j}$$

which is allometric in nature.

We may note also that if $W_{Q,1}$ has a linear allometric relationship with $W_{Q,i}$ $(i = 2, \ldots, q)$ (equation (6.13)), and if $W_{Q,1}$ is also related to W_j $(j = 1, \ldots, Q-1)$ (equation (6.14)); then all the $W_{Q,i}$ have linear allometric relationships with the W_j, and also that all the $W_{Q,i}$ have linear allometric relationships with all the $W_{Q,j}$ $(i \neq j)$ (Theorem 6.1).

Organ systems in which linear allometry may occur

The outstanding consequence of Theorems 6.2 and 6.3 is the demonstration of the 'precarious' nature of linear allometry from a mathematical viewpoint. While Theorem 6.1 shows that all N components of an entity can exhibit linear allometry amongst themselves, Theorem 6.2 shows that, in general, the concept of linear allometry cannot simultaneously apply among components and a component in relation to its entity at the same level. Several authors (e.g. Pearsall, 1927; Troughton, 1955, 1956, 1960, 1967; Brouwer, Jenneskens & Borggreve, 1961; Hunt, 1975; Stanhill, 1977a,b) have described linear allometric relationships between root and shoot weights. If this is true, then the relationship between either shoot or root and total plant cannot be linear.

The result of Theorem 6.3 can be applied to Levels 1 and 1a. Let the entity be the whole plant, W_N, the shoot which consists of two subcomponents – foliage and stem – linearly related to each other, and also let the foliage and stem be each linearly related to the root. Then, according to the result of Theorem 6.3, the root and shoot are not linearly related. Thus, if the observations of the above authors are correct, the second result of Theorem 6.3 shows that either the gradient of the linear allometric relationship between foliage and stem must be unity, or else there is no linear allometric relationship between these two organs, or between foliage, stem and root in any combination. The existence of linear allometric relationships between components of Level 1a have more rarely been described, but such relationships were postulated by Causton (1970) and Good & Good (1976).

Critical evaluations of the existence of particular linear allometric relationships before their application never appear to have been undertaken, but it must be admitted that an attempt to distinguish between whether (say) a root and a shoot had a linear relationship, or whether each had a linear relationship with the whole plant would be difficult, because of the small degree of curvature that would exist in a curvilinear allometric relationship. If the result of Theorem 6.2 is applied to this situation, we have

$$R = \alpha C^{\beta}$$

where C is shoot dry weight, and R is root dry weight (note again the distinction between italic R for root dry weight and non-italic R for whole plant relative growth rate); and so

$$\log_e W = \log_e C + \log_e(1 + \alpha C^{(\beta - 1)}) \qquad (6.17)$$

Differentiating (6.17), we have

$$\frac{d(\log_e W)}{d(\log_e C)} = \frac{1 + \beta\alpha C^{(\beta - 1)}}{1 + \alpha C^{(\beta - 1)}} \qquad (6.18)$$

Since $0.5 \leqslant \beta \leqslant 1.5$ usually in a vegetatively growing plant, it can be seen that the change in gradient with increasing C is not very great.

The existence of linear allometric relationships at other levels has scarcely been investigated, apart from in some biochemical features. A distinction has already been drawn between entities showing determinate and indeterminate growth patterns (Chapter 4, pages 86–87), but for the study of relationships between plant parts, the alternative concept of organs showing sequential or simultaneous growth is more appropriate. Thus, the components of Levels 1, 1a, and 2 grow simultaneously, for example foliage grows while the root grows; but even at this level there is sometimes more of a sequential pattern, for example, in a rosette plant much of the foliage growth occurs before that of the stem. Components at Level 3 and below grow sequentially, for example individual leaves on the main stem or a particular branch are formed sequentially at the apex and so go through their growth pattern in a sequential manner; early formed leaves will have ceased growth while leaf initiation still occurs at the apex, and at any one time there are leaves present in between the extremes which are at a whole range of different stages of development. The interrelationships between a group of components, i.e. whether they grow sequentially or simultaneously, is often correlated with each individual component's growth pattern; thus, individual leaves on the main stem or on a particular branch show determinate growth and grow sequentially, whereas individual stems can show indeterminate growth and grow simultaneously. A moment's reflection will show that a group of components exhibiting sequential *and* determinate growth cannot have linear allometric relationships between themselves except over very short time periods.

Allometry and the Richards function

Theorem 6.4 *If the growth of two components of an entity can be described by Richards functions,* $W_i = A_i(1 \pm e^{(\beta_i - \kappa_i t)})^{1-\nu_i}$ $i = 1, 2$; *then if* $\beta_1 = \beta_2$ and $\kappa_1 = \kappa_2$, *the two components have a linear allometric relationship.*

Let $u = (1 \pm e^{(\beta - \kappa t)})$ where $\beta = \beta_1 = \beta_2$ and $\kappa = \kappa_1 = \kappa_2$, then

$$W_i = A_i u^{-1/\nu_i} \qquad i = 1, 2 \tag{6. 19}$$

On eliminating u between the two components of (6. 19) and rearranging, we obtain

$$W_1 = A_1 A_2^{-\nu_2/\nu_1} . W_2^{\nu_2/\nu_1} \tag{6.20}$$

Equation (6.20) is linearly allometric, and so the theorem is proved; it can easily be extended to N components. The main feature of interest here is that the gradient of the allometric line is equal to the ratio ν_2/ν_1.

Linear allometry: physiological aspects

Undoubtedly the most interesting biological feature of linear allometry is obtained by differentiating the second equation of (6.4) with respect to time:

$$\frac{d(\log_e W_i)}{dt} = \beta . \frac{d(\log_e W_j)}{dt}$$

i.e.

$$\beta = R_i/R_j \tag{6.21}$$

Thus, the gradient of the line is equal to the ratio of the relative growth rates of the two components, and so can provide direct information on the partitioning of assimilates between plant parts. In terms of sources and sinks for assimilates, Warren Wilson (1972) has suggested that for actively growing tissues, relative growth rate could be regarded as a measure of sink activity; thus, β may be regarded as a ratio of sink activities for two different parts of a plant where there is a linear allometric relationship between them.

Earlier than Warren Wilson's suggestion, Nelder (1963) proposed a very simple supply and demand model for assimilates, which was subsequently amplified by Causton (1970). Nelder's hypothesis was that if β is not equal to unity for two parts of a plant, then the growth of the parts is limited by their relative demands for assimilates; whereas if $\beta = 1$ growth is most probably supply limited, but *could* also be demand limited. This may be clarified by writing (6.21) in the form

$$\frac{\delta W_1}{\delta W_2} = \beta\left(\frac{W_1}{W_2}\right)$$

i.e. in a small time interval, δt, the ratio of the increments in the dry weights of the two parts are proportional to the ratio of their sizes. Consider a rosette plant, comprising root and leaves only. Now if assimilates are limited in supply, and there is no 'demand control' by either of the two parts, then the ratio of increments of the two components would be equal to the ratio of their sizes ($\beta = 1$). On the other hand, if the parts are exerting a differential demand for assimilates, then $\beta \neq 1$ unless the two demands happen to be the same. Since values of β significantly different from unity appear to be more common than otherwise, one can infer that demand limited growth situations in plants are commoner than supply limited ones. It must be emphasized, however, that these ideas are strictly tenable only in very young plants in which the whole of each part is actively growing, unless the proportion of growing tissue to total tissue stays the same in each plant part.

Demand for assimilates by particular plant parts inevitably raises the question of hormone directed transport (see e.g. Lovell, 1971; Bowen and Wareing, 1971), so that we might infer that the slope of a linear allometric relationship is determined hormonally. However, morphology may also play a part. Again consider a rosette plant, now in the reproductive stage; now the stem is elongating and reproductive structures are growing thereon, while the possible addition of a few new, typically small, leaves on the stem will result in a low foliage relative growth rate. The root, however, may continue growing with what is now a relatively high relative growth rate, giving a β-value for root and leaves very different from unity. This situation could be regarded as one of 'negative' demand by the leaves, in that by this stage of the plant's development, the only leaves capable of growth are few in number and have a low potential growth rate.

Abrupt changes in the gradient of allometric relationships are commonly observed, and can be equated with sudden changes in the partitioning pattern (e.g. Troughton, 1956). Such changes may be associated with metabolic and/or morphological switches, notably the change from vegetative to reproductive growth, and a possible pattern for this kind of change has been outlined in the previous paragraph. A curvilinear allometric relationship, on the other hand, may indicate a decrease in the proportion of actively growing tissue in one component compared to that in the other. This is assuming that a curvilinear relationship is a true one, and not an artifact arising out of one of the many causes discussed in the previous section.

Linear allometry: statistical aspects

We shall leave aside the difficult question of assessing the existence of a single linear function, a single curvilinear function, or a combination of these in a set of data for later discussion. Assuming now that we have a set of log-log data to which the fitting of a single linear function is reasonable, the underlying philosophy of the estimation procedure will first be discussed, followed by the derivation of the procedure itself, using x–y notation.

Although not appearing explicitly in allometric functions, time is implicitly involved (page 174) since a progression along the curve of the function implies a progression in time. Also, the data for allometric relationships are obtained at each of a series of harvest times, although these distinct times are not usually apparent on a graph. However, at each harvest time there is a population of both x- and y-values, which can be visualized as in Fig. 6.1a; the three dimensional diagram can be reduced to a two dimensional one in the usual way for bivariate populations (Fig. 6.1b), thus depicting the populations on the x–y plane. If the growth of both x and y is sufficiently rapid and/or the harvests are sufficiently far apart, the data at each harvest will appear as separate clusters of points, as shown in Fig. 6.1c; otherwise the data points coalesce and appear as a band, as shown in Fig. 6.1d, the underlying population structure is then obscured but still exists.

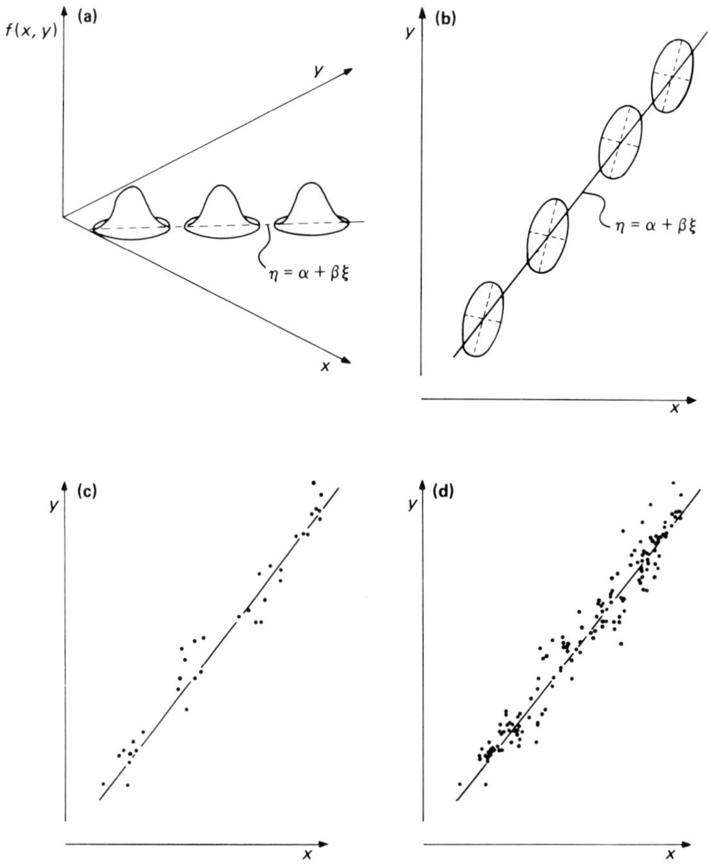

Fig. 6.1 (**a**) Population structure for linear functional relationship. (**b**) As (**a**) but projected onto the *x*-*y* plane. (**c**) Sample observations from four populations which are sufficiently distant from one another for each population to be distinguishable. (**d**) As (**c**) but with 16 populations which are no longer distinguishable.

Hence, we have the situation that both the *x*- and *y*-axes represent random variables, in contrast with the situation described in Chapter 3 where the *y*-axis only represented a random variable while *x* was a mathematical variable not subject to random variation. The distinction is vitally important, since it affects the whole fitting procedure; using the wrong fitting method will result in inefficient parameter estimation and, at least equally important, can also lead to wrong ideas about the variability structure of the data, and so in turn to erroneous statements about magnitudes of variation of the data.

There are various schemes of nomenclature for the two situations: the one we shall adopt refers to the line in which x is a mathematical variable as a regression line, and when x is a variate we have a functional relationship. Thus, our present problem is to fit a linear functional relationship to x–y data. Two review papers on the subject of functional relationships are Madansky (1959) and Moran (1971).

When fitting a regression line it was logical to start with the method of least squares, since it could be illustrated graphically, and then to demonstrate the similarity between this method and the more general one of maximum likelihood. Graphical illustration of least squares in connection with the linear functional relationship is, however, much less easy, and so the conceptual simplicity of least squares is lost. Hence, we shall use the method of maximum likelihood from the start.

Estimating a linear functional relationship by maximum likelihood

We start with the assumptions of the model as they apply to our log-log plant data.

1. At each of the h harvest times there is a bivariate population of x- and y-values, and we specify that the populations are bivariate normally distributed. Time, while not entering explicitly into the x-y relationship, defines the bivariate populations used to formulate the relationship; thus, time is called an instrumental variable.

2. The means of each population lie on the straight line $\eta_i = \alpha + \beta\xi_i$ (Fig. 6.1a,b), where ξ_i and η_i are the population means of the x- and y-values, respectively, at the ith harvest.

3. Observed x-y pairs of observations (one at each harvest for notational and conceptual simplicity) 'arise' from the underlying populations in the following manner:

$$\left.\begin{aligned} x_i &= \xi_i + \delta_i \\ y_i &= \eta_i + \epsilon_i = (\alpha + \beta\xi_i) + \epsilon_i \end{aligned}\right\} \quad i = 1, \ldots, h \tag{6.22}$$

   The δ_i and ϵ_i may thus be regarded as sample observations from bivariate normal distributions in which the mean of each variate is zero, i.e. $\mathscr{E}(\delta_i) = \mathscr{E}(\xi_i) = 0$, for all i.

4. Each of the h bivariate normal distributions has the same form. This implies $\mathscr{V}(\delta_i) = \sigma_{\delta\delta}$, $\mathscr{V}(\epsilon_i) = \sigma_{\varepsilon\varepsilon}$, $\mathscr{C}(\delta_i, \epsilon_i) = \sigma_{\delta\varepsilon}$, for all i; that is, the variances of the δ_i and ϵ_i are the same at each harvest, so also is the covariance between the two variates. The variance–covariance notation, in which there is a double subscript and no power of 2 in the case of variances, is convenient and also standard in multivariate work.

5. Each of the h observations are independent of one another.

The above five assumptions are analogous to the regression situation although, because of the bivariate as opposed to the univariate nature of the data, more assumptions have to be incorporated into the model. It should also be pointed out at this stage that whereas in the regression situation we needed to estimate 3 parameters: α, β, σ^2; now we need to estimate $h+5$ parameters: $\alpha, \beta, \sigma_{\delta\delta}, \sigma_{\varepsilon\varepsilon}, \sigma_{\delta\varepsilon}, \xi_i i = 1, \ldots, h$; although the ξ_i are usually of little or no interest, and have been termed incidental parameters (Neyman & Scott, 1951). The estimation of a linear functional relationship is obviously much more complicated than is the regression fitting of a straight line; in fact, without making additional assumptions, there is insufficient information in the above model to enable a consistent estimation to be made.

Lindley (1947) was the first to systemmatically study the linear functional relationship, although aspects of the problem had been examined earlier, for example Wald (1940). Lindley made the additional assumption that $\sigma_{\delta\varepsilon} = 0$ and showed that a maximum likelihood solution was then possible; however, it gave the nonsensical result that $\hat{\beta}^2 = \hat{\sigma}_{\varepsilon\varepsilon}/\hat{\sigma}_{\delta\delta}$, i.e. that the square of the estimated gradient was equal to the estimated variance of the ϵ_i divided by the estimated variance of the δ_i. This, now famous, result was originally taken to mean that the maximum likelihood method of estimation breaks down in the case of β for the linear functional relationship. However, it is shown in Sprent (1969) and Kendall & Stuart (1977) that, in effect, there is still insufficient information in the model for a consistent estimate of β to be obtained and, therefore, no maximum likelihood solution exists with these assumptions (Solari, 1969); evidently, more must be known or assumed about the structure of the underlying bivariate normal distributions.

If the magnitudes of $\sigma_{\delta\delta}$, $\sigma_{\varepsilon\varepsilon}$, and $\sigma_{\delta\varepsilon}$ are either known (most unlikely in practice), assumed, or can be estimated from the data, then Sprent (1966, 1968, 1969) has shown that the $h+2$ parameters, α, β, ξ_i, of the linear functional relationship can be estimated. Earlier, and less stringently, Lindley (1947) also showed that if $\sigma_{\delta\varepsilon} = 0$ and $\lambda = \sigma_{\varepsilon\varepsilon}/\sigma_{\delta\delta}$ was known or could be assumed, then the $h+3$ parameters, $\alpha, \beta, \sigma_{\delta\delta}, \xi_i$, could be estimated. Thus, in addition to the 5 assumptions listed earlier, we need a 6th assumption, consisting of the following two alternatives.

6a. The covariance of the bivariate normal distributions is zero, and the ratio of the variance is equal to λ: $\sigma_{\delta\varepsilon} = 0$ and $\lambda = \sigma_{\varepsilon\varepsilon}/\sigma_{\delta\delta}$.

6b. The variances and covariance of the bivariate normal distributions are known (or can be estimated) beforehand.

The derivation of the maximum likelihood solution using assumption 6a is relatively straightforward and will be demonstrated below; the derivation involving assumption 6b is more complicated and the result will merely be given, those interested in the proof can find it in Sprent (1969, pages 40–43).

With assumptions 1 to 5 and 6a, the likelihood of a single observation is

$$\frac{1}{2\pi(\sigma_{\delta\delta})^{1/2}(\lambda\sigma_{\delta\delta})^{1/2}}\cdot\exp\left[-\frac{1}{2}\left\{\frac{(x_i-\xi_i)^2}{\sigma_{\delta\delta}}+\frac{(y_i-\alpha-\beta\xi_i)^2}{\lambda\sigma_{\delta\delta}}\right\}\right]$$

and so the likelihood of h observations is given by

$$L=\frac{1}{(2\pi\lambda^{1/2}.\sigma_{\delta\delta})^h}\cdot\prod_{i=1}^{h}\exp\left[-\frac{1}{2\sigma_{\delta\delta}}\left\{(x_i-\xi_i)^2+\frac{1}{\lambda}(y_i-\alpha-\beta\xi_i)^2\right\}\right]$$

with

$$\log_e L=-h.\log_e(2\pi\lambda^{1/2})-h.\log_e\sigma_{\delta\delta}-\frac{1}{2\sigma_{\delta\delta}}\cdot\sum_{i=1}^{h}\left\{(x_i-\xi_i)^2+\frac{1}{\lambda}(y_i-\alpha-\beta\xi_i)^2\right\} \tag{6.23}$$

Estimation of the parameters

Differentiating (6.23) successively with respect to α, β, $\sigma_{\delta\delta}$, ξ_i, gives:

$$\frac{\partial(\log_e L)}{\partial\alpha}=\frac{1}{\lambda\sigma_{\delta\delta}}\cdot\sum(y_i-\alpha-\beta\xi_i) \tag{6.24}$$

$$\frac{\partial(\log_e L)}{\partial\beta}=\frac{1}{\lambda\sigma_{\delta\delta}}\cdot\sum(y_i-\alpha-\beta\xi_i)\xi_i \tag{6.25}$$

$$\frac{\partial(\log_e L)}{\partial\sigma_{\delta\delta}}=-\frac{h}{\sigma_{\delta\delta}}+\frac{1}{2\sigma_{\delta\delta}^2}\cdot\sum\left\{(x_i-\xi_i)^2+\frac{1}{\lambda}(y_i-\alpha-\beta\xi_i)^2\right\} \tag{6.26}$$

$$\frac{\partial(\log_e L)}{\partial\xi_i}=\frac{1}{\sigma_{\delta\delta}}\left\{(x_i-\xi_i)+\frac{\beta}{\lambda}(y_i-\alpha-\beta\xi_i)\right\}\quad i=1,\ldots,h \tag{6.27}$$

The resemblance of equations (6.24) and (6.25) to their regression counterparts (page 75) is evident, except that the unknown ξ_i replaces the known x_i.

On equating (6.27) to zero and rearranging, we obtain

$$\hat{\xi}_i=\frac{\lambda x_i+\hat{\beta}y_i-\hat{\alpha}\hat{\beta}}{\lambda+\hat{\beta}^2}\qquad i=1,\ldots,h \tag{6.28}$$

Putting (6.24) equal to zero, substituting for ξ_i using (6.28), and doing a preliminary rearrangement yields

$$\sum\{y_i(\lambda+\hat{\beta}^2)-\hat{\alpha}(\lambda+\hat{\beta}^2)-\hat{\beta}(\lambda x_i+\hat{\beta}y_i-\hat{\alpha}\hat{\beta}\}=0$$

which gives the result

$$\hat{\alpha} = \bar{y} - \hat{\beta}\bar{x} \tag{6.29}$$

where $\bar{x} = (1/h).\Sigma x_i$ and $\bar{y} = (1/h).\Sigma y_i$. Equating (6.25) to zero, substituting for ξ_i using (6.28), and effecting a preliminary rearrangement gives

$$\sum\{y_i(\lambda+\hat{\beta}^2) - \hat{\alpha}(\lambda+\hat{\beta}^2) - \hat{\beta}(\lambda x_i + \hat{\beta}y_i - \hat{\alpha}\hat{\beta})\}\{\lambda x_i + \hat{\beta}y_i - \hat{\alpha}\hat{\beta}\} = 0 \tag{6.30}$$

Since, from (6.29) $\Sigma(y_i - \hat{\alpha} - \hat{\beta}x_i) = 0$, $\hat{\beta}.\Sigma y_i - h\hat{\alpha}\hat{\beta} = \hat{\beta}^2\Sigma x_i$ (third term in the first pair of braces in (6.30)): (6.30) then reduces to

$$\sum\{y_i - \hat{\alpha} - \hat{\beta}x_i\}\{\lambda x_i + \hat{\beta}y_i - \hat{\alpha}\hat{\beta}\} = 0$$

Substituting for $\hat{\alpha}$, using (6.29), gives

$$\sum\{(y_i - \bar{y}) - \hat{\beta}(x_i - \bar{x})\}\{\hat{\beta}(y_i - \bar{y}) + \lambda x_i + \hat{\beta}^2\bar{x}\} = 0$$

which eventually yields the following quadratic in $\hat{\beta}$:

$$\hat{\beta}^2\sum(x_i - \bar{x})(y_i - \bar{y}) - \hat{\beta}\{\sum(y_i - \bar{y})^2 - \lambda(x_i - \bar{x})^2\} \\ - \lambda\cdot\sum(x_i - \bar{x})(y_i - \bar{y}) = 0 \tag{6.31}$$

Solution of (6.31) is

$$\hat{\beta} = \frac{\{\Sigma(y_i - \bar{y})^2 - \lambda\cdot\sum(x_i - \bar{x})^2\} \pm \sqrt{[\{\sum(y_i - \bar{y})^2 - \lambda.\sum(x_i - \bar{x})^2\}^2 - 4\lambda\cdot\sum^2(x_i - \bar{x})(y_i - \bar{y})]}}{2.\sum(x_i - \bar{x})(y_i - \bar{y})} \tag{6.32}$$

If the positive square root is used, $\hat{\beta}$ has the same sign as the sum of cross products, $\Sigma(x_i - \bar{x})(y_i - \bar{y})$, which is sensible.

It will be noticed that the right-hand side of (6.32) consists only of terms involving the observed data, together with the assumed ratio of variances, λ; hence $\hat{\beta}$ can be evaluated directly from the data. Substitution in (6.29) then yields $\hat{\alpha}$, and substitution into the h equations of (6.28) enables the ξ_i to be evaluated, if required.

From the third of the partial derivatives (6.26) equated to zero, we immediately obtain

$$\hat{\sigma}_{\delta\delta} = \frac{1}{2h}\cdot\sum\left\{(x_i - \hat{\xi}_i)^2 + \frac{1}{\lambda}(y_i - \hat{\alpha} - \hat{\beta}\hat{\xi}_i)^2\right\} \tag{6.33}$$

and so the magnitude of $\hat{\sigma}_{\delta\delta}$ can be found by substituting for $\hat{\alpha}$, $\hat{\beta}$, and $\hat{\xi}_i$, already obtained. However, by substituting for $\hat{\xi}_i$ in (6.33), and after some tedious algebra, we find that

$$\hat{\sigma}_{\delta\delta} = \frac{\lambda^2 + \hat{\beta}^2}{2h(\lambda + \hat{\beta}^2)}\cdot\sum\{(y_i - \bar{y}) - \hat{\beta}(x_i - \bar{x})\}^2 \tag{6.34}$$

a result that does not require explicit estimates of the ξ_i.

Variances and covariances of the parameter estimates

From equations (6.24), (6.25), and (6.27), we have the following second order partial derivatives:

$$\frac{\partial^2(\log_e L)}{\partial\alpha^2} = -\frac{h}{\lambda\sigma_{\delta\delta}} \tag{6.35}$$

$$\frac{\partial^2(\log_e L)}{\partial\alpha.\partial\beta} = -\frac{1}{\lambda\sigma_{\delta\delta}}\cdot\sum\xi_i \tag{6.36}$$

$$\frac{\partial^2(\log_e L)}{\partial\beta^2} = -\frac{1}{\lambda\sigma_{\delta\delta}}\cdot\sum\xi_i^2 \tag{6.37}$$

$$\frac{\partial^2(\log_e L)}{\partial\alpha.\partial\xi_i} = -\frac{\beta}{\lambda\sigma_{\delta\delta}} \tag{6.38}$$

$$\frac{\partial^2(\log_e L)}{\partial\beta.\partial\xi_i} = \frac{1}{\lambda\sigma_{\delta\delta}}(y_i-\alpha-2\beta\xi_i) \tag{6.39}$$

$$\frac{\partial^2(\log_e L)}{\partial\xi_i^2} = -\frac{1}{\lambda\sigma_{\delta\delta}}(\lambda+\beta^2) \tag{6.40}$$

$$\frac{\partial^2(\log_e L)}{\partial\xi_i.\partial\xi_j} = 0 \tag{6.41}$$

Equations (6.38) to (6.40) are for $i = 1, \ldots, h$; (6.41) is for $i \neq j$, and this result follows because of the independence of the observations. Hence, the following information matrix can be defined:

$$\mathbf{I} = \left[\begin{array}{cc|ccccc}
\frac{\partial^2(\log_e L)}{\partial\alpha^2} & \frac{\partial^2(\log_e L)}{\partial\alpha.\partial\beta} & \frac{\partial^2(\log_e L)}{\partial\alpha.\partial\xi_1} & \cdots & \frac{\partial^2(\log_e L)}{\partial\alpha.\partial\xi_i} & \cdots & \frac{\partial^2(\log_e L)}{\partial\alpha.\partial\xi_h} \\
\frac{\partial^2(\log_e L)}{\partial\alpha.\partial\beta} & \frac{\partial^2(\log_e L)}{\partial\beta^2} & \frac{\partial^2(\log_e L)}{\partial\beta.\partial\xi_1} & \cdots & \frac{\partial^2(\log_e L)}{\partial\beta.\partial\xi_i} & \cdots & \frac{\partial^2(\log_e L)}{\partial\beta.\partial\xi_h} \\
\hline
\frac{\partial^2(\log_e L)}{\partial\alpha.\partial\xi_1} & \frac{\partial^2(\log_e L)}{\partial\beta.\partial\xi_1} & \frac{\partial^2(\log_e L)}{\partial\xi_1^2} & \cdots & \frac{\partial^2(\log_e L)}{\partial\xi_1.\partial\xi_i} & \cdots & \frac{\partial^2(\log_e L)}{\partial\xi_1.\partial\xi_h} \\
\vdots & \vdots & \vdots & & \vdots & & \vdots \\
\frac{\partial^2(\log_e L)}{\partial\alpha.\partial\xi_i} & \frac{\partial^2(\log_e L)}{\partial\beta.\partial\xi_i} & \frac{\partial^2(\log_e L)}{\partial\xi_i.\partial\xi_1} & \cdots & \frac{\partial^2(\log_e L)}{\partial\xi_i^2} & \cdots & \frac{\partial^2(\log_e L)}{\partial\xi_i.\partial\xi_h} \\
\vdots & \vdots & \vdots & & \vdots & & \vdots \\
\frac{\partial^2(\log_e L)}{\partial\alpha.\partial\xi_h} & \frac{\partial^2(\log_e L)}{\partial\beta.\partial\xi_h} & \frac{\partial^2(\log_e L)}{\partial\xi_h.\partial\xi_1} & & \frac{\partial^2(\log_e L)}{\partial\xi_h.\partial\xi_i} & & \frac{\partial^2(\log_e L)}{\partial\xi_h^2}
\end{array}\right] \tag{6.42}$$

Strictly, when dealing with the variances and covariances of the parameters,

I should be inverted as a whole. This may be cumbersome, however, and it would be impossible to give an analytical solution.

An alternative approach, which may provide a good approximation, is to partition **I** as shown in (6.42), which can then be written more succinctly as

$$\mathbf{I} = \begin{bmatrix} \mathbf{B} & \boldsymbol{\Omega}' \\ \boldsymbol{\Omega} & \boldsymbol{\Xi} \end{bmatrix} \tag{6.43}$$

where **B** is a (2 × 2) symmetric matrix involving the parameters α and β, $\boldsymbol{\Xi}$ is an $(h \times h)$ diagonal matrix involving the incidental parameters ξ_i, and $\boldsymbol{\Omega}$ is an $(h \times 2)$ matrix involving both sets of parameters. Specifically, the elements of these matrices are as follows:

$$\mathbf{B} = -\frac{1}{\lambda\sigma_{\delta\delta}} \begin{bmatrix} h & \Sigma\xi_i \\ \sum\xi_i & \Sigma\xi_i^2 \end{bmatrix} \tag{6.44}$$

$$\Xi = -\frac{\lambda+\beta^2}{\lambda\sigma_{\delta\delta}} \begin{bmatrix} 1 \ldots\ldots 0 \ldots\ldots 0 \\ \vdots \qquad \vdots \qquad \vdots \\ 0 \ldots\ldots 1 \ldots\ldots 0 \\ \vdots \qquad \vdots \qquad \vdots \\ 0 \ldots\ldots 0 \ldots\ldots 1 \end{bmatrix} \tag{6.45}$$

$$\boldsymbol{\Omega} = \frac{1}{\lambda\sigma_{\delta\delta}} \begin{bmatrix} -\beta & (y_1-\alpha-2\beta\xi_1) \\ \vdots & \vdots \\ -\beta & (y_i-\alpha-2\beta\xi_i) \\ \vdots & \vdots \\ -\beta & (y_h-\alpha-2\beta\xi_h) \end{bmatrix} \tag{6.46}$$

For the function parameters,

$$|B| = -\frac{1}{(\lambda\sigma_{\delta\delta})^2}(h.\sum\xi_i^2 - \sum{}^2\xi_i)$$

and so

$$-\mathbf{B}^{-1} = \frac{\lambda\hat{\sigma}_{\delta\delta}/h}{\sum(\hat{\xi}_i-\bar{\xi})^2} \begin{bmatrix} \sum\xi_i^2 & -\sum\xi_i \\ -\sum\xi_i & h \end{bmatrix} \tag{6.47}$$

where $\bar{\xi} = (1/h).\Sigma\hat{\xi}_i$

Hence, the variance of $\hat{\beta}$ is given by

$$\mathscr{V}(\hat{\beta}) = \frac{\lambda\hat{\sigma}_{\delta\delta}}{\sum(\hat{\xi}_i-\bar{\xi})^2} \tag{6.48}$$

Now, from (6.28)

$$\sum\hat{\xi}_i = \frac{1}{\lambda+\hat{\beta}^2}\cdot\sum(\lambda x_i + \hat{\beta}y_i - \hat{\alpha}\hat{\beta})$$

and, substituting for $\hat{\alpha}$ from (6.29) gives

$$\sum\hat{\xi}_i = \frac{1}{\lambda+\hat{\beta}^2}\cdot\sum\{\lambda x_i + \hat{\beta}(y_i-\overline{y}) + \hat{\beta}^2\overline{x}\}$$

and so

$$\sum\hat{\xi}_i = \frac{1}{\lambda+\hat{\beta}^2}(\lambda\,.\sum x_i + h\hat{\beta}^2\overline{x})$$

as $\Sigma(y_i-\overline{y}) = \hat{\beta}\,.\,\Sigma(y_i-\overline{y}) = 0$. Hence, since $\overline{x} = (1/h)\,.\,\Sigma x_i$ and so $h\hat{\beta}^2\overline{x} = \hat{\beta}^2\Sigma x_i$,

$$\sum\hat{\xi}_i = \sum x_i$$

and

$$\overline{\xi} = \overline{x} \tag{6.49}$$

Again, using (6.28) and (6.29)

$$\sum(\hat{\xi}_i - \overline{\xi})^2 = \sum\left\{\frac{\hat{\beta}(y_i-\overline{y}) + \lambda x_i + \hat{\beta}^2\overline{x}_i}{\lambda+\hat{\beta}^2} - \overline{x}\right\}^2$$

which reduces to

$$\sum(\hat{\xi}_i - \overline{\xi})^2 = \frac{1}{(\lambda+\hat{\beta}^2)^2}\cdot\sum\{\hat{\beta}(y_i-\overline{y}) + \lambda(x_i-\overline{x})\}^2 \tag{6.50}$$

Substituting (6.50) into (6.48) gives

$$\mathscr{V}(\hat{\beta}) = \frac{\lambda\hat{\sigma}_{\delta\delta}(\lambda+\hat{\beta}^2)^2}{\sum\{\hat{\beta}(y_i-\overline{y}) + \lambda(x_i-\overline{x})\}^2} \tag{6.51}$$

Reverting to (6.47), the variance of $\hat{\alpha}$ is given by

$$\mathscr{V}(\hat{\alpha}) = \frac{\lambda\hat{\sigma}_{\delta\delta}(\sum\hat{\xi}_i^2)/h}{\sum(\hat{\xi}_i - \hat{\xi})^2} \tag{6.52}$$

Now, from (6.28) and (6.29)

$$\sum\xi_i^2 = \frac{1}{(\lambda+\hat{\beta}^2)^2}\cdot\sum\{\hat{\beta}(y_i-\overline{y}) + \lambda x_i + \hat{\beta}^2\overline{x}\}^2 \tag{6.53}$$

and substitution of (6.50) and (6.53) into (6.52) gives

$$\mathscr{V}(\hat{\alpha}) = \frac{\lambda\hat{\sigma}_{\delta\delta}\cdot\sum\{\hat{\beta}(y_i-\overline{y}) + \lambda x_i + \hat{\beta}^2\overline{x}\}^2}{h\,.\sum\{\hat{\beta}(y_i-\overline{y}) + \lambda(x_i-\overline{x})\}^2} \tag{6.54}$$

Finally from (6.47), the covariance of $\hat{\alpha}$ and $\hat{\beta}$ is given by

$$\mathscr{C}(\hat{\alpha}, \hat{\beta}) = -\frac{\lambda \hat{\sigma}_{\delta\delta}(\sum \hat{\xi}_i)/h}{\sum(\hat{\xi}_i - \bar{\hat{\xi}})^2}$$

which, in consequence of (6.49) and (6.50), is the same as

$$\mathscr{C}(\hat{\alpha}, \hat{\beta}) = \frac{\lambda \hat{\sigma}_{\delta\delta}(\lambda + \hat{\beta}^2)^2 \bar{x}}{\sum\{\hat{\beta}(y_i - \bar{y}) + \lambda(x_i - \bar{x})\}^2} \tag{6.55}$$

From (6.45), the variance of each ξ_i is given by

$$\mathscr{V}(\hat{\xi}_i) = \frac{\lambda \hat{\sigma}_{\delta\delta}}{\lambda + \hat{\beta}^2} \tag{6.56}$$

The maximum likelihood solution with Assumption 6b

First define the two matrices

$$\mathbf{X} = \begin{bmatrix} \Sigma(x_i - \bar{x})^2 & \Sigma(x_i - \bar{x})(y_i - \bar{y}) \\ \Sigma(x_i - \bar{x})(y_i - \bar{y}) & \Sigma(y_i - \bar{y})^2 \end{bmatrix}$$

and

$$\mathbf{\Sigma} = \begin{bmatrix} \sigma_{\delta\delta} & \sigma_{\delta\varepsilon} \\ \sigma_{\delta\varepsilon} & \sigma_{\varepsilon\varepsilon} \end{bmatrix}$$

where the elements of $\mathbf{\Sigma}$ are either known, assumed, or estimated prior to fitting the line. Where there are several (N) replicate values of x-y data from each population (as there will be in a plant growth experiment), obvious estimates of $\sigma_{\delta\delta}$, $\sigma_{\varepsilon\varepsilon}$, and $\sigma_{\delta\varepsilon}$, respectively, are

$$\left.\begin{aligned} s_{\delta\delta} &= (1/h) \,.\, \sum_{i=1}^{h} s_{\delta\delta}^{(i)} \\ s_{\varepsilon\varepsilon} &= (1/h) \,.\, \sum_{i=1}^{h} s_{\varepsilon\varepsilon}^{(i)} \\ s_{\delta\varepsilon} &= (1/h) \,.\, \sum_{i=1}^{h} s_{\delta\varepsilon}^{(i)} \end{aligned}\right\} \tag{6.57}$$

where for example $s_{\delta\varepsilon}^{(i)} = \dfrac{1}{N-1} \,.\, \sum_{j=1}^{N} (x_{ij} - \bar{x}_i)(y_{ij} - \bar{y}_i)$

and $\qquad \bar{x}_i = (1/N) \,.\, \sum_{j=1}^{N} x_{ij} \qquad \bar{y}_i = (1/N) \,.\, \sum_{j=1}^{N} y_{ij}$

Then

$$\hat{\beta} = \frac{\Sigma(x - \bar{x})(y - \bar{y}) - \phi\sigma_{\delta\varepsilon}}{\Sigma(x - \bar{x})^2 - \phi\sigma_{\delta\delta}} \tag{6.58}$$

where $$\phi = \frac{F - \sqrt{(F^2 - 4|\mathbf{X}||\mathbf{\Sigma}|)}}{2|\mathbf{\Sigma}|} \tag{6.59}$$

and $$F = \sigma_{\delta\delta}.\Sigma(y-\bar{y})^2 + \sigma_{\varepsilon\varepsilon}.\Sigma(x-\bar{x})^2 - 2\sigma_{\delta\varepsilon}.\Sigma(x-\bar{x})(y-\bar{y}) \tag{6.60}$$

All summations in (6.58) and (6.60) are for all observations.

As before $$\hat{\alpha} = \bar{y} - \hat{\beta}\bar{x} \tag{6.61}$$

If required, the $\hat{\xi}_i$ can be obtained from

$$\hat{\xi}_i = (x_i - \bar{x}) - \frac{(y_i - \hat{\alpha} - \hat{\beta}x_i)(\sigma_{\delta\varepsilon} - \hat{\beta}\sigma_{\delta\delta})}{\sigma_{\varepsilon\varepsilon} - 2\hat{\beta}\sigma_{\delta\varepsilon} + \hat{\beta}^2\sigma_{\delta\delta}} \tag{6.62}$$

Variances and covariances of the parameter estimates are not given by Sprent, neither shall we derive them here since the estimation procedure using Assumption 6a will be employed for the applications which follow.

Confidence intervals for the line

The underlying linear relationship has been specified as

$$\eta = \alpha + \beta\xi \tag{6.63}$$

(page 184). Suppose we now select a particular value of ξ, for which we require an estimate of η with its variance. Specifying a *particular* value of ξ implies that this quantity is no longer a variable, let alone a random one, and so we have

$$\hat{\eta} = \hat{\alpha} + \hat{\beta}\xi \tag{6.64}$$

and $$\mathscr{V}(\hat{\eta}) = \mathscr{V}(\hat{\alpha}) + \xi^2.\mathscr{V}(\hat{\beta}) + 2\xi.\mathscr{C}(\hat{\alpha}, \hat{\beta}) \tag{6.65}$$

This is, in fact, an ordinary linear regression situation, and the estimates of α and β should be obtained by the methods of Chapter 3.

The related problem of estimating η and its variance at a given *time* is more complicated, because now all the quantities in (6.63) are variates. The method will be described in Chapter 7, where it is utilized.

The maximum likelihood model in relation to plant growth data

Population specification and structure

As in the univariate cases (the growth of one plant part with respect to time) we suppose there to be a whole population of growth measurements at each harvest time, but now we have two distinct growth measurements which engender bivariate populations. Since we are working with the logarithms of the growth measurements, the bivariate normal distribution assumption is reasonable.

When data of the kind shown in Fig. 6.1d are observed, in which the population structure is obscured due to coalescence of points from neighbouring harvests, it might be thought that the data comprise sample observations from a single bivariate normally distributed population; an advantage of this line of thought would be that univariate regression would be adequate, so avoiding the complications inherent in linear functional relationships. It is also true that, in relation to plant growth, the populations (one at each harvest time) are artificially defined. Plant growth is a continuous process, and cannot naturally be partitioned in this way. Furthermore, if partitioning has to be done, it would be better to do it on the basis of some kind of plant chronology rather than time itself, since at any one harvest time the replicate plants will be at different developmental stages (page 133). When dealing with one plant part against time, the graph of the data demonstrates very obvious populations in the statistical sense, although one could replace the time-axis by a plant chronological axis (e.g. plastochron index; Lamoreaux, Chaney and Brown, 1978). This is a refinement which could be applied to the present, bivariate, case also. In either situation, the procedure would be to keep a plastochron age record of each harvested plant as well as its other growth measurements; then at the end of the experiment the plants would be arranged so that those with the same plastochron index (when rounded to the nearest whole number) would occur in one group, and so on. This would be a more natural way to define statistical populations, both in univariate and bivariate growth studies, and is a suggested future development.

For the present, however, we shall continue to assume that at each harvest time the two sets of growth measurements constitute a bivariate normally distributed population; the question then arises as to whether the populations defined in this way conform to the model assumptions listed on pages 184 and 185.

Although we shall adopt Assumption 6a in our examples of application, both alternatives of the sixth assumption demand homogeneity of variances and covariance along the length of a single line. When dealing with individual leaves, which show a determinate and sequential growth pattern, homogeneity of variance along a growth curve was not found (Table 4.7, page 120). Here, we are dealing with components at Level 1a which exhibit indeterminate and simultaneous growth, and there would seem to be no *a priori* reason why there should be variance heterogeneity along the length of a *single* allometric line. Differences in variability may be observed, however, between different allometric lines of a sequence for one set of plants as they grow; possible reasons for such differences will be discussed later (page 204).

Turning now more specifically to Assumption 6a there are, in fact, two assumptions for examination: zero covariances, and a value for the ratio of variances. With the populations as defined, in each of which there are plants of different developmental stages, the covariance is very unlikely to be truly zero; this is simply because there will tend to be a positive correlation between the parts of plants which are at different developmental stages. It is possible, however, that if populations were defined on a plant chronology basis so that only structural variability was present at each 'time', then correlation between the Level la components might be zero, but this is pure speculation at present.

With regard to the ratio of the two variances, if the hypothesis of constancy of variance is accepted for each component for a single allometric relationship, then a constant value for the ratio of two such variances follows. As to the value of this ratio, there is no fundamental reason why the coefficients of variation of the different parts, foliage, stem, and root, should differ from one another. In practice, root variability might be consistently greater than that of foliage and stem simply because of the difficulty of harvesting roots intact unless they are growing in solution culture. Even if this is true, however, the extra variability of roots would be due to experimental circumstances rather than to something inherent in the roots themselves. With these ideas in mind, a variance ratio of unity has been initially assumed.

Apart from population specification, all the postulates discussed in this sub-section are examined in relation to actual plant growth data in the applications section at the end of this chapter.

Replicate observations from each population

The derivation of the formulae for estimating the parameters of a linear functional relationship, and the variances and covariances of those estimates, has been given (pages 186–192) assuming only one sample observation from each harvest. Growth experiment data, however, always consist of several replicate values at each harvest, and the adjustments required to the formulae for N replicate observations at each harvest will now be given.

The logarithm of the likelihood is now

$$\log_e L = -Nh.\log_e(2\pi\lambda^{1/2}) - Nh.\log_e \sigma_{\delta\delta} - \frac{1}{2\sigma_{\delta\delta}}.\sum_{i=1}^{h}\sum_{j=1}^{N}\left\{(x_{ij}-\xi_i)^2+\frac{1}{\lambda}(y_{ij}-\alpha-\beta\xi_i)^2\right\} \qquad (6.66)$$

and so the partial derivatives with respect to α and β are the same as (6.24) and (6.25), respectively, but with y_{ij} replacing y_i and with double instead of single summation signs on the right-hand sides. The partial derivative with respect to ξ_i now becomes

$$\frac{\partial(\log_e L)}{\partial \xi_i} = \frac{1}{\sigma_{\delta\delta}}.\sum_{j=1}^{N}\left\{(x_{ij}-\xi_i)+\frac{\beta}{\lambda}(y_{ij}-\alpha-\beta\xi_i)\right\} \qquad (6.67)$$

which, when equated to zero and rearranged, gives

$$\hat{\xi}_i = \frac{\lambda \bar{x}_i + \hat{\beta}\bar{y}_i - \hat{\alpha}\hat{\beta}}{\lambda + \hat{\beta}^2} \tag{6.68}$$

where $\bar{x}_i = (1/N) . \sum_{j=1}^{N} x_{ij}$ and $\bar{y}_i = (1/N) . \sum_{j=1}^{N} y_{ij}$. By substituting for $\hat{\xi}_i$ in the partial derivatives with respect to α and β and equating to zero, we find that, as before,

$$\hat{\alpha} = \bar{y} - \hat{\beta}\bar{x} \tag{6.69}$$

where $\bar{x} = (Nh)^{-1} . \sum_{i=1}^{h} \sum_{j=1}^{N} x_{ij}$ and $\bar{y} = (Nh)^{-1} . \sum_{i=1}^{h} \sum_{j=1}^{N} y_{ij}$; and $\hat{\beta}$ is given by (6.32) with $\bar{x}_i$ substituted for every x_i, and $\bar{y}_i$ substituted for every y_i. Thus $\hat{\alpha}$, $\hat{\beta}$, and the $\hat{\xi}_i$ are calculated using the harvest sample mean values of the two growth attributes. The estimate of $\sigma_{\delta\delta}$ is given either by

$$\hat{\sigma}_{\delta\delta} = \frac{1}{2Nh} . \sum_{i=1}^{h} \sum_{j=1}^{N} \left\{ (x_{ij} - \hat{\xi}_i)^2 + \frac{1}{\lambda}(y_{ij} - \hat{\alpha} - \hat{\beta}\hat{\xi}_i)^2 \right\} \tag{6.70}$$

or

$$\hat{\sigma}_{\delta\delta} = \frac{\lambda^2 + \beta^2}{2Nh(\lambda + \beta^2)} . \sum_{i=1}^{h} \sum_{j=1}^{N} \{ (y_{ij} - \bar{y}) - \hat{\beta}(x_{ij} - \bar{x}) \}^2 \tag{6.71}$$

analogously to (6.33) and (6.34).

With regard to the variances and covariances of the parameter estimates, each of the second order derivatives (equations (6.35) to (6.40)) is now multiplied by N; additionally the y_i term in (6.39) is replaced by $\bar{y}_i$. Hence, all variance and covariance formulae are N^{-1} times the previous expressions ((6.48), (6.51), (6.52), (6.54), (6.55), and (6.56)), with $\bar{x}_i$ and $\bar{y}_i$ replacing x_i and y_i, respectively, where appropriate.

Linear segment and curvilinear allometry

Assessment of the situation

At present, it is not possible to suggest a rigorous set of methods to distinguish between a single straight line, a single curve, or a combination of straight lines and/or curves as being most appropriate for a single set of data. With sufficient observations, particularly where there are a large number of harvests, a visual inspection of both a harvest mean graph and an all individuals graph will usually give a reliable clue to the situation. Assistance in the graphical enquiry will be given by fitting a single straight line and examining for deviations. In this way, it should be relatively easy to assess whether a single straight line is adequate.

Curvilinear or multiphase linear segments?

If the idea of a single straight line is rejected for a given set of data, the question of whether a single curve or two or more straight lines is the more appropriate can be very difficult unless the trend is fairly obvious. Again, one can fit a seemingly appropriate relationship or relationships and graphically examine the result, but by this stage, if the data are segmented first, then fewer data are available in each segment, and the situation can become unsatisfactorily vague.

At this point, a recourse to physiological considerations is useful. Bearing in mind that the gradient of an allometric relationship between two parts of a plant at any point is a measure of the ratio of the relative growth rates of the two parts, with the implications already discussed in terms of the partitioning of dry matter during growth (page 181), we can say that not only are there several documented clear cut cases of abrupt changes in the gradient of an allometric relationship (e.g. Troughton, 1956; Causton, 1970), but also that the idea is quite feasible from the physiological viewpoint. The change from vegetative to reproductive growth is abrupt rather than gradual in many species, and this may be clearly shown using allometric relationships (Troughton, 1956; Stanhill, 1977a). Even during vegetative growth, definite abrupt changes in the gradient of an allometric line may occur (Causton, 1970), and this could be induced by a sudden change in the hormonal status of the plant. On the other hand, a gradual change in the ratio of relative growth rates, consequent upon a gradual change in the partitioning pattern, is also feasible; the whole question is very much an open one at present, and possibly varies from one species to another, and is affected by different environmental conditions.

In the present state of knowledge,* it would seem reasonable to advocate that every effort should be made to describe allometric relationships by straight lines rather than by curves. The two reasons for this recommendation can be summarized as statistical expediency and previous experience. As will be shown later (p. 199), fitting even a quadratic functional relationship to data introduces a problem which does not arise with a straight line. Also, although abrupt changes in linear allometric relationships have been reported in the literature on more than one occasion, only one instance of a curvilinear relationship is known to us (Causton, 1970), and the cause of this can be identified as an artificial pruning treatment on the plants concerned.

Linear segments

Two phase regression lines have been discussed by Sprent (1961) and

* (Added just prior to printing.) Currah & Barnes (1979) clearly showed that the allometric relationship between the shoot and storage root of carrot is curvilinear. Subsequently, Barnes (1979) developed a mathematical model demonstrating a possible physiological basis for this curvilinearity. The model has the added advantage that linear allometry is included as an alternative outcome.

Hinkley (1969). Basically, one of two situations may arise in a given situation: either there is a particular value of x where an hypothesis predicts there should be a change in the gradient of linear relationships; or else there is no prior hypothesis, and the data merely indicate that two straight lines, with differing gradients and intersecting at an unknown x-value, are required for an adequate description. The latter situation is relevant to allometry, and we shall confine the discussion to this case.

The main problems with two phase regression lines are to decide which data points concur with each line, thus deciding on a range of x-values in which the transition will occur, and then ensuring that the fitted lines do indeed intersect within the range of x selected. It may well be found that, having initially partitioned the data into two groups and fitted a straight line to each group, the point of intersection does not then occur within the expected range of x; this would give the anomolous situation that one or more particular observations had been used as data for one regression line but, because of the position of the resulting intersection point, those observations may ultimately lie in the domain of the other regression line (the word 'domain' is not used here in its strict mathematical sense). In this circumstance, it may be possible to rectify the anomoly by regrouping the data and refitting the lines; more than one attempt may be required, and there is no guarantee of ultimate success.

Similar principles are involved for two phase linear functional relationships, but here, in the absence of a non-random variable axis, there may be no *a priori* reason for selecting one variate axis for defining a range within which the transition point may occur more than the other variate axis. Such is the case with allometric relationships, and all that can be said in the absence of additional evidence is that provided the point of intersection occurs correctly with respect to one of the two axes, then the two lines should be accepted. If at some time in the future a change in gradient of an allometric relationship in a particular species, growing under a set of defined environmental conditions, can be identified with a particular stage of development, then this could form the basis for establishing the approximate position of the point of intersection. Although one example of this nature would be scarcely useful, a number of such examples might show that helpful generalizations could be made.

Curvilinear allometry

In all data sets examined by us so far, any curvature present in a log-log plot of data is small and unidirectional. Hence, in the absence of any definite biological model to account for this curvature (assuming that curvature does not exist because of conditions described in Theorems 6.2 and 6.3), we shall assume that a quadratic functional relationship will adequately describe any curvature that may arise in studies of allometry, but the method to be described below is directly extendable to higher order polynomials.

Maximum likelihood estimation of a quadratic functional relationship

The method is a direct extension of the linear case given earlier in this chapter. Again, we use Assumptions 1 to 5 and 6a (pages 184 and 185), except that Assumption 2 now reads that the means of each population lie on the curve represented by

$$\eta_i = \alpha + \beta_1 \xi_i + \beta_2 \xi_i^2 \tag{6.72}$$

The logarithm of the likelihood of h observations, one from each population, is given by

$$\log_e L = -h.\log_e (2\pi\lambda^{1/2}) - \mathrm{h}.\log_e \sigma_{\delta\delta} - \frac{1}{2\sigma_{\delta\delta}} \cdot \sum_{i=1}^{h} \left\{ (x_i - \xi_i)^2 + \frac{1}{\lambda}(y_i - \alpha - \beta_1 \xi_i - \beta_2 \xi_i^2)^2 \right\} \tag{6.73}$$

Partally differentiating (6.73) successively with respect to α, β_1, β_2, $\sigma_{\delta\delta}$, ξ_i, gives

$$\frac{\partial (\log_e L)}{\partial \alpha} = \frac{1}{\lambda \sigma_{\delta\delta}} \cdot \sum (y_i - \alpha - \beta_1 \xi_i - \beta_2 \xi_i^2) \tag{6.74}$$

$$\frac{\partial (\log_e L)}{\partial \beta_1} = \frac{1}{\lambda \sigma_{\delta\delta}} \cdot \sum (y_i - \alpha - \beta_1 \xi_i - \beta_2 \xi_i^2) \xi_i \tag{6.75}$$

$$\frac{\partial (\log_e L)}{\partial \beta_2} = \frac{1}{\lambda \sigma_{\delta\delta}} \cdot \sum (y_i - \alpha - \beta_1 \xi_i - \beta_2 \xi_i^2) \xi_i^2 \tag{6.76}$$

$$\frac{\partial (\log_e L)}{\partial \sigma_{\delta\delta}} = -\frac{h}{\sigma_{\delta\delta}} + \frac{1}{2\sigma_{\delta\delta}^2} \cdot \sum \left\{ x_i - \xi_i)^2 + \frac{1}{\lambda}(y_i - \alpha - \beta_1 \xi_i - \beta_2 \xi_i^2)^2 \right\} \tag{6.77}$$

$$\frac{\partial (\log_e L)}{\partial \xi_i} = \frac{1}{\sigma_{\delta\delta}}(x_i - \xi_i) + \frac{1}{\lambda}(\beta_1 + 2\beta_2 \xi_i)(y_1 - \alpha - \beta_1 \xi_i - \beta_2 \xi_i^2) \quad i = 1, \ldots, h \tag{6.78}$$

From (6.78) equated to zero and rearrnged, we have

$$2\hat{\beta}_2^2 \hat{\xi}_i^3 + 3\hat{\beta}_1 \hat{\beta}_2 \hat{\xi}_i^2 + (\hat{\beta}_1^2 + 2\hat{\alpha}\hat{\beta}_2 - 2\hat{\beta}_2 y_i + \lambda)\xi_i + (\hat{\alpha}\hat{\beta}_1 - \hat{\beta}_1 y_i - \lambda x_i) = 0 \tag{6.79}$$

which is cubic in $\hat{\xi}_i$. Thus one or three possible values of $\hat{\xi}_i$ exist corresponding to an observation (x_i, y_i). It is at this point that the difficulty in fitting anything but a linear functional relationship arises; in the linear case, there is only one $\hat{\xi}_i$ for each (x_i, y_i), given by (6.28), and this may then be substituted back into the other relationships such that explicit estimates can be obtained for the other parameters. For the present, quadratic, case there is no explicit analytical solution for $\hat{\xi}_i$ in (6.79), although numerical solutions may be obtained (e.g. Heading, 1963).

An iterative method is required, and again the Newton–Raphson procedure is suitable; indeed, O'Neill, Sinclair & Smith (1969) found this method to be the most successful for polynomial functional relationships. The second order partial derivatives are:

$$\frac{\partial^2(\log_e L)}{\partial\alpha^2} = -\frac{h}{\lambda\sigma_{\delta\delta}} \tag{6.80}$$

$$\frac{\partial^2(\log_e L)}{\partial\alpha\,\partial\beta_1} = -\frac{1}{\lambda\sigma_{\delta\delta}}.\Sigma\xi_i \tag{6.81}$$

$$\frac{\partial^2(\log_e L)}{\partial\alpha\,\partial\beta_2} = -\frac{1}{\lambda\sigma_{\delta\delta}}.\Sigma\xi_i^2 \tag{6.82}$$

$$\frac{\partial^2(\log_e L)}{\partial\beta_1^2} = -\frac{1}{\lambda\sigma_{\delta\delta}}.\Sigma\xi_i^2 \tag{6.83}$$

$$\frac{\partial^2(\log_e L)_1}{\partial\beta_1\,\partial\beta_2} = -\frac{1}{\lambda\sigma_{\delta\delta}}.\Sigma\xi_i^3 \tag{6.84}$$

$$\frac{\partial^2(\log_e L)}{\partial\beta_2^2} = -\frac{1}{\lambda\sigma_{\delta\delta}}.\Sigma\xi_i^4 \tag{6.85}$$

$$\frac{\partial^2(\log_e L)}{\partial\alpha\,\partial\xi_i} = -\frac{1}{\lambda\sigma_{\delta\delta}}(\beta_1+2\beta_2\xi_i) \tag{6.86}$$

$$\frac{\partial^2(\log_e L)}{\partial\beta_1\,\partial\xi_i} = -\frac{1}{\lambda\sigma_{\delta\delta}}(\alpha-y_i+2\beta_1\xi_i+3\beta_2\xi_i^2) \tag{6.87}$$

$$\frac{\partial^2(\log_e L)}{\partial\beta_2\,\partial\xi_i} = -\frac{\xi_i}{\lambda\sigma_{\delta\delta}}(2\alpha-2y_i+3\beta_1\xi_i+4\beta_2\xi_i^2) \tag{6.88}$$

$$\frac{\partial^2(\log_e L)}{\partial\xi_i^2} = -\frac{1}{\lambda\sigma_{\delta\delta}}(\lambda-2\beta_2y_i+2\alpha\beta_2+\beta_1^2+6\beta_1\beta_2\xi_i+6\beta_2^2\xi_i^2)$$
$$i=1,\ldots,h \tag{6.89}$$

and the information matrix is similar to (6.42) for the linear functional relationship, except that it is now of order $(h+3)$. In terms of the partitioned matrix (6.43), **B** has an extra row and column for the quadratic coefficient, and **Ω** has likewise an extra column. From this information and equations (6.80) to (6.89), the three matrices, **B**, **Ξ**, and **Ω** may be written out by the reader if required. Once again, we reiterate the advantage of the Newton-Raphson method that the parameter estimates, together with their variances and covariances, are produced in one operation.

With many harvests, however, there could be problems in using the information matrix as a whole, not because of inversion difficulties, but through convergence problems occasioned by employing the Newton–Raphson method in a large number of dimensions. In an attempt to simplify the situation, O'Neill *et al.* (1969) employed orthogonal polynomials, which renders **B** diagonal in addition to **Ξ** which is diagonal in any case. They also argued that the elements of **Ω** should be numerically small in relation to those in **B** and **Ξ**, and so

$$\mathbf{I}^{-1} \simeq \begin{bmatrix} \mathbf{B}^{-1} & \mathbf{O} \\ \mathbf{O} & \boldsymbol{\Xi}^{-1} \end{bmatrix} \tag{6.90}$$

O'Neill *et al.* then fitted the orthogonal polynomial functional relationship by iterations. Each iteration consisted of two parts: the calculation of the $\hat{\xi}_i$ by inverting **Ξ**, followed by the fitting of a polynomial curve to the points $(\hat{\xi}_i, y_i)$ by ordinary regression. The rationale behind the second part of an iteration for the quadratic function can be appreciated by comparing equations (6.74) to (6.76) with (3.53) to (3.55), where it will be observed that the unknown ξ_i in the former set of equations replace the known x_i in the latter; an identical situation exists for higher order polynomials. Starting values of the coefficients of the polynomial functional relationship for the first iteration were provided by regression on the data points (x_i, y_i).

The method was said to yield reasonable results, but often very numerous iterations were required if convergence was slow. From our present results with a linear functional relationship, we know that the off-diagonal elements in $\mathbf{I}^{-1}$ are not negligibly small (Tables 6.5 and 6.7, pages 208 and 209), and herein probably lies the root of the problem.

For our present purpose, we require a direct fit of ordinary (non-orthogonal) polynomials, and this means that the approximation of (6.90) will be even more crude since **B** is no longer a diagonal matrix. On attempting what was essentially the same method as O'Neill *et al.* for a quadratic functional relationship, we found by a parallel least squares method that $\mathcal{S}^2$ first decreased, and then increased and converged with successive iterations; but the converged value of $\mathcal{S}^2$ was higher than the minimum value it attained in early iterations. We are quite unable to explain this strange behaviour.

Among the data presented in this book, however, there are only two sub-sets of data, out of a total of 20, which cannot be considered linear, and for these we have found the Newton-Raphson method applied to **I** as a whole to be satisfactory.

Finally, we should mention that where there are N replicate observations per harvest, all y_i are replaced by $\bar{y}_i$ in all the partial derivatives, and the second order partial derivatives are each multiplied by N.

Applications

Before describing and discussing the allometric relationships found in the four species experimental data, it is relevant to examine some of the properties of the data in relation to the assumptions of the estimation model being used. We have already discussed these aspects briefly from a theoretical viewpoint; now, by using the birch and wheat data as examples, the assumptions of the maximum likelihood model can be more realistically assessed. Also, the effects of three changes in the estimation method will be examined.

Properties of the data

The data and results required for the whole discussion are given in Tables 6.1 (birch) and 6.2 (wheat). The normality of the data has not been investigated, neither has the appropriateness of a linear or quadratic functional relationship, where each has been respectively used; it would require very marked deviations of the data from any of these three assumptions to show as significant differences, and there is no reason to suspect such marked anomolies in our data.

The harvests are grouped in Tables 6.1 and 6.2 to correspond with the segmentation of the data employed for estimating foliage-stem allometric relationships (Table 6.8, Figs. 6.4a,b and 6.5a,b); there is, however, some correspondence between these groupings and those for the foliage-root allometric relationships, but the latter have more segments.

Variance homogeneity of a single component

For a group of harvests for a particular component, the homogeneity of the variances has been assessed by Bartlett's test (page 117), and the results are shown in the first three columns of Tables 6.1 and 6.2. Only two results out of 15 are significant: birch foliage, harvests 1 and 2, due to an exceptionally high variance of the harvest 2 data; and wheat stem, harvests 2 to 11, due to moderately high variabilities at harvests 2 and 5. The groups comprising birch foliage, harvests 10 to 24, and birch stem over the same period almost show a significant heterogeneity (but the correction factor, which would reduce the value of χ^2, has not been calculated), but in no case is there a *systematic* change in variance with time. Thus, variance homogeneity would appear, in general, to be a realistic assumption (but see Table 6.9 for sunflower).

Table 6.1 Variance of foliage, stem and root data for birch, together with their correlation coefficients. The significance of the correlations for individual and groups of harvests, the *F*-ratio for variance homogeneity between pairs of parts at each harvest and groups of harvests, and the χ^2 result for Bartlett's test of homogeneity of a group of variances for each organ, are also given. The groups of harvests are according to the segmentation of the data for fitting allometric relationships to leaves and stem. The segmentation of the data for foliage-root allometric relationships is very similar, except that harvest 9 belongs to the final group.

Harvest	s_l^2	s_s^2	s_r^2	$\hat{\rho}_{ls[8]}$	$\hat{\rho}_{lr[8]}$	$F_{ls[9,9]}$	$F_{lr[9,9]}$
1	0.0206	0.1118	0.1156	0.71*	0.21	5.43*	5.61
2	0.3167	0.2003	0.2452	0.83**	0.39	1.58	1.29
Mean	0.1687	0.1561	0.1804	0.75**	0.61*	1.08	1.07
$\chi^2_{[1]}$	13.25***	0.75	1.24				
3	0.0499	0.3883	0.1607	0.31	0.05	7.79**	3.22
4	0.0581	0.0893	0.4353	0.57	0.47	1.54	7.50**
5	0.1315	0.1683	0.0864	0.66*	−0.03	1.28	1.52
6	0.1800	0.1988	0.1992	0.79**	−0.35	1.10	1.10
7	0.1136	0.1053	0.1460	0.96***	0.42	1.08	1.29
8	0.1139	0.1533	0.1192	0.93***	0.72*	1.35	1.05
9	0.1365	0.1726	0.1745	0.95***	0.95***	1.27	1.28
Mean	0.1119	0.1823	0.1888	0.71**	0.29	1.63	1.69*
$\chi^2_{[6]}$	5.37	6.54	7.63				
10	0.0624	0.1039	0.0512	0.93***	0.83**	1.67	1.22
11	0.1212	0.1341	0.1279	0.97***	0.45	1.11	1.06
12	0.899	0.0914	0.0918	0.94***	0.91***	1.02	1.02
13	0.1960	0.2204	0.2570	0.97***	0.86**	1.13	1.31
14	0.1424	0.2271	0.1192	0.97***	0.85**	1.59	1.20
15	0.1091	0.2082	0.1165	0.93***	0.90***	1.79	1.07
16	0.0201	0.0505	0.0307	0.96***	0.63	2.52	1.53
17	0.1143	0.2049	0.1814	0.97***	0.87***	1.79	1.59
18	0.0742	0.1137	0.1586	0.94***	0.68	1.53	2.14
19	0.0501	0.0864	0.0987	0.89***	0.75	1.72	1.97
20	0.0939	0.1295	0.1096	0.98***	0.87***	1.38	1.17
21	0.1164	0.1210	0.0681	0.82**	0.37	1.04	1.71
22	0.0362	0.0302	0.0552	0.76*	0.37	1.20	1.52
23	0.0273	0.0342	0.0801	0.59	0.20	1.25	2.93
24	0.0449	0.0593	0.0803	0.84**	0.52	1.37	1.37
Mean	0.0866	0.1210	0.1084	0.92***	0.72**	1.40*	1.25
$\chi^2_{[14]}$	23.35	23.41	17.21				

Correlation between components

In contrast, the assumption of no correlation between components in individual harvests would appear to be untenable, but the significance of the correlations appears to vary widely. The most obvious correlations are between foliage and stem in birch between harvests 7 and 20, but earlier and later than this group of harvests the correlation is weaker. Correlations between leaves and stem of wheat are generally high up to the 10th harvest, but not as consistently so as in birch. On the other hand, correlations

Table 6.2 Variances of foliage, stem, and root data for wheat, together with their correlation coefficients. The significance of the correlations for individual and groups of harvests, the *F*-ratio for variance homogeneity between pairs of parts at each harvest and groups of harvests, and the χ^2 result for Bartlett's test of homogeneity of a group of variances for each organ are also given. The groups of harvests are according to the segmentation of the data for fitting allometric relationships to leaves and stem. The segmentation of the data for foliage-root allometric relationships has an extra transition between harvests 7 and 8.

Harvest	s_l^2	s_s^2	s_r^2	$\hat{\rho}_{ls[4]}$	$\hat{\rho}_{lr[4]}$	$F_{ls[5,5]}$	$F_{lr[5,5]}$
2	0.0229	0.1719	0.0674	0.92**	0.65	7.51*	2.94
3	0.0148	0.0670	0.0204	0.93**	0.71	4.53	1.38
4	0.0522	0.0819	0.0273	0.86*	0.83*	1.57	1.91
5	0.0554	0.1373	0.0677	0.99***	0.85*	2.48	1.22
6	0.0107	0.0515	0.0448	0.93**	0.63	4.81	4.19
7	0.0205	0.0348	0.0213	0.81*	0.91*	1.70	1.04
8	0.0091	0.0099	0.1456	0.72	0.75	0.09	16.00*
9	0.0166	0.0224	0.0378	0.85*	0.67	1.35	2.28
10	0.0334	0.0282	0.0459	0.98***	0.75	1.18	1.37
11	0.0135	0.0124	0.0283	0.46	0.59	1.09	2.10
Mean	0.0249	0.0617	0.0507	0.85**	0.67	2.48**	2.04*
$\chi^2_{[9]}$	9.17	18.97*	9.34				
12	0.0326	0.1042	0.1178	0.35	−0.01	3.20	3.61
13	0.0208	0.0462	0.0348	0.77	0.53	2.22	1.67
14	0.0168	0.0266	0.538	0.71	0.52	1.58	3.20
15	0.0070	0.0491	0.0594	−0.22	−0.34	7.01	8.49*
Mean	0.0193	0.0565	0.0665	0.42	0.16	2.93*	3.45*
$\chi^2_{[3]}$	2.77	2.42	2.00				

between leaves and root are less marked; only 3 correlation coefficients out of 14 are significant in wheat, and a few very low values are found in both species.

Earlier, it was suggested that if plants in a single harvest were at the same developmental stage, there might be no correlation between the sizes of the Level la components; that is, a significant correlation found between components of plants at different stages of development at a given harvest might arise simply because of this developmental variability, and would not be a feature of structural variability. From the data examined, this might be true for leaves and root, and maybe also for stem and root; however, the correlations seem to be so high between leaves and stem in the vegetative phase of wheat, and in birch (whose growth is vegetative throughout the experimental period), that the idea of zero correlation between foliage and stem associated with structural variability may have to be discarded. Much more work is necessary to resolve this question.

Ratios of variances of pairs of components

On page 194, it was stated that there was no fundamental reason why the three components – leaves, stem, and root – should have markedly different variances, but that the roots might show a higher variability due to the difficulty of harvesting them intact. The data in Tables 6.1 and 6.2 partly refute this suggestion; the variance of the foliage is consistently lower than that of the other two components, and on average, the variance of the roots is about equal to that of the stem. True, the F-ratios are rarely significantly different from unity in individual harvests, but when harvests are aggregated the lower foliage variance usually shows as a significant effect. Evidently, the variance of the roots is not affected by harvesting difficulties unless, of course, the variability of roots is basically less than that of the stems; but we have no evidence of this.

Why the foliage should have a lower variance than the other Level Ia components is purely a matter of conjecture at this stage. The lower variance shown by the foliage as a whole may be due in some way to the determinate growth pattern shown by the individual leaves; the individual parts of the root system are probably indeterminate in their growth habit, while the individual internodes of birch would also show an indeterminate growth pattern due to secondary thickening after the cessation of extension growth. On the other hand, there should be no secondary thickening in the internodes of wheat. In this species, however, the experimental period extended over the whole duration of vegetative growth and the beginning of the reproductive phase. During vegetative growth, individual leaves were produced in quick succession from a shoot apex (about 3 per week on the main stem), and the expansion of the resulting individuals was rapid. The growth of the foliage system therefore proceeded in a very orderly fashion allowing, perhaps, restricted scope for variation. Stem growth at this time, although relatively high, was low in absolute terms because of the very small size of the structure, and this growth pattern may allow for greater variability than in the foliage.

Some effects of changes in the estimation method

Having shown that the correlations between the three Level Ia components may not necessarily be zero, even if developmental variability could be ignored, and also that foliage appears to have a lower variability than stem and root, it is necessary to examine the effect of different assumptions on the parameter estimates for a few sets of data. This has been done, again for birch and wheat, where linear rather than curvilinear relationships seem appropriate, and the following four methods have been compared:

(a) linear functional relationship method, assuming $\sigma_{\delta\varepsilon} = 0$ and $\lambda = 1$ (our preferred method, see page 185);
(b) linear functional relationship method, again assuming $\sigma_{\delta\varepsilon} = 0$, but using values of λ more in keeping with the results given in Tables 6.1 and 6.2 $-\lambda = 1.5$ for birch, and $\lambda = 2.5$ for wheat;

(c) linear functional relationship method (Sprent's generalized least squares, 1966, 1969), using data estimates for $\sigma_{\delta\delta}$, $\sigma_{\varepsilon\varepsilon}$, $\sigma_{\delta\varepsilon}$;
(d) ordinary linear regression, to see how parameter estimates using methods (b) and (c) compare with (a), the last perhaps containing suspect assumptions, and (d) which is definitely an inappropriate method.

Results are given in Table 6.3. In the two cases where only two harvests are involved, the results are, of course, identical for each method. Elsewhere, in every case but one (wheat $s(l)$, harvests 12 to 15, where there is a reversal of the trend between methods (b) and (c)) there is a decline in the value of the gradient of the estimated line from method (a) through to (d). In some cases, the decline through the methods is steady (birch $s(l)$, harvests 3 to 9; wheat $s(l)$, harvests 2 to 11), but more often there is a larger discontinuity between gradient values of methods (c) and (d) than elsewhere, thus indicating that there was less of a difference among the three functional relationship methods than there was between them as a whole and regression. It does seem though that the model in which $\sigma_{\delta\varepsilon} = 0$ and $\lambda = 1$ 'over-corrects' the depressed value of the gradient occasioned by the use of linear regression. Similar progressions are seen in the results of α-estimates, but the trends are more erratic, and may go in either direction depending on the placement of the data in relation to the intercept.

Table 6.3 Comparison of the estimates of parameters α and β by four different methods: (a) linear functional relationship, assuming $\sigma_{\delta\varepsilon} = 0$ and $\lambda = 1$; (b) linear functional relationship, assuming $\sigma_{\delta\varepsilon} = 0$ and $\lambda = 1{\cdot}5$ for birch and $2{\cdot}5$ for wheat; (c) linear functional relationship, using data estimates of $\sigma_{\delta\delta}$, $\sigma_{\varepsilon}^{\varepsilon}$, $\sigma_{\delta\varepsilon}$; (d) ordinary linear regression.

Method	(a)		(b)		(c)		(d)	
	$\hat{\alpha}$	$\hat{\beta}$	$\hat{\alpha}$	$\hat{\beta}$	$\hat{\alpha}$	$\hat{\beta}$	$\hat{\alpha}$	$\hat{\beta}$
BIRCH								
$s(l)$ H1–2	−5.5852	0.4629	−5.5852	0.4629	−5.5852	0.4629	−5.5852	0.4629
$s(l)$ H3–9	−1.3866	1.0726	−1.3872	1.0725	−1.3884	1.0722	−1.3900	1.0719
$s(l)$ H10–24	−0.8169	1.2734	−0.8170	1.2732	−0.8170	1.2732	−0.8168	1.2720
$r(l)$ H1–2	−6.9179	0.2897	−6.9179	0.2897	−6.9179	0.2897	−6.9179	0.2897
$r(l)$ H3–8	−2.5885	0.7584	−2.5908	0.7580	−2.5928	0.7576	−2.5984	0.7565
WHEAT								
$s(l)$ H2–11	−0.4065	1.3409	−0.4105	1.3395	−0.4180	1.3369	−0.4186	1.3368
$s(l)$ H12–15	2.1401	3.2709	2.1289	3.2601	2.1420	3.2728	2.0448	3.1781
$r(l)$ H12–15	1.6495	2.7993	1.6464	2.7964	1.6457	2.7957	1.6300	2.7794

Despite the fact that method (c) would seem to be the more appropriate model for the present data, we have not utilized it for the examples presented in this book; firstly, because undoubtedly more complicated expressions would be required for the variances and covariances of the parameter estimates, and secondly, more importantly, we feel that more should be known about the nature of the variances and covariances of the plant components at each harvest before advocating the use of a more complicated model. There is also the reason that the model in which $\sigma_{\delta\varepsilon} = 0$ would undoubtedly extend more easily to curvilinear relationships; hence, selecting this model throughout is desirable on the grounds of uniformity.

Comparison of methods using $-\mathbf{I}^{-1}$
with $-\mathbf{B}^{-1}$ *and* $-\mathbf{\Xi}^{-1}: \lambda = 1$

In the derivation of the method for obtaining the variances and covariances of the parameter estimates, it was shown how the whole information matrix, of order $(h + 2)$, could be partitioned (equations (6.42) and (6.43)) in order to simplify the computation of the required quantities. Obviously, the correct way is to invert **I** as a whole, and strictly, $\sigma_{\delta\delta}$ should be involved as well. However, the latter is a rather different kind of parameter than the others (α, β, ξ_i) and, moreover, it does not normally figure in the maximum likelihood estimation of a linear regression; hence, we have omitted terms involving $\sigma_{\delta\delta}$ from **I**. Barnett (1970), however, in a linear functional relationship model involving a much more complicated set of assumptions regarding the data than we need to make here, included not only $\sigma_{\delta\delta}$ in his information matrix, but also λ as a parameter to be estimated.

The inversion of **B** and $\mathbf{\Xi}$ separately would give the same result as the inversion of **I** only if the off-diagonal elements of the latter were all zero; this would imply that the off-diagonal elements of **B** and $\mathbf{\Xi}$ would be zero (false for **B**, true for $\mathbf{\Xi}$), similarly, all the elements of $\mathbf{\Omega}$ would be zero (false) (see equations (6.44), (6.45), and (6.46)). The question now is, how do the results of the two inversion methods compare? We have contrasted the two methods for most of the segmented sets of data in the four species where linear relationships apply. The trends of the results are identical in every case, although they differ in detail. Accordingly, a comparison of the variances and correlation coefficient of $\hat{\alpha}$ and $\hat{\beta}$ are given for wheat and most segments of the birch data (Table 6.4), and a more detailed comparison for one set of data (birch $s(l)$, harvests 3 to 9) (Tables 6.5 and 6.6).

In Table 6.4, it is evident that the variances of $\hat{\alpha}$ and $\hat{\beta}$ are lower where matrix **B** alone has been inverted than those obtained by inversion of **I**, and also that the factor by which the variances are lowered increases with the number of harvests involved. This is to be expected because the order of **I** is $(h + 2)$ as against 2 for the order of **B**. The correlations of $\hat{\alpha}$ and $\hat{\beta}$ are unaffected by the method used. The reason for the low correlation coefficient for birch $s(l)$, harvests 10 to 24, is that the point $(\overline{l}, \overline{s})$ is very near the intercept $(0, \hat{\alpha})$, and so a change of gradient of this line would occasion a relatively small change in the intercept. It should, however, be pointed out

Table 6.4 Comparison of the variances and correlation coefficient of $\hat{\alpha}$ and $\hat{\beta}$ of linear allometric relationships for birch and wheat data, obtained by inversion of **I** and of **B** and **Ξ** separately.

	Inversion of **I**			Inversion of **B**		
	$\mathscr{V}(\hat{\alpha})$	$\mathscr{V}(\hat{\beta})$	ρ	$\mathscr{V}(\hat{\alpha})$	$\mathscr{V}(\hat{\beta})$	ρ
BIRCH						
$s(l)$ H1–2	7.73678	0.12389	1.00	6.37164	0.10203	1.00
$s(l)$ H3–9	0.08381	0.00321	0.98	0.03897	0.00149	0.98
$s(l)$ H10–24	0.00170	0.00070	0.16	0.000647	0.00027	0.16
$r(l)$ H1–2	6.18527	0.09905	1.00	5.70634	0.09138	1.00
$r(l)$ H3–8	0.09121	0.00322	0.98	0.05782	0.00204	0.98
WHEAT						
$s(l)$ H2–11	0.02165	0.00240	0.96	0.00773	0.00086	0.96
$s(l)$ H12–15	1.09049	0.99849	0.99	0.09299	0.08514	0.99
$r(l)$ H12–15	0.90663	0.82989	0.99	0.10253	0.09385	0.99

that direct comparisons of these approximate variances do not provide a firm foundation for the drawing of conclusions; a proper simulation study is required.

In Tables 6.5 and 6.6, values of $\hat{\alpha}$, $\hat{\beta}$, and the $\hat{\xi}_i$ are given, together with their variances. Table 6.6 also restates the correlation coefficient of $\hat{\alpha}$ and $\hat{\beta}$, while Table 6.5 gives the entire correlation matrix, calculated from the variance-covariance matrix, $-\mathbf{I}^{-1}$. The parameter estimates are, of course, unaffected by the two methods, but the variance of the $\hat{\xi}_i$ shows a minimum in the mid-range of the data and increases towards either end of the line (Table 6.5). Since Ξ is a diagonal matrix with all elements equal (equation (6.45)), $\mathscr{V}(\hat{\xi}_i)$ are also equal to one another and are given by (6.56); the actual value, 0.00617 (Table 6.6), is lower than the lowest variance, $\mathscr{V}(\hat{\xi}_6)$.

The correlation matrix in Table 6.5 shows interesting features. The closer a pair of $\hat{\xi}_i$ are to one another, the higher are their correlations, and this is particularly true at the ends of the line. As the separation between two $\hat{\xi}_i$s increases, the correlation coefficients decrease in absolute value ultimately becoming negative. Bearing in mind that the straight line is constrained to pass through the point $(\bar{l}, \bar{s})$, a moment's thought will reveal how this pattern of correlations occurs, and also how the patterns of correlation between $\hat{\alpha}$ and $\hat{\beta}$ with the $\hat{\xi}_i$ are determined. This pattern would always be the same with regard to $\hat{\beta}$ and the $\hat{\xi}_i$, namely, positive correlations where $\hat{\xi}_i < \bar{l}$ and *vice versa*; but the pattern found between $\hat{\alpha}$ and $\hat{\xi}_i$ in Table 6.5, where $\bar{l} < 0$, would be reversed in data where $\bar{l} > 0$.

As a final summing up, we may say that the off-diagonal elements of $-\mathbf{I}^{-1}$, while often not large, are certainly not negligible, and therefore inversion of **B** and **Ξ** separately cannot be recommended, except in special

Table 6.5 Parameter estimates, their variances and correlation matrix, for the linear allometric relationship fitted to the birch stem leaves data, harvests 3 to 9, calculated by inverting matrix **I** as a whole; $\lambda = 1$. The hatched lines correspond with the partitioning of as shown in equation (6.42).

Parameter	Estimate	Variance	Correlation matrix								
α	−1.387	0.08381	1.00								
β	1.073	0.00321	0.98	1.00							
ξ_3	−6.926	0.01020	0.46	0.54	1.00						
ξ_4	−6.074	0.00811	0.25	0.34	0.30	1.00					
ξ_5	−5.403	0.00732	0.05	0.13	0.19	0.18	1.00				
ξ_6	−4.820	0.00720	−0.13	−0.04	0.10	0.12	0.13	1.00			
ξ_7	−4.211	0.00768	−0.33	−0.25	−0.02	0.04	0.10	0.15	1.00		
ξ_8	−3.887	0.00819	−0.42	−0.35	−0.08	0.01	0.09	0.15	0.22	1.00	
ξ_9	−3.589	0.00873	−0.48	−0.42	−0.12	−0.02	0.07	0.15	0.23	0.27	1.00

Table 6.6 Parameter estimates, their variances and covariance, for the linear allometric relationship fitted to the birch stem-leaves data, harvests 3 to 9; calculated by inversion of matrices **B** and **Ξ** separately.

Parameter	Estimate	Variance	Correlation coefficient of $\hat{\alpha}$ and $\hat{\beta}$
α	−1.387	0.03897	0.98
β	1.073	0.00149	
ξ_3	−6.926	0.00617 (for ξ_3 to ξ_9)	
ξ_4	−6.074		
ξ_5	−5.403		
ξ_6	−4.820		
ξ_7	−4.211		
ξ_8	−3.887		
ξ_9	−3.589		

circumstances (e.g. where computing facilities are unavailable) when formulae (6.51), (6.54), (6.55) and (6.56) may be used as approximate estimators.

The effect of changing λ

Table 6.3 showed that the estimates of α and β were affected only in the fourth place of decimals in birch where λ was changed from 1 to 1.5, but that the third or even second place of decimals was affected in wheat where λ was changed from 1 to 2.5. We have therefore studied the effect of changing λ on the other estimated parameters and their variances and covariances, and present the results for birch $s(l)$, harvests 3 to 9, with $\lambda = 1.5$, in Table 6.7. These data are the same as those given in Table 6.5 for $\lambda = 1$ and the two tables are of identical format, so facilitating comparisons. It can be seen

Table 6.7 Parameter estimates, their variances and correlation matrix, for the linear allometric relationship fitted to the birch stem-leaves data harvests 3 to 9; calculated by inverting matrix **I** as a whole. $\lambda = 1.5$. The hatched lines correspond with the partitioning of **I** as shown in equation (6.42).

Parameter	Estimate	Variance	Correlation matrix								
α	−1.387	0.08196	1.00								
β	1.073	0.00313	0.97	1.00							
ξ_3	−6.927	0.00857	0.40	0.48	1.00						
ξ_4	−6.071	0.00720	0.21	0.21	0.22	1.00					
ξ_5	−5.401	0.00670	0.04	0.11	0.14	0.13	1.00				
ξ_6	−4.827	0.00662	−0.10	−0.03	0.07	0.09	0.10	1.00			
ξ_7	−4.209	0.00693	−0.28	−0.21	−0.02	0.01	0.07	0.10	1.00		
ξ_8	−3.883	0.00727	−0.36	−0.30	−0.06	0.01	0.06	0.10	0.16	1.00	
ξ_9	−3.591	0.00760	−0.42	−0.36	−0.09	0.01	0.05	0.10	0.17	0.20	1.00

that, while $\hat{\alpha}$ and $\hat{\beta}$ are unchanged to the third place of decimals, the $\hat{\xi}_i$ are altered slightly but with no pattern of direction of change. All variances and correlation coefficients are reduced in numerical value, and the largest decreases are found among the $\hat{\xi}_i$. Whether these changes are due to using a more appropriate value of λ for the data in hand, or whether they depend on the numerical value of λ, is impossible to say without further investigation, but it is probable that the latter suggestion is the true one.

Species results

At the outset, it is necessary to reiterate that the estimation assumptions used in this section are $\sigma_{\delta\varepsilon} = 0$, $\lambda = 1$, and that $-\mathbf{I}^{-1}$ is used for the variances and covariances of the parameter estimates. Although many, if not most, of our data sets show properties which do not accord with these assumptions, we are employing this consistent method for all the experimental data, partly because of the speculations given in the discussions on pages 192 to 194, and also because of the much greater mathematical difficulty of the maximum likelihood method if $\sigma_{\delta\varepsilon} \neq 0$. Changing the value of λ from unity does not, however, introduce extra mathematical difficulties, but we would recommend that the biological reality of any particular λ-value suggested by the data should be ascertained first, before using it as an assumption in the estimation procedure of an allometric relationship.

Results are given in Table 6.8 and in Figs. 6.2 (sunflower), 6.3 (maize), 6.4 (wheat), and 6.5 (birch). In each figure, four graphs are given: (a) stem-leaves, all observations; (b) stem-leaves, harvest means; (c) root-leaves, all observations; (d) root-leaves, harvest means; and on all graphs the fitted allometric relationships are shown.

Segmentation into two or more linear or curvilinear phases has been effected purely on a visual basis, using the harvest mean graphs shown in this chapter together with graphs of the harvest means against time, and assuming as far as possible that linear segments underlie the observed data; only in two extreme cases have curvilinear segments been fitted because there seemed to be no sensible alternative. In two instances a single straight line appeared to adequately define the data over the entire experimental period. The visual criterion for segmentation was on the basis of the harvest mean graphs because the data trends are easier to see on such a graph, and also because the results on page 195 show that it is the harvest mean values which appear in the parameter estimate equations rather than the individual observations. The reasons for striving to have linear, rather than curvilinear, segments have been given on page 196.

Sunflower

Of the four species, this one's allometric relationships are the most straightforward to interpret. We have already demonstrated the stem

Table 6.8 Species results: the parameter estimates (excluding the ξ_i), their variances, and $\sigma^{\delta}_{\delta\delta}$, the estimated variance of the foliage at each harvest. Assumptions are that $\sigma_{\delta\varepsilon} = 0$ and $\lambda = 1$; $-\mathbf{I}^{-1}$ provides the variance and covariances of the parameter estimates.

	$\hat{\alpha}$	$\hat{\beta}$ or $\hat{\beta}_1$	$\hat{\beta}_2$	$\mathscr{V}(\hat{\alpha})$	$\mathscr{V}(\hat{\beta})$ or $\mathscr{V}(\hat{\beta}_1)$	$\mathscr{V}(\hat{\beta}_2)$	$\sigma_{\delta\delta}$
SUNFLOWER							
s(*l*) H1–4	–1.0159	1.0086	–	0.25478	0.02383	–	0.0296
s(*l*) H5–8	–0.2556	1.2843	s(*l*)	0.09077	0.02187	–	0.0612
s(*l*) H9–16	–0.2545	1.6590	–	0.00346	0.01281	–	0.0618
r(*l*) H1–16	–0.2033	1.2568	–	0.00401	0.00105	–	0.1125
MAIZE							
s(*l*) H1–16	–0.9861	1.1584	–	0.01706	0.00243	–	0.2887
r(*l*) H1–11	0.6668	0.7617	–	0.06396	0.00629	–	0.4057
r(*l*) H12–16	–0.463	1.8224	–	0.02191	0.16316	–	0.1504
WHEAT							
s(*l*) H2–11	–0.4065	1.3409	–	0.02165	0.00240	–	0.0372
s(*l*) H12–15	2.1401	3.2709	–	1.09049	0.99849	–	0.0318
r(*l*) H2–7	–4.8899	1.3358	–0.2896	3.28	1.18	0.0259	0.0593
r(*l*) H8–11	–3.2519	–2.3514	–1.0017	17.1	20.6	1.48	0.1191
r(*l*) H12–15	1.6495	2.7993	–	0.90663	0.82989	–	0.0358
BIRCH							
s(*l*) H1–2	–5.5852	0.4629	–	7.73678	0.12389	–	0.1462
s(*l*) H3–9	–1.3866	1.0726	–	0.08381	0.00321	–	0.1328
s(*l*) H10–24	–0.8169	1.2734	–	0.00170	0.00070	–	0.0945
r(*l*) H1–2	–6.9179	0.2897	–	6.18527	0.09905	–	0.1309
r(*l*) H3–8	–2.5885	0.7584	–	0.09121	0.00322	–	0.1359
s(*l*) H9–16	–0.7820	0.9600	–	0.01242	0.00263	–	0.1064
r(*l*) H17–20	–0.4502	1.0304	–	0.01113	0.02847	–	0.0995
r(*l*) H21–24	–1.5935	2.1374	–	0.33086	0.13015	–	0.0572

growth strategy of this species in Chapter 2, and the present way of viewing growth data also highlights this strategy.

The two graphs showing the individual plant values (Fig. 6.2a,c) indicate that the variability of the root is greater than that of the stem in this species, since both components are plotted against leaves, and the root-leaves graph shows a greater scatter of points than the stem-leaves graph. This supposition is confirmed by the results shown in Table 6.9; evidently, the pattern of variances of the Level 1a components of sunflower differs from that shown in birch and wheat, and is more in line with our original suppositions (page 194). The reason for the lower stem variability, in

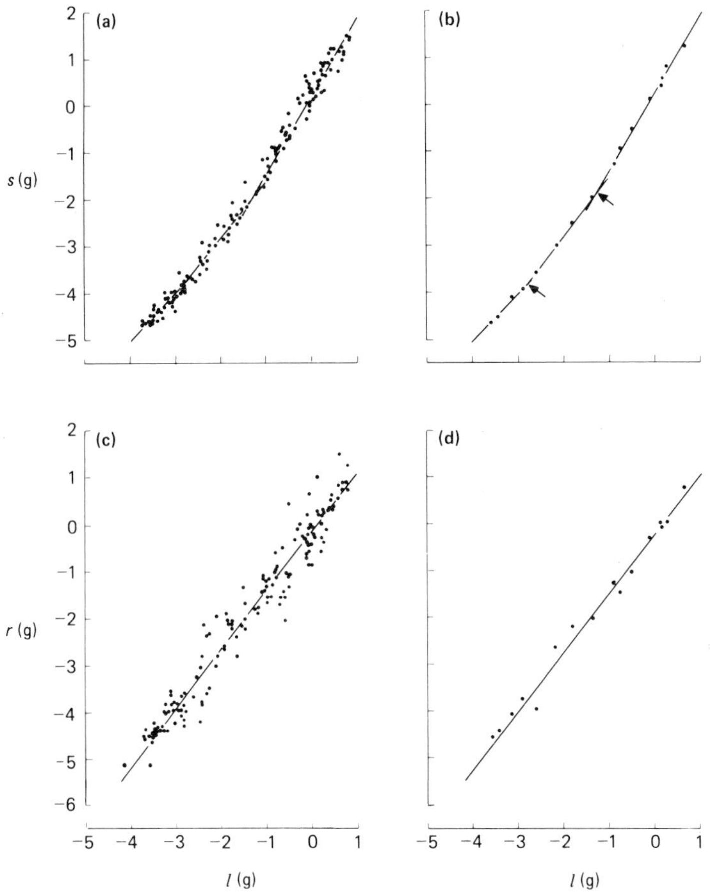

Fig. 6.2 Allometric relationships for sunflower; between stem weight and leaf weight, (**a**) and (**b**), and root weight and leaf weight, (**c**) and (**d**); with all points, (**a**) and (**c**), and harvest means, (**b**) and (**d**). Intersections of allometric segments are indicated →.

comparison with foliage variability in sunflower as compared with birch and wheat, may again be attributed to growth pattern. The stem of sunflower is a large structure from early on, there is an orderly sequence of internode extension without secondary thickening at this stage, and it is evident that internode growth is very much synchronized with individual leaf (or leaf pair) growth. Although the use of $\lambda = 1$ for the root-leaves relationship instead of λ-values suggested by the data would have little effect on the values of $\hat{\alpha}$ and $\hat{\beta}$, we do have the anomalous result that $\hat{\sigma}_{\delta\delta}$ is given as 0.1125 for the root-leaves relationship but much less than this for the stem-leaves relationships, even though it is estimating the same quantity – foliage variability at a single harvest. This is because the variance of the stem is lower than the variance of the root (Table 6.9), and with fixed λ the variances of the stem and root influence the magnitude of the estimate of $\sigma_{\delta\delta}$. Evidently, if the estimate of $\sigma_{\delta\delta}$ is of particular interest, great care must be exercised over the choice of a value for λ.

Despite the higher variability in the root-leaves graph, it is quite clear that a single linear relationship adequately describes these data for the whole experimental duration. The gradient of the line is 1.2568, and is very significantly greater than unity. Thus, there is a higher demand for assimilates by the root than by the foliage throughout this experiment.

Examination of the stem-leaves graph shows that a single linear relationship will not adequately describe these data. Changes of slope appeared to occur between harvests 4 and 5, and between harvests 8 and 9; straight lines fitted subsequently confirm these positions of change, and there is no evidence of curvilinearity in any of the three segments. The results show an increasing demand by the stem in relation to the leaves.

Looking at all the relationships for the experimental duration, over the period covered by the first 4 harvests the gradient of the stem-foliage line is essentially unity; thus we may picture a demand by the root over the shoot, with the growth of the stem and leaves being supply limited (page 181). Over the second phase, the root-stem relationship is essentially unity (1.2568/1.2843 = 0.9786), whereas the leaves show a lower relative growth rate than either stem or root. This may be interpreted as a negative demand situation by the leaves (page 182) which could be brought about by a low rate of leaf production so that there is a limited amount of foliage tissue available for expansion; the stem and root might then be supply limited, neither organ having a higher relative growth rate over the other, but under these circumstances equal demands seem more likely. Over the third phase, growth of the stem is paramount; its demand for assimilates exceeds that of the root (1.6590/1.2568 = 1.3200) and, again, the foliage exerts only a small demand.

Table 6.9 Variances of leaves, stem, and root for sunflower, averaged across the indicated harvests. The figures in brackets are not relevant to the relationships fitted.

Harvests	Foliage	Stem	Root
1 to 4	0.0428	0.0229	(0.0773)
5 to 8	0.0662	0.0714	(0.1951)
9 to 16	0.0616	0.0762	(0.2159)
1 to 16	0.0569	(0.0568)	0.1628

Besides the strategy of high stem growth, sunflower also seems to have a pattern of low leaf growth. To meet the demand for rapid growth, characteristic of this species, unit leaf rate must be high, and this is indeed shown in this experiment (Fig. 2.7b), and by others (Warren Wilson, 1966c). Sunflower is a C_3 plant; it would not, therefore, be expected to show a particularly high light saturated photosynthetic rate, although there is evidence that this rate is higher than for most other C_3 species. Two other explanations for the high unit leaf rate may be offered: the photosynthetic rate may be high at relatively low light intensities (unlikely in this species), or that photosynthetic rate is less limited by a low demand for photosynthate in sunflower as compared with other species; this latter hypothesis would seem to be very credible for sunflower.

Maize

One straight line appears to adequately describe the stem-leaves relationship, and its gradient, while not high (1.1584), differs significantly from unity. Thus, the relative growth rate of the stem is higher than that of the foliage right from the beginning; this is, perhaps, not surprising when it is considered that the stem forms a large part of the mature plant, and that at the time of the first harvest in this experiment the stem was only one quarter the size of the leaves in terms of dry weight.

For the root-leaves relationship, two straight lines are required, intersecting between harvests 11 and 12. The first one has a gradient of 0.7617, which is significantly below unity and indicates that root demand for assimilates is low during early growth in this species under the prevailing experimental conditions but the root weight ratio is high at the beginning of the experiment (Fig. 2.5c). The situation is reversed later, but examination of the primary data (not shown) reveals that the change in allometric gradient was occasioned by a slowing down of leaf growth, which was correlated with the cessation of leaf production, rather than by an increase in root growth. However, this still means that the root now has the highest relative growth rate of the three parts, a situation that probably does not persist into the reproductive phase to any great extent.

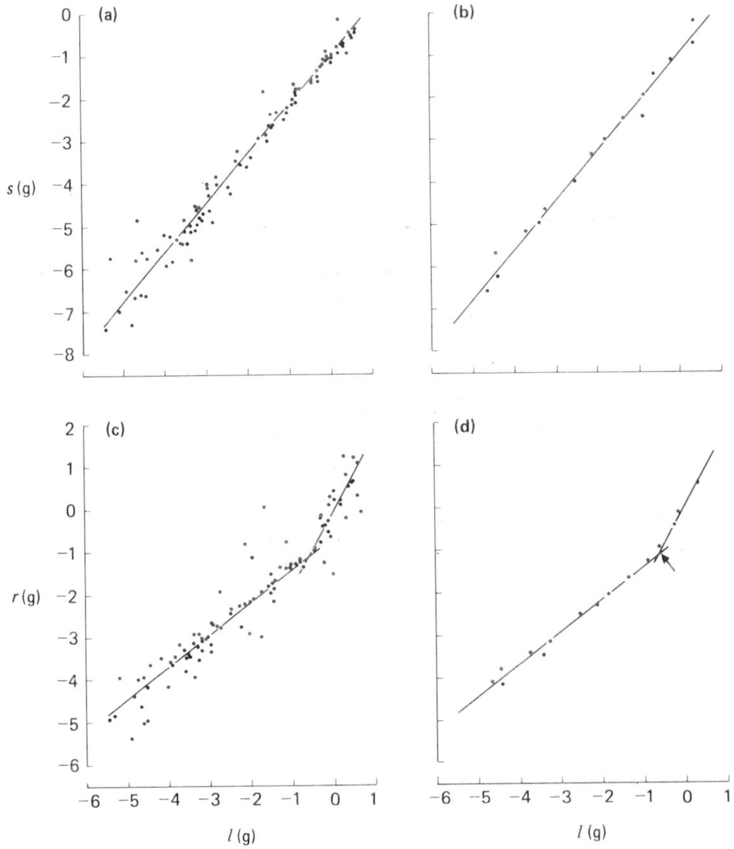

Fig. 6.3 Allometric relationships for maize between stem weight and leaf weight, (**a**) and (**b**), and root weight and leaf weight, (**c**) and (**d**); with all points, (**a**) and (**c**), and harvest means, (**b**) and (**d**). Intersections of allometric segments are indicated →.

Wheat

In this species, harvest 1 seems to have given anomalous results, and so was discarded; it is conceivable that the allometric relationship might change between harvests 1 and 2. The remaining stem-foliage data showed two distinct phases, with the junction between harvests 11 and 12. As in maize, the slope of the allometric line is significantly greater than unity even in the early stages of growth, very significantly so in the case of wheat. At the first harvest, the stem is less than 1/7 the size of the foliage, whereas the ratio is larger (1/4) in maize; thus it appears that the sink activity of the very small stem in the early stages of wheat growth is high. For the last part of the

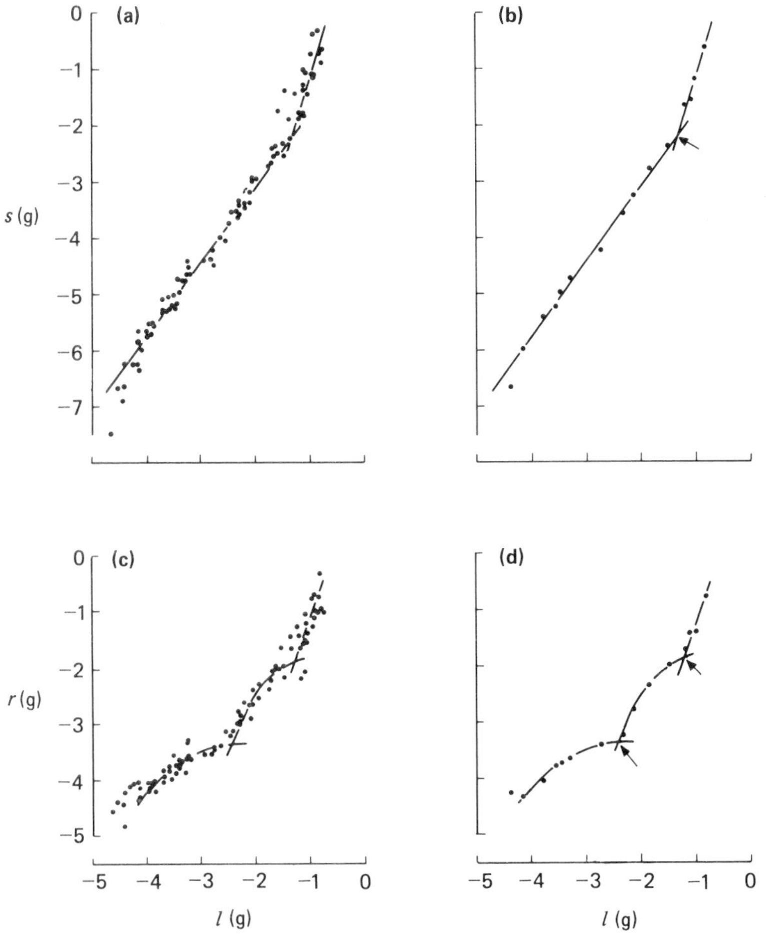

Fig. 6.4 Allometric relationships for wheat between stem weight and leaf weight, (**a**) and (**b**), and root weight and leaf weight, (**c**) and (**d**); with all points, (**a**) and (**c**), and harvest means, (**b**) and (**d**). Intersections of allometric segments are indicated →.

experiment, growth of the foliage almost ceases, but the stem continues to grow at only a slightly reduced rate, allowing the gradient of the allometric line to rise to the very high level of 3.27. This last phase is associated with reproductive growth; no further leaves are initiated, at least on the main apex, after harvest 11.

The root-leaves data are singular, in that between harvests 2 to 11 there appear to be two distinct curvilinear phases; quadratic functional relationships were, therefore, used to describe the data. The first phase, from harvest 2 to 7, has an initial gradient of 1.0900 declining to 0.0485, with a mean of 0.5692; the second phase, from harvest 8 to 11, has an initial gradient of 2.4367 decreasing to 0.1729, with a mean of 1.3048. Inspection of the primary data shows that over the first phase, root growth is decidedly curvilinear with respect to foliage growth, but the cause of curvilinearity of the allometric relationship in the second phase is obscure.

Birch

Because the growth of birch in its first year is wholly vegetative, and because the duration of this experiment covers so small a part of the total life span of an individual tree, this species might have been expected to show particularly simple allometric relationships. Such was not the case, however, but the stem-leaves relationships show a fairly straightforward picture with three phases, showing an increasing propensity toward stem growth. The root-leaves graph, however, seems to indicate a series of discontinuities. Examination of the primary data (Venus, 1978) shows that the discontinuity between harvests 2 and 3 is occasioned by a large difference between mean root weights at the two harvests not matched by a similar difference in leaf weights; between harvests 8 and 9 there is again a large difference in mean root weights, but only a small difference between mean foliage weights; and similarly between harvests 16 and 17. It thus appears that in each instance the discontinuity, if real, is caused by a sudden increase in root growth which is either not matched by that of the remainder of the plant or is correlated with a decrease in the growth of foliage and stem. In a previous experiment (Causton, 1970), involving the growth of first year plants of birch and *Acer pseudoplatanus* (sycamore), smooth transitions were found in the gradients of linear allometric relationships of birch which involved the root, but one discontinuity was found in sycamore, which again could be correlated with an apparent sudden increase in root growth coupled with a decrease in stem and foliage growth. The number of discontinuities in the root-leaves allometric relationships of the present experiment is curious; the only explanation we can offer concerns the unusual way in which the plants were grown as alternative to potting on as the root systems enlarged (page 43).

Looking at the plant as a whole, between the first two harvests the leaves are exerting a high demand for assimilates compared with stem and root; indeed the slope of the root-leaves line is so low (0.2897) that one is led to suppose that the root has a negative demand at this time. This hypothesis could be carried through to the next phase, between harvests 3 and 8 or 9, where either the root has a negative demand, or the shoot exerts a positive demand; but if, because of a gradient of essentially unity between stem and leaves at this time, we infer that the growth of these two parts relative to one another is supply limited, then the suggestion that shoot growth exerts a

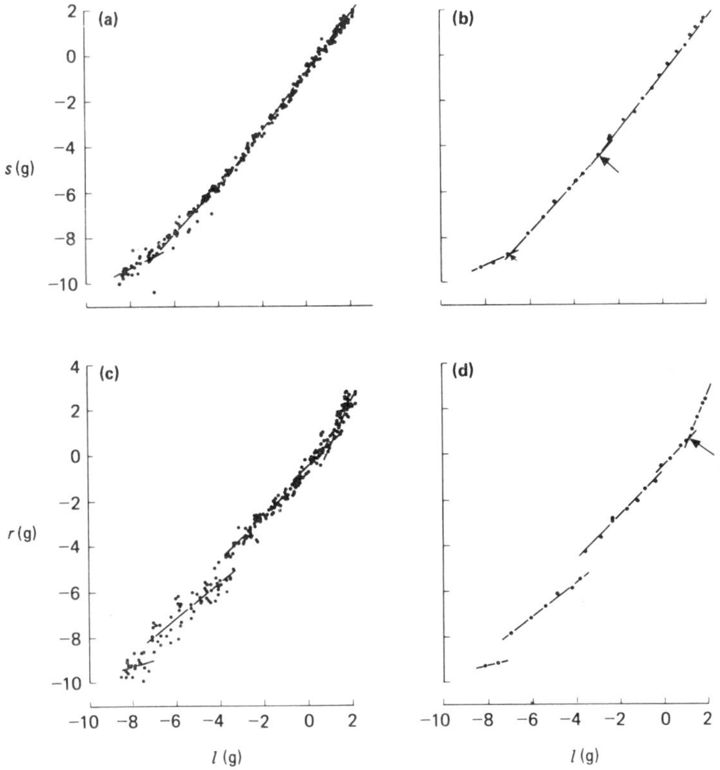

Fig. 6.5 Allometric relationships for birch between stem weight and leaf weight, (**a**) and (**b**), and root weight and leaf weight, (**c**) and (**d**); with all points, (**a**) and (**c**), and harvest means, (**b**) and (**d**). Intersections of allometric segments are indicated →.

positive demand is more likely. The next two phases, harvests 9 or 10 to 16 and 17 to 20, may be taken together, since it is covered by one stem-leaves relationship, and the two root-leaves relationships have essentially a slope of unity; only the discontinuity in the root-leaves data induces us to recognize two phases, which are characterized by a clear stem demand over the other parts. The final phase, harvests 21 to 24 shows a preponderance of root growth.

An explanation of these changes, in terms of plant strategy, can be made in relation to the fact that birch is a pioneer woody species (Grime, 1979). Birch species all have small, winged, seeds for wide dispersal, but consequently have only a small food reserve. In the earliest phase, leaf growth is thus of paramount importance, but relatively early stem growth is important too, in order that the seedlings may successfully compete with the surrounding herbaceous vegetation. For some considerable time, therefore, shoot growth takes precedence over that of the root, but ultimately the root must 'catch up' and this is seen to be happening in the final phase of this experiment.

The physiological significance of allometric data

In the above sub-sections we have presented allometric data on four species, and interpreted the results in terms of plant strategies. The word 'strategy' is, perhaps, an unfortunate one with its connotations of active decision making. What is implied, however, is the evolution of certain properties, both morphological and physiological, in response to environmental pressures, and the nature of these pressures can be both physical and biotic. The whole subject has been reviewed recently in an interesting book by Grime (1979).

The physiological bases of different strategies is a field of study that has scarcely been broached. Of particular interest would be the physiological cause of changes in the gradient of allometric lines, and as a prelude to this, the provision of an accurate specification of the stage of development at which a change occurs is necessary. To be reliable, the estimate of the position of change must be obtained from well-replicated experiments analysed by the most appropriate statistical methods, and the final result should be repeatable under conditions which are as near as possible the same in each experiment. Eventually, after some considerable experience of correlating allometric relationships with physiological processes, it might be possible to directly infer some aspects of the physiological status of a plant by analysis of allometric and other growth data. This would have the advantage of being a relatively simple and inexpensive way of making deductions of physiological relevance in the growing plant.

7

The whole plant: a synthetic growth model

In this chapter we shall consider the growth of the whole plant in terms of the growth of its component parts. A model of whole plant vegetative growth will be built in three stages: firstly, the fitted Richards functions for each leaf will be added together to produce a foliage growth curve; secondly, this foliage growth curve will be used in conjunction with the allometric functions to produce stem and root growth curves; and finally, foliage, stem and root growth curves will be added to give a growth curve for the whole plant. At each stage, component-entity ratios and relative growth rate curves are derived, and a unit leaf rate curve is finally obtained for the whole plant. Full attention is given to the statistical aspects: thus for example, the addition of estimated individual leaf weights (given by the fitted Richards functions), at any time, to give a foliage weight is not simply a mathematical addition but a statistical one in which the variances and covariances of the weight estimates are taken into account; confidence intervals will also be given for each curve. Every model curve will be compared with the directly fitted curve (usually the Richards function) in Chapter 2 in the case of components and entities, or with the directly derived curve in the case of component-entity ratios, relative growth and unit leaf rates. This model is more ambitious than that of Hunt & Bazzaz (1980) because even though component and entity curves were presented simultaneously, they estimated variances of the entity curves directly and not by utilizing the variances of the components.

A model of this type, assuming that the forms of interdependence are correct, would make possible the examination of the effects of changes to single leaf growth upon the whole foliage system and the total plant. This approach is relevant to plants growing in a fluctuating environment to assist in deducing the effects of short term environmental changes on the long term growth of the plant. Additionally, the model allows for the assessment of interrelationships at various component entity levels, and could be useful in the theoretical investigation of control and feedback mechanisms within the plant.

Before dealing with the model itself and its application to specific sets of data, we shall consider the growing plant as a mathematical system. Apart from its intrinsic interest, one reason for doing this is to elucidate the mathematical constraints on growth patterns, because any physiological and morphogenetical constraints on these patterns must lie within the mathematical confines.

Some general mathematical relationships involved in plant growth

The method of presentation will be similar to that employed in the previous chapter on the mathematical aspects of allometry. A general notation will be used, the component entity level is unspecified, and the primary results will be given in the form of two theorems; then the theorems will be discussed with more specific reference to particular plant parts, using the conventional notation adopted throughout the book.

For the theorems, at time t the dry weight of the entity is W, and the entity is considered to be made up of Q components whose dry weights are W_i, $i = 1, \ldots, Q$. Then we have the obvious and fundamental relationships:

$$\left.\begin{aligned} W &= \sum_{i=1}^{Q} W_i \\ \frac{\mathrm{d}W}{\mathrm{d}t} &= \sum_{i=1}^{Q} \frac{\mathrm{d}W_i}{\mathrm{d}t} \end{aligned}\right\} \qquad (7.1)$$

namely, that the entity is composed of the sum of its parts, and that the absolute growth rate of the entity is equal to the sum of the absolute growth rates of its parts. Also, at time t, we define Q component entity ratios, x_i, $i = 1, \ldots, Q$; i.e.

$$x_i = W_i / W \qquad (7.2)$$

and so

$$\sum_{i=1}^{Q} x_i = 1 \qquad (7.3)$$

Finally, the relative growth rate of the entity and of the ith component at time t are defined as usual:

$$\left.\begin{aligned} \mathrm{R} &= \frac{1}{W} \cdot \frac{\mathrm{d}W}{\mathrm{d}t} \\ \mathrm{R}_i &= \frac{1}{W_i} \cdot \frac{\mathrm{d}W_i}{\mathrm{d}t} \end{aligned}\right\} \qquad (7.4)$$

Theorems 7.1 and 7.2

Theorem 7.1 *In an entity consisting of Q components, the relative growth rate of the entity, in terms of the relative growth rate of the components, is given at any instant by the sum of the products of each component relative* *growth rate and component-entity ratio; i.e. by*

$$\mathrm{R} = \sum_{i=1}^{Q} \mathrm{R}_i \mathrm{x}_i \tag{7.5}$$

Multiply both sides of the second equation in (7.1) by $1/W$, giving

$$\frac{1}{W}\cdot\frac{\mathrm{d}W}{\mathrm{d}t} = \frac{1}{W}\cdot\sum_{i=1}^{Q}\frac{\mathrm{d}W_i}{\mathrm{d}t}$$

Then, on multiplying the ith term on the right-hand side by W_i/W_i, the required result is obtained:

$$\mathrm{R} = \sum_{i=1}^{Q}\frac{W_i}{W}\cdot\frac{1}{W_i}\cdot\frac{\mathrm{d}W_i}{\mathrm{d}t} = \sum_{i=1}^{Q}\mathrm{R}_i\mathrm{x}_i$$

Theorem 7.2 *The rate of change of the ith component entity ratio at any instant, is given by the product of the ith component entity ratio and the difference between the relative growth rate of the entity and that of the ith component; i.e. by*

$$\frac{\mathrm{dx}_i}{\mathrm{d}t} = \mathrm{x}_i(\mathrm{R}_i - \mathrm{R}) \tag{7.6}$$

By the quotient rule for differentiation, we have

$$\frac{\mathrm{d}(W_i/W)}{\mathrm{d}t} = \frac{W(\mathrm{d}W_i/\mathrm{d}t) - W_i(\mathrm{d}W/\mathrm{d}t)}{W^2}$$

$$= \frac{1}{W}\left(\frac{\mathrm{d}W_i}{\mathrm{d}t} - \frac{W_i}{W}\frac{\mathrm{d}W}{\mathrm{d}t}\right)$$

On multiplying the right-hand side by W_i/W_i, we obtain the required result:

$$\frac{\mathrm{dx}_i}{\mathrm{d}t} = \frac{W_i}{W}\left(\frac{1}{W_i}\cdot\frac{\mathrm{d}W_i}{\mathrm{d}t} - \frac{1}{W}\cdot\frac{\mathrm{d}W}{\mathrm{d}t}\right) = \mathrm{x}_i(\mathrm{R}_i - \mathrm{R})$$

Corollary *The rate of change of the ith component entity ratio at any instant, in terms of the relative growth rates of the components, is given by*

$$\frac{\mathrm{dx}_i}{\mathrm{d}t} = \mathrm{x}_i\left\{(1-\mathrm{x}_i)(\mathrm{R}_i - \mathrm{R}_Q) + \sum_{\substack{j=1\\ j\neq i}}^{Q-1}(\mathrm{R}_Q - \mathrm{R}_j)\mathrm{x}_j\right\} \quad i \neq Q \tag{7.7}$$

where R_Q *is the relative growth rate of the Qth component.*

First substitute (7.5) for R in (7.6), giving

$$\frac{dx_i}{dt} = x_i(R_i - \sum_{j=1}^{Q} R_j x_j)$$

i.e. $$\frac{dx_i}{dt} = x_i\left\{R_i - \sum_{j=1}^{Q-1} R_j x_j - R_Q\left(1 - \sum_{j=1}^{Q-1} x_j\right)\right\} \qquad i \neq Q$$

since $x_Q = 1 - \sum_{j=1}^{Q-1} x_j$.

Hence

$$\frac{dx_i}{dt} = x_i\left\{R_i - R_Q + \sum_{j=1}^{Q-1} (R_Q - R_j)x_j\right\} \qquad i \neq Q$$

$$= x_i\left\{R_i - R_Q + (R_Q - R_i)x_i + \sum_{\substack{j=1 \\ j \neq i}}^{Q-1} (R_Q - R_j)x_j\right\} \qquad i \neq Q$$

Therefore

$$\frac{dx_j}{dt} = x_i\left\{(1 - x_i)(R_i - R_Q) + \sum_{\substack{j=1 \\ j \neq i}}^{Q-1} (R_Q - R_j)x_j\right\} \qquad i \neq Q$$

as required.

Relative growth rates and component entity ratios

The result of Theorem 7.1 shows how the relative growth rates of the components are related to that of the entity; it is the relative growth rate equivalent of the summation formula for absolute growth rate, given as the second equation of (7.1). For example, at Level la, we have

$$R = R_L\left(\frac{L}{W}\right) + R_S\left(\frac{S}{W}\right) + R_R\left(\frac{R}{W}\right) \qquad (7.8)$$

One consequence of Theorem 7.1 is that if the relative growth rates of all the components are equal to one another, the relative growth rate of the entity is the same as that for the components. For if $R_1 = R_2 = \ldots\ldots = R_Q = R$ (say), then from (7.5) and (7.3)

$$Rx_1 + Rx_2 + \ldots\ldots + Rx_Q = R(x_1 + x_2 + \ldots\ldots + x_Q) = R \qquad (7.9)$$

Evidently, component relative growth rates and component-entity ratios are intimately connected, and this association again appears in Theorem 7.2 which relates the rate of change of a component entity ratio to the relative growth rates of the component and the entity. For example, for the leaf weight ratio we have

$$\frac{d(L_W/W)}{dt} = \frac{L_W}{W}(R_L - R) \tag{7.10}$$

which shows that if $R_L > R$ the leaf weight ratio is increasing, and conversely. If we consider a plant in vegetative growth having leaves and root only (a rosette plant), then the corollary to Theorem 7.2 shows that

$$\frac{d(L_W/W)}{dt} = \frac{L_W}{W}\left(1 - \frac{L_W}{W}\right)(R_L - R_R) \tag{7.11}$$

If the plant has a stem as well, then

$$\frac{d(L_W/W)}{dt} = \frac{L_W}{W}\left\{\left(1 - \frac{L_W}{W}\right)(R_L - R_R) + \frac{S}{W}(R_R - R_S)\right\} \tag{7.12}$$

The mathematics of exponential growth

Exponential growth is said to occur in the early stages of vegetative growth after the seedling has ceased to be dependent on seed reserves, and it implies a constant relative growth rate. The equations describing exponential growth are:

$$\left.\begin{aligned} W &= ae^{Rt} \\ \text{and} \quad \log_e W &= \log_e a + Rt \end{aligned}\right\} \tag{7.13}$$

The properties of the exponential function are well known and will not be reiterated here. The purpose of this section is to demonstrate that even in a seemingly simple quantitative situation, as in exponential growth, there are complications owing to the separation of distinct growth centres. The term 'exponential growth', applied to a plant, is vague; does it mean that the plant as a whole is growing exponentially, or that its main organ systems are growing exponentially, or that both these conditions are occurring simultaneously? Theorem 7.3 and the ensuing discussion put these problems into perspective.

Theorem 7.3 *If all Q components of an entity are growing exponentially, then the entity is not growing exponentially unless the relative growth rates of all the components are equal to one another. When the relative growth rates of all the components are unequal, the relative growth rate of the entity always increases according to a logistic function.*

If W grows exponentially, then

$$W = \sum_{i=1}^{Q} a_i e^{R_i t} = a e^{Rt} \tag{7.14}$$

By a method which exactly follows the proof of the lemma in Chapter 6 (page 175), this implies that

$$R_1 = R_2 = \ldots . R_Q \tag{7.15}$$

which proves the first part of the theorem.

To prove the second part of the theorem we will start with a two-component plant, and then equation (7.7) yields

$$\frac{dx_1}{dt} = x_1(1 - x_1)(R_1 - R_2)$$

This is a logistic function (since all R_i are constants), which on integration gives

$$x_1 = (1 + c e^{(R_2 - R_1)t})^{-1} \tag{7.16}$$

where $c = e^C$, C being the constant of integration. From (7.5), we have

$$R = R_2 - x_1(R_2 - R_1) \tag{7.17}$$

Substituting for x_1 in (7.17) using (7.16) gives

$$R = R_2 - \frac{R_2 - R_1}{1 + c e^{(R_2 - R_1)t}} \tag{7.18}$$

The second term on the right-hand side of (7.18) is logistic in nature, but is of a special kind since the asymptotic maximum value in the numerator and the exponent in the denominator are identical. The term may be denoted by $Lf\{(R_2 - R_1):t\}$, since apart from c, $(R_2 - R_1)$ is the only parameter. Equation (7.18) is then written as

$$R = R_2 - Lf\{(R_2 - R_1):t\} \tag{7.19}$$

If $R_2 < R_1$, $Lf\{(R_2 - R_1):t\}$ is negative and decreases from 0 to $(R_2 - R_1)$; thus R is increasing from a lower asymptote, R_2, to an upper asymptote, R_1. If $R_2 > R_1$, $Lf\{(R_2 - R_1):t\}$ is positive and decreases from $(R_2 - R_1)$ to 0; thus R increases from a lower asymptote R_1 to an upper asymptote R_2. Hence the theorem is proved for two components.

In the general case, we apply (7.19) successively, as follows. Let S_Q be a 'compound' component consisting of R_1 up to R_{Q-1}. Then by (7.19)

$$R = S_Q - Lf\{(S_Q - R_Q):t\} \tag{7.20}$$

Now partition S_Q into R_Q and a 'compound' component, S_{Q-1}, consisting of

R_1 up to R_{Q-2}; then again by (7.19)

$$S_Q = S_{Q-1} - Lf\{(S_{Q-1} - R_{Q-1}):t\} \tag{7.21}$$

Substituting (7.21) into (7.20) gives

$$\begin{aligned} R = {} & S_{Q-1} - Lf\{(S_{Q-1} - R_{Q-1}):t\} \\ & - Lf\{[S_{Q-1} - Lf\{(S_{Q-1} - R_{Q-1}):t\} - R_Q]:t\} \end{aligned} \tag{7.22}$$

If we are dealing with a three-component plant ($Q = 3$), then $S_{Q-1} = R_1$ and $R_{Q-1} = R_2$; then (7.22) becomes

$$R = R_1 - Lf\{(R_1 - R_2):t\} - Lf\{[R_1 - Lf\{(R_1 - R_2):t\} - R_3]:t\}$$

It is not easy to deduce the form of the changes of the relative growth rate of the entity from this equation merely by inspection, but Table 7.1 shows how the asymptotic trend of this quantity may be investigated by resolving the equation into its elementary components. It is evident from the last column of Table 7.1 that the relative growth rate of the entity always increases with time within the bounds of the minimum and maximum component relative growth rates.

Growth of the plant and its parts at Levels 1 and 1a

Theorem 7.3 shows that only if the relative growth rates of the components are equal, which implies equality with the entity (equations (7.9) and (7.15)), can we say that the whole plant *and* its main parts are simultaneously growing exponentially. Otherwise, the theorem shows that if the parts are growing exponentially with different relative growth rates, the plant as a whole is growing with an increasing relative growth rate, or 'superexponentially'; conversely, it is reasonable to suppose that if the plant is growing exponentially then the components are growing 'subexponentially' with a declining relative growth rate. These two facets will now be examined for a rosette plant in vegetative growth, consisting of leaves and root (Level 1a) or, equivalently, shoot and root (Level 1). The results of Theorem 7.3 will not be used; a different approach will be adopted which makes only one assumption – that a linear allometric relationship exists between the two components – i.e.

$$r = \alpha + \beta l \tag{7.23}$$

and so

$$R_R = \beta R_L \tag{7.24}$$

Now, from (7.5)

$$R = R_L\left(\frac{L_W}{W}\right) + R_R\left(\frac{R}{W}\right)$$

Table 7.1 Resolution of equation (7.22) to show the behaviour of R for a three-component plant, in which each component is growing exponentially.

Relative size of component R_i	Asymptote $t \rightarrow$	$\frac{R_2 - R_1}{1 + ce^{(R_2 - R_1)t}}$	$R_2 - R_3 - R_3 - \frac{R_2 - R_1}{1 + ce^{(R_2 - R_1)t}}$	R (from (7.19))	= R
$R_1 < R_2 < R_3$	$-\infty$	$R_2 - R_1$	$R_2 - R_3 - (R_2 - R_1) = R_1 - R_3$	$R_2 - (R_2 - R_1) - 0$	R_1
	$+\infty$	0	$R_2 - R_3 - 0 = R_2 - R_3$	$R_2 - 0 - (R_2 - R_3)$	R_3
$R_1 < R_3 < R_2$	$-\infty$	$R_2 - R_1$	$R_2 - R_3 - (R_2 - R_1) = R_1 - R_3$	$R_2 - (R_2 - R_1) - 0$	R_1
	$+\infty$	0	$R_2 - R_3 - 0 = R_2 - R_3$	$R_2 - 0 - 0$	R_2
$R_3 < R_1 < R_2$	$-\infty$	$R_2 - R_1$	$R_2 - R_3 - (R_2 - R_1) = R_1 - R_3$	$R_2 - (R_2 - R_1) - (R_1 - R_3)$	R_3
	$+\infty$	0	$R_2 - R_3 - 0 = R_2 - R_3$	$R_2 - 0 - 0$	R_2

and since $L_w/W + R/W = 1$, we have

$$R = R_L(L_w/W) + R_R(1 - L_w/W) \tag{7.25}$$

Substituting for R_R in (7.25) using (7.24) gives

$$R = R_L\{\beta + (L_w/W)(1-\beta)\} \tag{7.26}$$

Differentiating both sides of (7.26) with respect to time, we have, using the product rule,

$$\frac{dR}{dt} = R_L . \frac{d\{\beta + (L_w/W)(1-\beta)\}}{dt} + \{\beta + (L_w/W)(1-\beta)\} . \frac{dR_L}{dt}$$

i.e.
$$\frac{dR}{dt} = R_L(1-\beta) . \frac{d(L_w/W)}{dt} + \{\beta + (L_w/W)(1-\beta)\} . \frac{dR_L}{dt} \tag{7.27}$$

Now, substituting for $d(L_w/W)/dt$ in (7.27) using (7.10), we have

$$\frac{dR}{dt} = R_L\left(\frac{L_w}{W}\right)(R_L - R)(1-\beta) + \{\beta + \frac{L_w}{W} 1 - \beta)\} . \frac{dR_L}{dt} \tag{7.28}$$

From (7.26)

$$\frac{L_w}{W} = \frac{R - \beta R_L}{R_L(1-\beta)} \tag{7.29}$$

and, substituting (7.29) in (7.28) and rearranging finally yields

$$\frac{dR}{dt} = (R - \beta R_L)(R_L - R) + \frac{R}{R_L} \cdot \frac{dR_L}{dt} \tag{7.30}$$

Equation (7.30) is the fundamental relationship required. Either dR_L/dt can be equated to zero, that is the leaves (and hence the root, because of the linear allometric relationship) have a constant relative growth rate (exponential growth); or $dR/dt = 0$, which is equivalent to the plant having a constant relative growth rate. First assume that the leaves and root are each growing exponentially, then (7.30) becomes

$$\frac{dR}{dt} = (R - \beta R_L)(R_L - R) \tag{7.31}$$

Solution of this differential equation, in which the variables are separable, is effected by partial fractions; so

$$\frac{dR}{(R - \beta R_L)(R_L - R)} = dt$$

which leads to

$$\frac{dR}{R - \beta R_L} + \frac{dR}{R_L - R} = R_L(1-\beta)dt$$

Integrating:

$$\log_e(R - \beta R_L) - \log_e(R_L - R) = R_L(1-\beta)t + c$$

where c is the constant of integration. Taking antilogarithms of both sides gives

$$\frac{R - \beta R_L}{R_L - R} = \exp\{c + R_L(1-\beta)t\}$$

and adding 1 to both sides and rearranging gives

$$\frac{R_L(1-\beta)}{R_L - R} = 1 + \exp\{c + R_L(1-\beta)t\}$$

from which

$$R = R_L - \frac{R_L(1-\beta)}{1 + \exp\{c + R_L(1-\beta)t\}} \tag{7.32}$$

Equation (7.32) is identical in form to equation (7.18) in Theorem 7.3. Equation (7.31) is graphed in Fig. 7.1a, assuming $\beta > 1$; R increases from R_L to $\beta R_L(R_R)$, with dR/dt increasing first before decreasing. If $\beta < 1$, βR_L is nearer the origin, and R increases from βR_L to R_L.

Now assume that the plant grows exponentially, and (7.30) becomes

$$\frac{dR_L}{dt} = -\frac{R_L}{R}(R - \beta R_L)(R_L - R) \tag{7.33}$$

Again, the variables are separable in this differential equation,

$$\frac{dR_L}{R_L(R - \beta R_L)(R_L - R)} = -\frac{dt}{R}$$

and resolving the left-hand side into partial fractions yields

$$\frac{\beta^2}{1-\beta}\left(\frac{dR_L}{R - \beta R_L}\right) + \frac{1}{1-\beta}\left(\frac{dR_L}{R_L - R}\right) - \frac{dR_L}{R_L} = -R\,dt$$

Integrating:

$$\frac{1}{1-\beta}\cdot\log_e(R_L - R) - \frac{\beta}{1-\beta}\cdot\log_e(R - \beta R_L) - \log_e R_L = -Rt + c$$

where c is the constant of integration, and multiplying throughout by $(1-\beta)$ gives

$$\log_e(R_L - R) - \beta\,.\log_e(R - \beta R_L) - (1-\beta)\,.\log_e R_L = k - R(1-\beta)t$$

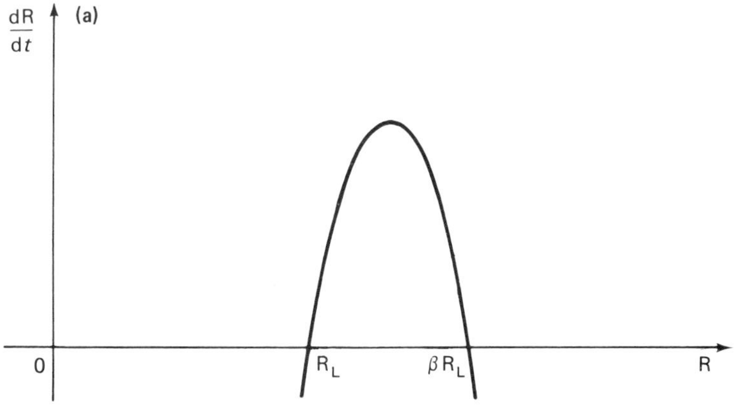

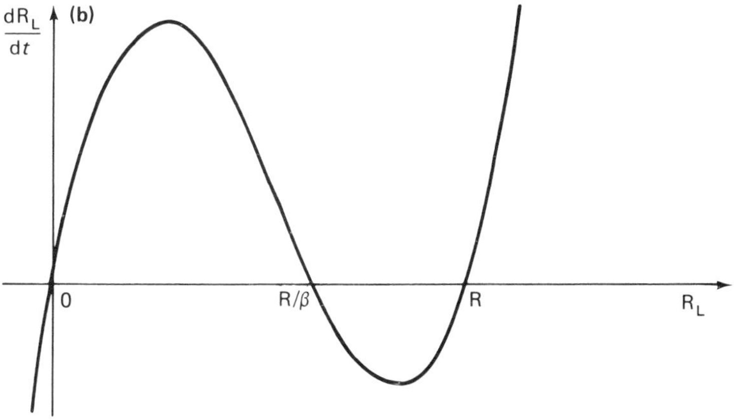

Fig. 7.1 (**a**) Graph of equation (7.31). (**b**) Graph of equation (7.33).

where $k = (1-\beta)c$. Taking antilogarithms of both sides gives

$$\frac{R_L^{(\beta-1)}(R_L - R)}{(R - \beta R_L)^\beta} = \exp\{k - R(1-\beta)t\} \tag{7.34}$$

The solution of (7.33) cannot be simplified any further, and examination of the trend in R_L is best made from (7.33) itself, in which dR_L/dt is a third degree polynomial in R_L with roots $R_L = 0$, $R_L = R/\beta$, $R_L = R$. The coefficient of R_L^3 is positive, and so the general trend of the curve is from bottom left to top right of the graph (Fig. 7.1b), and again we assume $\beta > 1$. Since dR/dt is positive when $dR_L/dt = 0$ (the case already considered), it is reasonable to assume that $\mathrm{dR}_L/\mathrm{d}t$ is negative when $\mathrm{dR}/\mathrm{d}t = 0$. Hence R_L decreases from R to R/β over the range where the curve is below the R_L-axis. If $\beta < 1$, R and R/β are reversed in position, but otherwise the situation is unchanged, with R_L decreasing from R/β to R. Similar conclusions can be drawn from equation (7.34) by allowing t to tend to $\pm\infty$.

The model: methods

Foliage growth

The growth in dry weight of the single leaves on a plant can be represented by a series of estimated Richards functions; so that for leaf k,

$$\hat{l}_k = a_k + \log_e (1 \pm e^{(b_k + K_k t)})^{-1/n_k} \qquad (7.35)$$

and from this function can be generated a series of estimated dry weight values, $\hat{l}_{ki}$, $i = t_1, \ldots, t_h$, and the variances, $\mathscr{V}(l_{ki})$. Throughout this chapter terms such as $\mathscr{V}(l_{ki})$ should be read as $\mathscr{V}(l_{ki})$, see p. 28. The times t_1 and t_h are, as before, the initial and final harvest times in the experiment providing data for the model; but in calculating the model results, estimates and variances of the dry weight of the kth leaf may be obtained at times other than the experimental harvesting times, and so the value of h need be no longer equal to the number of harvests actually taken. The model results to be presented in the next part of this chapter were, in fact, calculated on a daily basis. Similarly, a series of estimated relative growth rate values, $\hat{\mathrm{R}}_{Lki}$, can be obtained together with variance estimates, $\mathscr{V}(\mathrm{R}_{Lki})$, for the kth leaf (pages 108–109).

In order to obtain an estimate of foliage dry weight at a particular time, a simple summation might at first sight appear to be all that is required; however, the dry weight functions (7.35) are in logarithmic form, and the summation is also required in this form. Mathematically, the logarithm of foliage dry weight at the ith time is given by

$$l_i = \log_e \left(\sum_{k=1}^{q} e^{l_{ki}} \right) \qquad (7.36)$$

where q represents the number of leaves present at that time. In the interests of clarity from here onward, we shall dispense with the hat-notation ($\hat{\ }$) for estimates involving single leaves, for which the estimation methods have already been described in Chapter 4; and the hat will only be employed for the estimates derived by methods explained in this chapter. Thus, in equation (7.36) l_{ki} is no longer hatted. The expected values and variances cannot be found by simple linear statistical theorems, as would be the case if this was a simple summation, but approximation formulae (Appendix, Page 261) are needed. Thus.

$$\hat{l}_i \simeq \log_e\left(\sum_{k=1}^{q} e^{l_{ki}}\right) + \frac{1}{2}\cdot\sum_{k=1}^{q}\frac{\partial^2 l_i}{\partial l_{ki}^2}\cdot\mathscr{V}(l_{ki}) + $$
$$+\frac{1}{2}\cdot\sum_{k=1}^{q}\sum_{f=1}^{k-1}\frac{\partial^2 l_i}{\partial l_{ki}\partial l_{fi}}\cdot\mathscr{C}(l_{ki}, l_{fi}) \tag{7.37}$$

$$\mathscr{V}(l_i) \simeq \sum_{k=1}^{q}\left\{\frac{\partial l_i}{\partial l_{ki}}\right\}^2\cdot\mathscr{V}(l_{ki}) + \sum_{k=1}^{q}\sum_{f=1}^{k-1}\frac{\partial l_i}{\partial l_{ki}}\cdot\frac{\partial l_i}{\partial l_{ki}}\cdot\mathscr{C}(l_{ki}, l_{fi}) \tag{7.38}$$

where

$$\frac{\partial l_i}{\partial l_{ki}} = e^{l_{ki}}\left(\sum_{k=1}^{q} e^{l_{ki}}\right)^{-1} \qquad \frac{\partial^2 l_i}{\partial l_{ki}\partial l_{fi}} = -e^{l_{ki}}e^{l_{fi}}\left(\sum_{k=1}^{q} e^{l_{ki}}\right)^{-2}$$

$$\frac{\partial^2 l_i}{\partial l_{ki}^2} = e^{l_{ki}}\left(\sum_{k=1}^{q} e^{l_{ki}} - e^{l_{fi}}\right)\left(\sum_{k=1}^{q} e^{l_{ki}}\right)^{-2}$$

at time i. The total number of leaves, q, will vary from one time to another, depending upon the number of leaves that have been initiated. At first sight the estimate $\hat{L}_i$ can be obtained as $\exp(\hat{l}_i)$; however, this is not correct because we require an estimate of the geometric mean rather than the arithmetic and we can use the assumption that the $\hat{l}_{ij}$ are approximately normally distributed to obtain a more realistic estimate. The estimated foliage dry weight at the ith time, and the variance, are given by

$$\hat{L}_i \simeq \exp\left\{\hat{l}_i + \frac{1}{2}\cdot\mathscr{V}(l_i)\right\} \tag{7.39}$$

$$\mathscr{V}(L_i) \simeq \exp\left[2\left\{\hat{l}_i + \frac{1}{2}\cdot\mathscr{V}(l_i)\right\}\right]\cdot[\exp\{\mathscr{V}(l_i)\} - 1] \tag{7.40}$$

Formulae of this type will be used throughout the pages that follow in similar circumstances to those described above without further discussion.

Leaf foliage ratios

The second facet of interest concerning foliage growth is the ratios of single leaves to total foliage: the component entity ratio at Level 3. The calculations to find estimated values (and their variances) of these ratios are similar to those used to find the ratios involved in whole plant growth analysis (p. 23). Thus, for the kth leaf at time i,

$$\log_e(L_{ki}/L_i) = l_{ki} - \hat{l}_i \tag{7.41}$$

and so

$$\mathscr{E}\{\log_e(L_{ki}/L_i)\} = l_{ki} - \hat{l}_i \tag{7.42}$$

and
$$\mathscr{V}\{\log_e(L_{ki}/L_i)\} = \mathscr{V}(l_{ki}) + \mathscr{V}(l_i) - 2.\mathscr{C}(l_{ki}, l_i) \tag{7.43}$$

where, for typographical convenience, the symbol $\mathscr{E}$ is to be read as the estimate of the expected value rather than the expected value itself. Since L_{kij} and L_{ij} are approximately log-normally distributed, so (L_{ki}/L_{ij}) are also approximately log-normally distributed; therefore

$$\mathscr{E}(L_{ki}/L_i) \simeq \exp[\mathscr{E}\{\log_e(L_{ki}/L_i)\} + \frac{1}{2}.\mathscr{V}\{\log_e(L_{ki}/L_i)\}] \tag{7.44}$$

$$\mathscr{V}(L_{ki}/L_i) \simeq \exp[2.\mathscr{E}\{\log_e(L_{ki}/L_i)\} + \frac{1}{2}\cdot\mathscr{V}\{\log_e(L_{ki}/L_i)\}] \times (\exp[\mathscr{V}\{\log_e(L_{ki}/L_i)\}] - 1) \tag{7.45}$$

Equations (7.42) to (7.45) are analogous to equations (2.11) to (2.14). The quantities required for the evaluation of equations (7.44) and (7.45) have all previously been found except for the covariance between a single leaf and the total foliage. This can be found from the estimates of variances of the single leaf and the foliage, together with the covariances between all single leaves present; this is possible because the foliage is the sum of the individual leaves. The derivation is given on page 237 for the case of the three Level 1a components in relation to the whole plant, but a generalization applies equally well to a foliage system consisting of q leaves. Thus, from equation (7.75), we have in our present terms for the kth leaf

$$\mathscr{C}(L_{ki}, L_i) = \mathscr{V}(L_{ki}) + \sum_{f=1}^{q} \mathscr{C}(L_{ki}, L_{fi}) \qquad f \neq k \tag{7.46}$$

and application of equation (7.76) finally gives

$$\mathscr{C}(l_{ki}, l_i) \simeq \mathscr{C}(L_{ki}, L_i)/(\hat{L}_{ki}, \hat{L}_i) \tag{7.47}$$

All the quantities are now available for the estimation of leaf foliage ratios and variances.

Foliage relative growth rate

The third aspect of foliage growth that can be examined is the foliage relative growth rate. Recasting equation (7.5) in our present symbols gives

$$\mathrm{R}_{Li} = \sum_{k=1}^{q} \mathrm{R}_{Lki}(L_{ki}/L_i) \tag{7.48}$$

at time i. Applying the same approximation formulae as before to equation (7.48), we have

$$\hat{R}_{Li} \simeq \sum_{k=1}^{q} \{\mathscr{E}(L_{ki}/L_i)R_{Lki}\} + \frac{1}{2}.\sum_{k=1}^{q} \frac{\partial^2 R_{Li}}{\partial(L_{ki}/L_i)^2}.\mathscr{V}(L_{ki}/L_i)$$

$$+\frac{1}{2}.\sum_{k=1}^{q} \frac{\partial^2 R_{Li}}{\partial R^2_{Lki}}.\mathscr{V}(R_{Lki}) + \sum_{k=1}^{q}\sum_{f=1}^{k} \frac{\partial^2 R_{Li}}{\partial R_{Lki}\partial(L_{fi}/L_i)}.\mathscr{C}\{(L_{fi}/L_i), R_{Lki}\} \tag{7.49}$$

and

$$\mathscr{V}(R_{Li}) \simeq \sum_{k=1}^{q} \left\{\frac{\partial R_{Li}}{\partial(L_{ki}/L_i)}\right\}^2.\mathscr{V}(L_{ki}/L_i) + \sum_{k=1}^{q} \left\{\frac{R_{Li}}{R_{Lki}}\right\}^2.\mathscr{V}(R_{Lki})$$

$$+2.\sum_{k=1}^{q}\sum_{f=1}^{k} \frac{\partial R_{Li}}{\partial(L_{fi}/L_i)}.\frac{\partial R_{Li}}{\partial R_{Lki}}.\mathscr{C}\{(L_{fi}/L_i), R_{Lki}\} \tag{7.50}$$

The partial derivatives are given by

$$\frac{\partial R_{Li}}{\partial R_{Lki}} = \frac{L_{ki}}{L_i} \qquad \frac{\partial R_{Li}}{\partial(L_{ki}/L_i)} = R_{Lki}$$

$$\frac{\partial^2 R_{Li}}{\partial R^2_{Lki}\partial(L_{fi}/L_i)} = 1 \quad (f = k) \qquad = 0 \quad (f \neq k)$$

$$\frac{\partial^2 R_{Li}}{\partial R^2_{Lki}} = \frac{\partial^2 R_{Li}}{\partial(L_{ki}/L_i)^2} = 0$$

Substituting these into (7.49) and (7.50) gives, respectively,

$$\hat{R}_{Li} \simeq \sum_{k=1}^{q} \{\mathscr{E}(L_{ki}/L_i)R_{Lki}\} + \sum_{k=1}^{q} \mathscr{C}\{L_{ki}/L_i), R_{Lki}\} \tag{7.51}$$

and

$$\mathscr{V}(R_{Li}) \simeq \sum_{k=1}^{q} R^2_{Lki}.\mathscr{V}(L_{ki}/L_i) + \sum_{k=1}^{q} (L_{ki}/L_i)^2.\mathscr{V}(R_{Lki})$$

$$+2.\sum_{k=1}^{q}\sum_{f=1}^{k} R_{Lki}(L_{ki}/L_i).\mathscr{C}\{L_{fi}/L_i), R_{Lki}\} \tag{7.52}$$

At present, there is no evidence to suggest what the correlation might be between the relative growth rate of a leaf and leaf foliage ratios; consequently, we have set the covariances in (7.51) and (7.52) to zero in the applications of the model which follow.

All the above methods can be employed starting with the areas of the individual leaves, instead of dry weights, and so an identical set of model results can be obtained in terms of leaf areas.

Stem and root growth

The modelling of the foliage generates four sets of values that are used to estimate the growth curves of stem and root. The four quantities are the estimate of foliage dry weight, $\hat{l}_i$, and its variance, $\mathscr{V}(l_i)$, together with the estimate of foliage relative growth rate, $\hat{R}_{Li}$, and its variance, $\mathscr{V}(R_{Li})$, at time i. The allometric relationships between the Level 1a components, described in Chapter 6, are then used to predict the growth of stem and root from the given foliage estimates. We shall demonstrate the method used to obtain stem dry weight and stem relative growth rate estimates at time i; an identical procedure then follows to obtain the corresponding estimates for the root.

The allometric relationship will be assumed to be a linear one. In the rare instances where a quadratic relationship more adequately describes the relationship between the Level 1a components, the extension of the methods is straightforward and will not be detailed here. Thus, at the ith time,

$$s_i = a + bl_i \tag{7.53}$$

where a and b are estimates of α and β, respectively. If we were merely required to estimate the stem size for a given foliage size, then we should simply have $\hat{s}_i = a + bl_i$ and $\mathscr{V}(s_i) = \mathscr{V}(a) + l_i^2 . \mathscr{V}(b) + 2l_i . \mathscr{C}(a,b)$. However, we require the estimate of stem size at a particular time, and not for a specified foliage size. Since the foliage dry weight at time t is only an estimate of a randomly variable quantity, we must take into account the variability of the foliage estimate as well as the variabilities of the parameter estimates in (7.53); thus, all three variables on the right-hand side of (7.53) are variates (random variates). We apply the usual formulae;

$$\hat{s}_i \simeq a + b\hat{l}_i + \frac{1}{2}\left\{\frac{\partial^2 s_i}{\partial a^2} \cdot \mathscr{V}(a) + \frac{\partial^2 s_i}{\partial b^2} \cdot \mathscr{V}(b) + \frac{\partial^2 s_i}{\partial l_1^2} \cdot \mathscr{V}(l_i)\right\}$$
$$+ \frac{\partial^2 s_i}{\partial a . \partial b} \cdot \mathscr{C}(a, b) + \frac{\partial^2 s_i}{\partial a . \partial l_i} \cdot \mathscr{C}(a, l_i) + \frac{\partial^2 s_i}{\partial b . \partial l_i} \cdot \mathscr{C}(b, l_i) \tag{7.54}$$

and

$$\mathscr{V}(s_i) \simeq \left\{\frac{\partial s_i}{\partial a}\right\}^2 \cdot \mathscr{V}(a) + \left\{\frac{\partial s_i}{\partial b}\right\}^2 \cdot \mathscr{V}(b) + \left\{\frac{\partial s_i}{\partial l_i}\right\}^2 \cdot \mathscr{V}(l_i)$$
$$+ 2\left\{\frac{\partial s_i}{\partial a} \cdot \frac{\partial s_i}{\partial b} \cdot \mathscr{C}(a, b) + \frac{\partial s_i}{\partial a} \cdot \frac{\partial s_i}{\partial l_i} \cdot \mathscr{C}(a, l_i) + \frac{\partial s_i}{\partial b} \cdot \frac{\partial s_i}{\partial l_i} \cdot \mathscr{C}(b, l_i)\right\}$$
$$\tag{7.55}$$

The partial derivatives are given by

$$\frac{\partial s_i}{\partial a} = 1 \qquad \frac{\partial s_i}{\partial b} = \hat{l}_i \qquad \frac{\partial s_i}{\partial l_i} = b$$

$$\frac{\partial^2 s_i}{\partial a^2} = \frac{\partial^2 s_i}{\partial b^2} = \frac{\partial^2 s_i}{\partial l_i^2} = \frac{\partial^2 s_i}{\partial a \, . \, \partial b} = \frac{\partial^2 s_i}{\partial a \, . \, \partial l_i} = 0$$

and

$$\frac{\partial^2 s_i}{\partial b \, . \, \partial l_i} = 1$$

Thus (7.54) and (7.55), respectively, reduce to

$$\hat{s}_i \simeq a + b\hat{l}_i + \mathscr{C}(b, \hat{l}_i) \tag{7.56}$$

$$\mathscr{V}(s_i) \simeq \mathscr{V}(a) + \hat{l}_i^2 \, . \, \mathscr{V}(b) + b^2 \, . \, \mathscr{V}(l_i) + 2\{\hat{l}_i \, . \, \mathscr{C}(a, b) + b \, . \, \mathscr{C}(a, l_i) + bl_i \, . \, \mathscr{C}(b, l_i)\} \tag{7.57}$$

The estimated values and their variances are all known, but the only covariance term that seems to be well-established here is that between a and b. In the estimation of the linear allometric relationship, covariance terms between a and $\hat{l}_i$ and between b and $\hat{l}_i$ emerge as an integral part of the method ($\hat{l}_i$ has the symbol ξ_i in Chapter 6). These $\hat{l}_i$, however, are obtained from the allometric fitting procedure and are not the same as the $\hat{l}_i$ obtained by summation in this chapter, page 231. If the model were only to be used to estimate s_i (and other quantities) at times when harvests were actually taken, then covariance-values involving $\hat{l}_i$ would be available. However, the results to be presented have been calculated at daily intervals over the experimental period, but only over parts of the durations of two experiments were daily harvests actually taken. Moreover, once it has been tested against experimental data, the model may be used to assess the effects of hypothetical changes in the growth of plant parts on the growth of other parts and of the whole plant. In this case, there would be no harvest data available at all, and so covariance-values involving $\hat{l}_i$ would be impossible to guess. The covariance between a and b would remain unchanged if these parameters of the allometric relationship were not altered, but the converse is also true, thus making the covariance of a and b difficult to quantify if the values of these parameters themselves are changed.

The other aspect of the covariance problem to consider is whether there really is a non-zero covariance between a and $\hat{l}_i$ and between b and $\hat{l}_i$ when using the already fitted allometric relationship to estimate the size of one plant part with that of another. A moment's thought will enable one to appreciate how such correlations arise when fitting allometric relationships to data, but once the estimated relationship has been formulated, then it is much less easy to see why the correlations under discussion should not be zero when the allometric relationship is being used for prediction purposes.

In view of all this, we shall retain $\mathscr{C}(a, b)$ as given in the allometric function estimation procedure, but make the assumption that $\mathscr{C}(a, l_i) = \mathscr{C}(b, l_i) = 0$. Then, equations (7.56) and (7.57) reduce to

$$\hat{s}_i \simeq a + b\hat{l}_i \tag{7.58}$$

$$\mathscr{V}(s_i) \simeq \mathscr{V}(a) + \hat{l}_i^2 \,.\, \mathscr{V}(b) + b^2 \,.\, \mathscr{V}(l_i) + 2l_i \,.\, \mathscr{C}(a, b) \tag{7.59}$$

Notice that (7.58) is the same as would be used if $\hat{l}_i$ were known without error, and that (7.59) differs from this situation only by the addition of the term $b^2 \,.\, \mathscr{V}(l_i)$.

Stem and root relative growth rates

An estimate of the stem relative growth rate and its variance are found from the relationship

$$\mathrm{R}_{Si} = b\mathrm{R}_{Li} \tag{7.60}$$

Without derivations, application of the same approximation formulae to (7.60) give

$$\hat{\mathrm{R}}_{Si} \simeq b\hat{\mathrm{R}}_{Li} + \mathscr{C}(b, \mathrm{R}_{Li}) \tag{7.61}$$

$$\mathscr{V}(\mathrm{R}_{Si}) \simeq \mathrm{R}_{Li}^2 \,.\, \mathscr{V}(b) + b^2 \,.\, \mathscr{V}(\mathrm{R}_{Li}) + 2b\mathrm{R}_{Li} \,.\, \mathscr{V}(b, \mathrm{R}_{Li}) \tag{7.62}$$

Again, we will assume that the covariance between b and R_{Li} is zero. Although at first sight these two quantities would appear to be directly related through the allometric relationship, this is only true from the viewpoint that b is the ratio between R_{Si} and R_{Li}. In absolute terms, R_{Li} can change its value independently of b, and so an assumption of zero correlation between these quantities seems realistic.

Plant growth

At this stage, at time i, we have available estimates of the logarithms of foliage, stem, and root dry weights, $\hat{l}_i$, $\hat{s}_i$, and $\hat{r}_i$, Mathematically, we then have

$$W_i = L_i + S_i + R_i$$

so

$$e^{w_i} = e^{l_i} + e^{s_i} + e^{r_i}$$

and hence

$$w_i = \log_e(e^{l_i} + e^{s_i} + e^{r_i}) \tag{7.63}$$

Therefore, we have

$$\hat{w}_i \simeq \log_e(e^{\hat{l}_i} + e^{\hat{s}_i} + e^{\hat{r}_i}) + \frac{1}{2}\left\{\frac{\partial^2 w}{\partial l_i^2} \cdot \mathscr{V}(l_i) + \frac{\partial^2 w}{\partial s_i^2} \cdot \mathscr{V}(s_i) + \frac{\partial^2 w}{\partial r_i^2} \cdot \mathscr{V}(r_i)\right\}$$

$$+ \frac{\partial^2 w}{\partial l_i \,.\, \partial s_i} \cdot \mathscr{C}(l_i, s_i) + \frac{\partial^2 w}{\partial l_i \,.\, \partial r_i} \cdot \mathscr{C}(l_1 \,.\, r_i) + \frac{\partial^2 w}{\partial s_i \,.\, \partial r_i} \cdot \mathscr{C}(s_i, r_i) \tag{7.64}$$

and

$$\mathscr{V}(w_i) \simeq \left\{\frac{\partial w}{\partial l_i}\right\}^2 \cdot \mathscr{V}(l_i) + \left\{\frac{\partial w}{\partial s_i}\right\}^2 \cdot \mathscr{V}(s_i) + \left\{\frac{\partial w}{\partial r_i}\right\}^2 \cdot \mathscr{V}(r_i)$$

$$+2\left\{\frac{\partial w}{\partial l_i} \cdot \frac{\partial w}{\partial s_i} \cdot \mathscr{C}(l_i, s_i) + \frac{\partial w}{\partial l_i} \cdot \frac{\partial w}{\partial r_i} \cdot \mathscr{C}(l_i, r_i) + \frac{\partial w}{\partial s_i} \cdot \frac{\partial w}{\partial r_i} \cdot \mathscr{C}(s_i, r_i)\right\} \tag{7.65}$$

where, for example

$$\frac{\partial w}{\partial l_i} = e^{\hat{l}_i}(e^{\hat{l}_i} + e^{\hat{s}_i} + e^{\hat{r}_i})^{-1}$$

$$\frac{\partial^2 w}{\partial l_i^2} = e^{\hat{l}_i}(e^{\hat{s}_i} + e^{\hat{r}_i})(e^{\hat{l}_i} + e^{\hat{s}_i} + e^{\hat{r}_i})^{-2}$$

$$\frac{\partial^2 w}{\partial l_i \cdot \partial s_i} = -e^{\hat{l}_i}e^{\hat{s}_i}(e^{\hat{l}_i} + e^{\hat{s}_i} + e^{\hat{r}_i})^{-2}$$

The other partial derivatives follow analogously. Since the stem and root weights have been estimated directly from foliage weight, we assume that there is a high correlation between these parts. Thus, we assume a correlation coefficient of 0.9 between any two of these components in order to find the corresponding covariance from the variances. For example, the covariance of stem and root is given by

$$\mathscr{C}(s_i, r_i) \simeq 0.9\{\mathscr{V}(s_i) . \mathscr{V}(r_i)\}^{\frac{1}{2}} \tag{7.66}$$

Leaf, stem, and root weight ratios

These ratios are obtained in an analogous way to the leaf foliage ratios on page 233; we illustrate the method for the stem weight ratio, and the other two follow analogously.

$$\mathscr{E}\{\log_e(S_i/W_i)\} = \hat{s}_i - \hat{w}_i \tag{7.67}$$

$$\mathscr{V}\{\log(S_i/W_i)\} = \mathscr{V}(s_i) + \mathscr{V}(w_i) - 2 . \mathscr{C}(s_i, w_i) \tag{7.68}$$

$$\mathscr{E}(S_i/W_i) \simeq \exp[\mathscr{E}\{\log_e(S_i/W_i)\} + \frac{1}{2} \cdot \mathscr{V}\{\log_e(S_i/W_i)\}] \tag{7.69}$$

$$\mathscr{V}(S_i/W_i) \simeq \exp[2 . \mathscr{E}\{\log_e(S_i/W_i)\} + \frac{1}{2} \cdot \mathscr{V}\{\log_e(S_i/W_i)\}] \times (\exp[\mathscr{V}\{\log_e(S_i/W_i)\}] - 1) \tag{7.70}$$

The covariance term, $\mathscr{C}(s_i, w_i)$, is derived as follows. Dispensing with the suffix i, we have at any time

$$W = L + S + R$$

$$L = W - S - R$$

$$S = W - L - R$$

and $$R = W - L - S$$

and so

$$\mathscr{V}(W) = \mathscr{V}(L) + \mathscr{V}(S) + \mathscr{V}(R) + 2.\mathscr{C}(L, S) + 2.\mathscr{C}(L, R) + 2.\mathscr{C}(S, R) \tag{7.71}$$

$$\mathscr{V}(L) = \mathscr{V}(W) + \mathscr{V}(S) + \mathscr{V}(R) - 2.\mathscr{C}(S, W) - 2.\mathscr{C}(R, W) + 2.\mathscr{C}(S, R) \tag{7.72}$$

$$\mathscr{V}(S) = \mathscr{V}(W) + \mathscr{V}(L) + \mathscr{V}(R) - 2.\mathscr{C}(L, W) - 2.\mathscr{C}(R, W) + 2.\mathscr{C}(L, R) \tag{7.73}$$

$$\mathscr{V}(R) = \mathscr{V}(W) + \mathscr{V}(L) + \mathscr{V}(S) - 2.\mathscr{C}(L, W) - 2.\mathscr{C}(S, W) + 2.\mathscr{C}(L, S) \tag{7.74}$$

On adding (7.72) to (7.74), subtracting (7.73), and substituting for $\mathscr{V}(W)$ in the result, using (7.71), we find that

$$\mathscr{C}(S, W) = \mathscr{V}(S) + \mathscr{C}(L, S) + \mathscr{C}(S, R) \tag{7.75}$$

To convert $\mathscr{C}(S, W)$ into $\mathscr{C}(s, w)$, we make use of a third approximation formula, given by Kendall & Stuart (1977) and in our appendix (equation (A.3)), where $k = 2, x_1 = S, x_2 = W, f(x_1) = \log_e S = s$, and $g(x_1) = \log_e W = w$. Hence, $\partial f/\partial S = 1/S$, $\partial f/\partial W = 0$, $\partial g/\partial S = 0$, $\partial g/\partial W = 1/W$, and substituting into (A.3), we have

$$\mathscr{C}(s, w) \simeq \frac{1}{SW} \cdot \mathscr{C}(S, W) \tag{7.76}$$

Substituting for $\mathscr{C}(S, W)$ in (7.76), using (7.75), and replacing the suffix i, we have

$$\mathscr{C}(s_i, w_i) \simeq \frac{1}{S_i, W_i} \cdot \{\mathscr{V}(S_i) + \mathscr{C}(L_i, S_i) + \mathscr{C}(S_i, R_i)\} \tag{7.77}$$

Finally, all quantities on the right-hand side of (7.77) need to be replaced by their logarithmic forms, and so for example

$$S_i = \exp\{\hat{s}_i + \tfrac{1}{2}.\mathscr{V}(s_i)\} \tag{7.78}$$

$$\mathscr{V}(S_i) = \exp[2\{\hat{s}_i + \tfrac{1}{2}.\mathscr{V}(s_i)\}][\exp\{\mathscr{V}(s_i)\} - 1] \tag{7.79}$$

$$\mathscr{C}(S_i, R_i) \simeq e^{(\hat{s}_i + \hat{r}_i)}.\ \mathscr{C}(s_i, r_i) \tag{7.80}$$

Equations (7.78) and (7.79) are the exact transformations associated with a log-normal distribution, and (7.80) is another application of (A.3) which is simply a reversal of (7.76).

Leaf area ratio and specific leaf area

These ratios can be estimated using equations analogous to (7.67) to (7.70). The covariances, between foliage area and total plant dry weight and between foliage area and foliage dry weight, respectively, may be estimated by the same method as used in the functional approach to whole plant growth analysis (page 55).

Plant relative growth rate

From (7.5), we have

$$\mathrm{R}_i = \mathrm{R}_{Li}(L_i/W_i) + \mathrm{R}_{Si}(S_i/W_i) + \mathrm{R}_{Ri}(\mathrm{R}_i/W_i) \tag{7.81}$$

and the estimate of plant relative growth rate and its variance are derived by equations analogous to (7.51) and (7.52). Again, we set the covariances to zero as we have no evidence of a correlation between a component entity ratio and a component relative growth rate; thus, using the symbols of (7.81) in (7.51) and (7.52), we obtain

$$\hat{\mathrm{R}}_i \simeq \hat{\mathrm{R}}_{Li} \cdot \mathscr{E}(L_i/W_i) + \hat{\mathrm{R}}_{Si} \cdot \mathscr{E}(S_i/W_i) + \hat{\mathrm{R}}_{Ri} \cdot \mathscr{E}(R_i/W_i) \tag{7.82}$$

and

$$\begin{aligned}\mathscr{V}(\mathrm{R}_i) \simeq{}& \hat{\mathrm{R}}^2_{Li} \cdot \mathscr{V}(L_i/W_i) + \hat{\mathrm{R}}^2_{Si} \cdot \mathscr{V}(S_i/W_i) + \hat{\mathrm{R}}^2_{Ri} \cdot \mathscr{V}(R_i/W_i) \\ &+ \{\mathscr{E}(L_i/W_i)\}^2 \cdot \mathscr{V}(\mathrm{R}_{Li}) + \{\mathscr{E}(S_i/W_i)\}^2 \cdot \mathscr{V}(\mathrm{R}_{Si}) + \{\mathscr{E}(R_i/W_i)\}^2 \cdot \mathscr{V}(\mathrm{R}_{Ri})\end{aligned} \tag{7.83}$$

where, again, the symbol $\mathscr{E}$ is to be read as 'estimate of'. All the quantities on the right-hand sides of (7.82) and (7.83) have previously been estimated, and so $\hat{\mathrm{R}}_i$ and $\mathscr{V}(\mathrm{R}_i)$ can be found.

Unit leaf rate

Finally, unit leaf rate may be estimated in precisely the same way as in the functional approach to whole plant growth analysis, as all model estimates of the required quantities have already been obtained.

Discussion on model formulation

Throughout the development of this model of plant growth it has only been possible, in many places, to obtain approximate estimates of parameters and variances, and the accumulated effect of so many approximations cannot go without comment. Approximations arise in a number of situations: (1) finding a quantity, and associated variance estimate, which is given by a non-linear combination of variates; (2) assuming that all growth attributes are log-normally distributed; and (3) estimating correlations between growth attributes where no precise information is available.

The first circumstance cannot be avoided because of the nature of the expressions involved, although in time derivations of exact results may be forthcoming; the second point involves what can usually be assumed to be a very close approximation. The third factor certainly needs more investigation to attempt to find ways of estimating the correlations accurately. However, even if all three contributing approximations were reduced to the minimum possible, we are still uncertain of the additive effect. Rigorous testing of the assumptions and estimation procedures is now required, using simulation studies, to investigate the validity of the model results. This work has not yet been done by us and perhaps we should not go on to apply the model to data without more preliminary work. We would justify the application of the model on the grounds that we are exploring its possibilities, and we compare the results obtained with those given by the classical and functional growth analysis methods of Chapter 2, into which we have a greater insight. The difficulty of this approach is that it is all too easy to fall back on the classical results as a standard with which to compare the model. This is tantamount to assuming that the model is a curve fitted to the classical results which is, of course, not the case. In general we use the classical results as guidelines with which to compare the model purely because we have no knowledge of the underlying relationship. When we are considering the growth curves of entities, and also the ratio trends, a comparison with the raw data and classical results can yield useful information, although day to day variation may obscure underlying trends of ratios to some extent. Comparisons of growth rate trends are more difficult to make, as model curves give, in effect, mean rate estimates for every 24 hour period, whereas classical results may be mean rates for periods of up to 7 days.

We believe the model holds promise for the biologist who is concerned with whole plant growth tends and who wants to interpret these in terms of the component growth of the plant. We hope that further statistical testing will enable us to apply the model with greater confidence.

The model: results

In this part of the chapter we shall discuss the results mainly from the viewpoint of the adequacy of the model in describing the various facets of plant growth, and so the description of the model results should be read in conjunction with the relevant results sections of Chapters 2 and 6. Although attempts are made to obtain the maximum biological information from the model (perhaps too much in some instances) the tentative nature of the methods must be remembered throughout. Four sets of graphs will be presented for each species. The first set shows leaf (foliage), stem, and root dry weights, together with foliage area, and each graph presents the harvest mean values of these quantities with the fitted Richards functions (Chapter 2 and Appendix tables) and the model curves superimposed. The second set of graphs shows the trends in leaf foliage ratios produced by the model: one graph for dry weights, the other for areas. The first graph of the third set shows whole plant dry weights, on the same basis as the graphs of the first set; while the second graph shows relative growth rates with the results given by: (i) classical growth analysis (as given in Chapter 2), (ii) derivation from the fitted Richards function, and (iii) the model, superimposed on one another. The final set of graphs presents leaf, stem, and root weight ratios, specific leaf areas, leaf area ratios, and unit leaf rates, depicted in the same way as for relative growth rate.

In the cases of the last formed, and occasionally the earliest, leaves, where the data did not extend over the major part of the growth curve, Richards function fittings were unobtainable. In order to obtain model results for the entire experimental period, polynomial functions were used in these instances, and the details are given in Appendix Table A.10.

Sunflower

The four sets of graphs for this species' results are Figs. 7.2 to 7.5. As regards foliage and stem dry weights (Fig. 7.2a,c), the model provides a good simulation of the growth of these organs – better than the fitted Richards functions do. However, in the case of foliage area (Fig. 7.2b) the Richards function is better at the ends of the growth curve, particularly at the top where the model trend seems to be completely wrong; but the irregular trend of foliage area increase in the middle of the experiment is accommodated by the model. For root weight (Fig. 7.2d), the model provides a superior fit at the beginning of the experiment, but there is little to choose between the model and the Richards function in the centre of the growth curve. At the upper end of the root weight growth curve, there is the interesting situation that the Richards function seems to fit the data better, but that the trend of the model curve appears to be more in accord with the trend of root growth at this stage. The model and Richards function curves have 95% confidence bands constructed around them and one would expect that approximately 95% of the mean dry weight values would be contained within these bands. This is not always the case, indicating an inadequacy of the $\mathscr{V}(l)$ expressions. In general the Richards function variances are too small and in a number of cases, for example (b) and (d), the model variances are also too small. This latter discrepancy may well be due

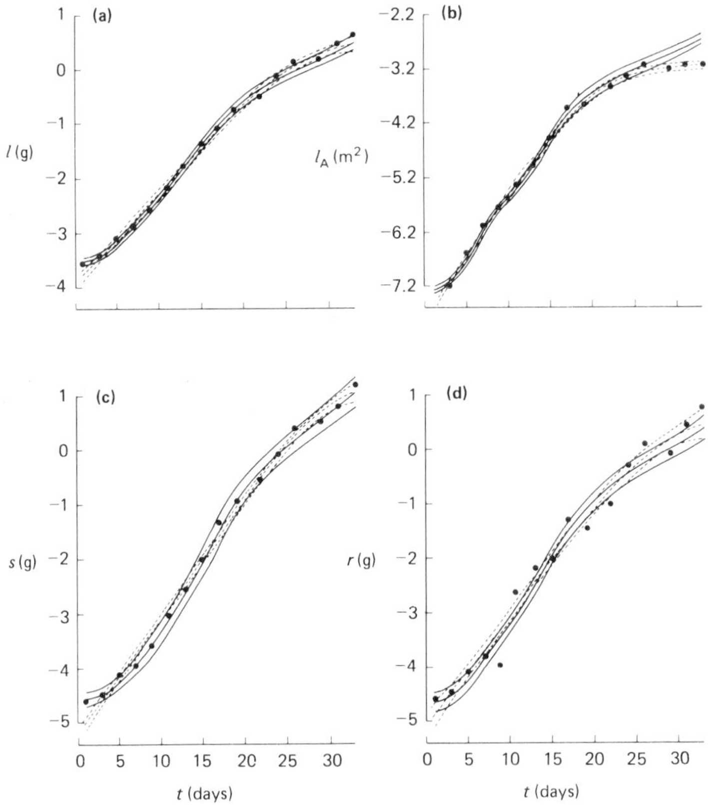

Fig. 7.2 Growth curves with 95% confidence bands for sunflower, of (**a**) foliage dry weight, (**b**) foliage area, (**c**) stem dry weight, and (**d**) root dry weight. Model curves ——, Richards function curves ---- and primary data mean values ●.

to us making incorrect assumptions about covariance terms and more work is required on this subject.

The leaf foliage curves (Fig. 7.3) are of similar form for all species, and so need only be commented on once. These curves show, at any selected time, the relative importance of the dry weight and area of a particular leaf in relation to the foliage as a whole. Naturally, successive leaves have progressively lower maxima as the foliage comprises an increasing number of fully expanded leaves, and the rapidity with which the curve of any particular leaf rises and falls is partly related to its duration of expansion (Chapter 5).

The model trend is markedly superior to that of the Richards function for whole plant growth (Fig. 7.4a), especially at the lower end of the

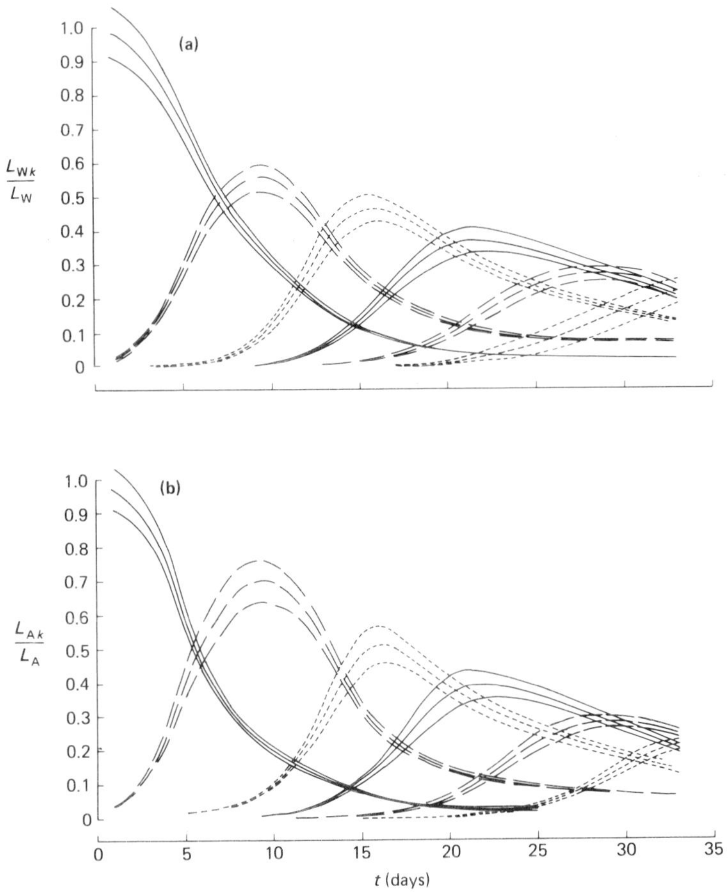

Fig. 7.3 Leaf/foliage ratio curves with 95% confidence bands for (**a**) dry weight and (**b**) leaf area, of cotyledons —— and leaf pairs 1 — —, 2 ----, 3 ——, 4 — —and 5 ---- of sunflower.

curve. Trends become somewhat ambiguous at the end of the experiment because of the irregularity of the primary data harvest means. With regard to relative growth rate (Fig. 7.4b), there is no question as to the superiority of the model over the derived Richards function during most of the experiment. It is difficult to have any confidence in the minor oscillations of the model relative growth rate curve, but the overall trends are encouraging when one remembers the large amount of approximation that has been incorporated into the model by this stage. The trend shown by the model at the end of the duration is somewhat disturbing but, as already noted, the trend in plant growth at the end of this experiment is not well defined

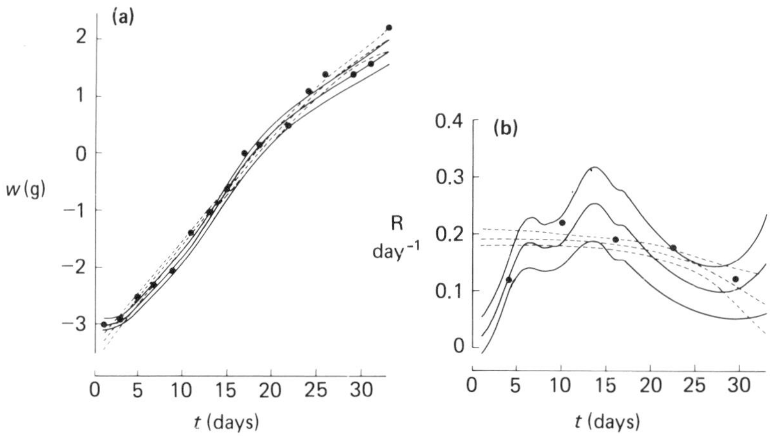

Fig. 7.4 Whole plant growth curves with 95% confidence bands for (**a**) dry weight and (**b**) relative growth rate, of sunflower. Model curves ——, Richards function curves ---- and primary data mean values and classical growth analysis values ●.

anyway. It is of interest to compare the model relative growth rate result with that derived from a polynomial function fitting. In this sunflower experiment, using an objective procedure for selecting the appropriate degree of polynomial (page 84), a fourth degree function was indicated for describing whole plant dry weight. The resulting relative growth rate function is thus a cubic polynomial, and the curve of this function bears some resemblance to the present model curve (Venus & Causton, 1979a, Fig. 2c).

The model curve provides a good description of the stem weight ratio (Fig. 7.5b), but owing to the irregularity of the directly calculated leaf and root weight ratios, neither the Richards function derived curve or the present model curve can be preferred (Fig. 7.5a,c). The abrupt broadening and narrowing of the confidence intervals over the first 17 days in the case of the stem weight ratio is due to two changes of allometric relationship between stem and leaves over this period. A cursory glance at the classical results for many of the ratios on Fig. 7.5 gives the impression that a straight line would describe the relationships as well as, if not better than, the model or Richards function derived curves. But, we are not concerned with the describing of these trends in isolation, we are aiming at a model that can also tell us something about how the growth of components and entities at Level 1a contributes to whole plant growth quantities.

The model trends in both specific leaf area (Fig. 7.5d) and leaf area ratio (Fig. 7.5e) suffer from the divergence of the model curve of leaf area from the corresponding data towards the end of the experimental period (Fig.

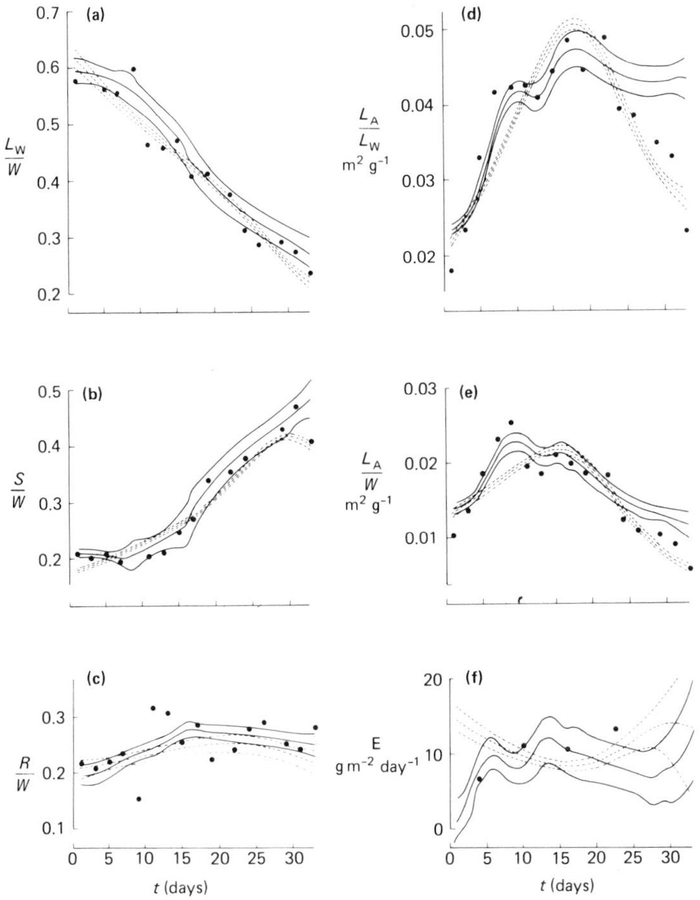

Fig. 7.5 Derived curves with 95 % confidence bands for sunflower of (**a**) leaf weight ratio, (**b**) stem weight ratio, (**c**) root weight ratio, (**d**) specific leaf area, (**e**) leaf area ratio and (**f**) unit leaf rate. Model curves ——, Richards function curves ----, and classical growth analysis values •.

7.2b). The resulting divergence over the last 5 days of the specific leaf area curve from the directly calculated points is very large, but is not as great in the case of leaf area ratio. Otherwise, in both these ratios, the model curves describe the data better than the curves derived from Richards functions. Similar remarks may be made about unit leaf rate (Fig. 7.5f) as for relative growth rate, but the model trend after about day 20 is unreliable because of the divergencies of model curves from actual and directly calculated data, already discussed.

Wheat

The wheat results (Figs. 7.6 to 7.9) are adversely affected by the complicated root-foliage allometric relationships (Fig. 6.4c, d), so that although trends in the basic and directly calculated data are reproduced well, the model confidence intervals are excessively wide. The only relationships not affected by the complicated root-foliage allometry are leaf dry weight (Fig. 7.6a) and leaf area (Fig. 7.6b) where, in the latter, the model confidence intervals are smaller than those given by the fitted Richards function. The model stem (Fig. 7.6c) and root (Fig. 7.6d) growth curves follow the data extremely well, except at the end of the experiment where an apparently premature cessation of growth is indicated. The wide confidence band is a consequence of the two curvilinear foliage-root allometric relationships whose parameter estimate variances are large, and the extra

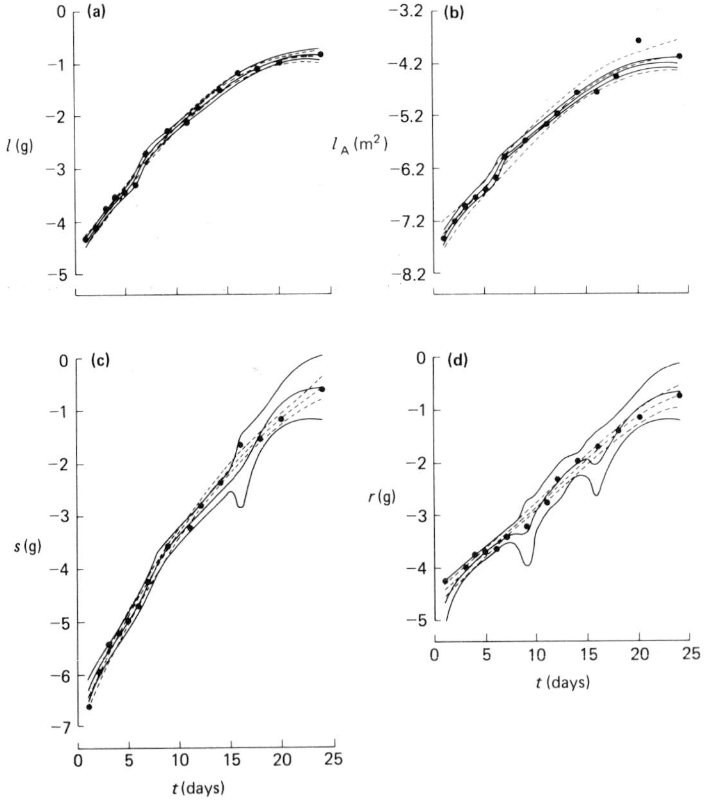

Fig. 7.6 Growth curves with 95% confidence bands for wheat, of (**a**) foliage dry weight, (**b**) foliage area, (**c**) stem dry weight, and (**d**) root dry weight. Model curves ——, Richards function curves---- and primary data mean values ●

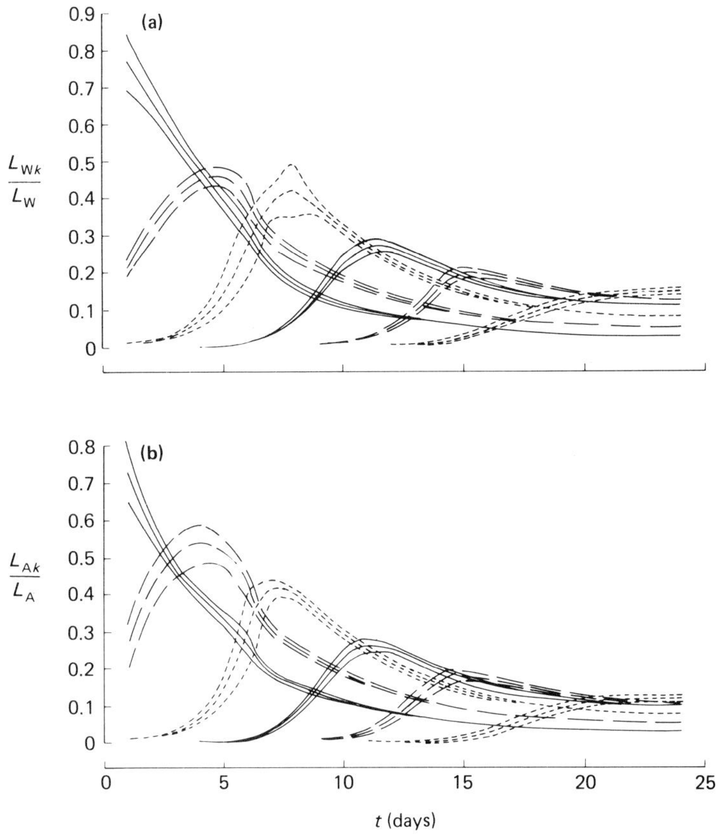

Fig. 7.7 Leaf/foliage ratio curves with 95% confidence bands for (**a**) dry weight and (**b**) leaf area, of leaves 1 ——, 2 — —, 3 ----, 4 ——, 5 — — and 6 ---- of wheat.

wide confidence intervals coincide with the changeover positions between allometric relationships where greater uncertainty prevails. Similar remarks apply to the plant growth curve in Fig. 7.8a.)

The oscillations in the model relative growth rate curve (Fig. 7.8b) over the first 11 days are doubtless associated with the two curvilinear allometric functions used to describe the foliage-root relationship over that time. The later oscillations are more difficult to explain but are likely to be artificially created, and obtaining a clue to their cause would require manipulation of the model components, i.e. slight changes could be made to the parameters of the allometric relationships and/or to the parameters of one or more of the Richards functions describing individual leaf growth. This model curve of root growth emphasizes a need for further work to investigate the

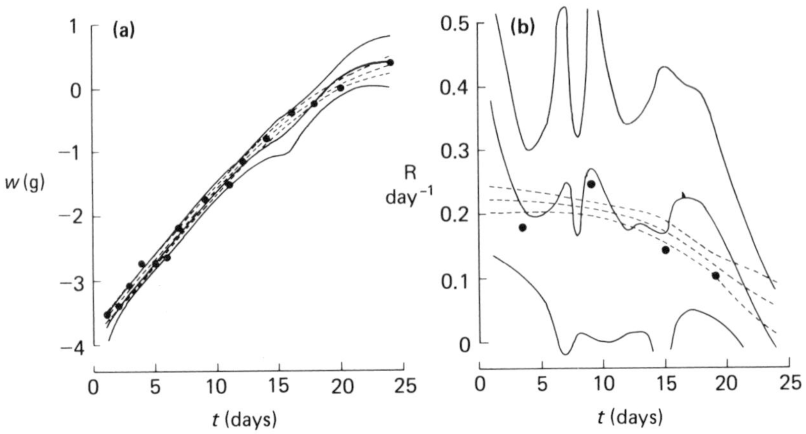

Fig. 7.8 Whole plant growth curves with 95% confidence bands for (**a**) dry weight and (**b**) relative growth rate, of wheat. Model curves ——, Richards function curves ---- and primary data mean values and classical growth analysis values ●.

variance structure of the relationships involved. It is well established that approximate combinations of large variances behave erratically and the problem is exacerbated here by the difficulty of defining covariances. It is evident, however, that the foliage-root allometry needs to be defined more precisely.

The leaf, stem, and root weight ratio model trends (Fig. 7.9a,b,c) all show a levelling off at the end of the experiment, which is almost certainly an artifact. At this stage the foliage virtually ceases to grow, but the stem at least continues growth and a different foliage-stem allometric relationship, having a very high gradient, would ensue. However, the experimental duration did not include a phase of this kind, and so the information at the end of this experiment is bound to be imprecise. Apart from this, there is not much to choose between the model curves and the derived Richards function curves for the component entity ratios as far as trends are concerned, but the model confidence intervals mostly include the harvest mean values directly calculated from the primary data, whereas the derived Richards function confidence intervals do not. In the case of the area ratios (Fig. 7.9d,e), the model confidence intervals are smaller than those given by the derived Richards functions. The harvest mean points are too scattered to enable one to favour the model curves or the derived Richards functions, but the former seem to describe better the specific leaf area trend in the latter part of the experiment. The unit leaf rate model curve (Fig. 7.9f) is very similar to that of relative growth rate, already discussed, and no further comment is necessary. Again, classical results, at first sight, appear linear, see page 47.

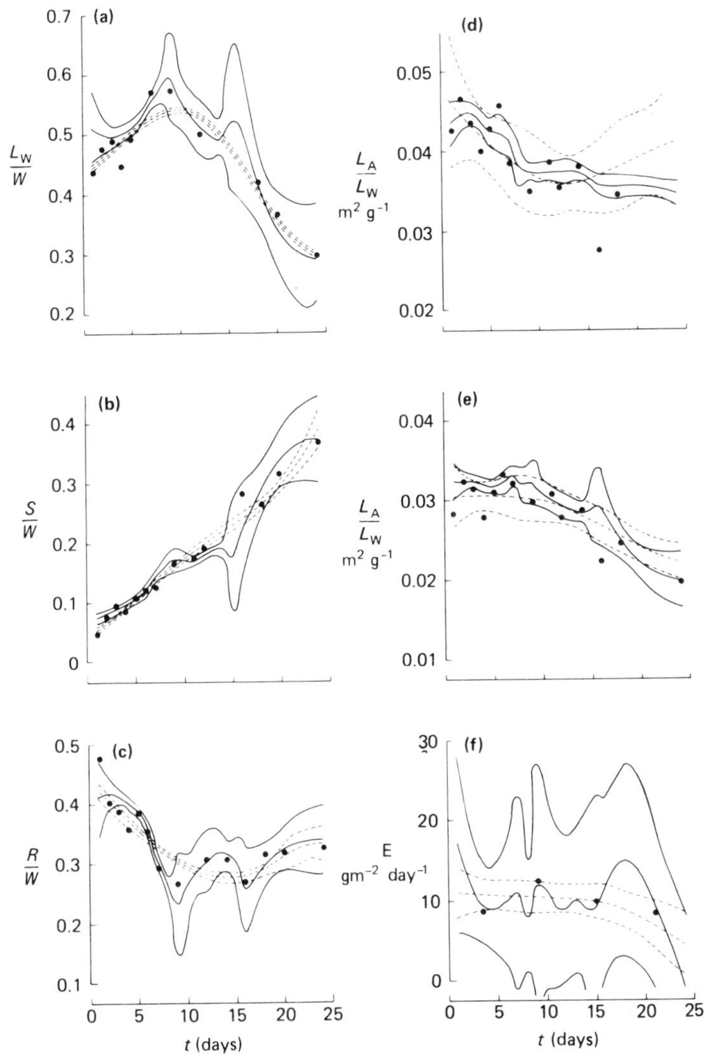

Fig. 7.9 Derived curves with 95% confidence bands for wheat of (**a**) leaf weight ratio, (**b**) stem weight ratio, (**c**) root weight ratio, (**d**) specific leaf area, (**e**) leaf area ratio and (**f**) unit leaf rate. Model curves ——, Richards function curves ----, and classical growth analysis values •.

Maize

The variability of the maize growth data is higher than for any of the other three species, and this feature of the primary data gives rise to the

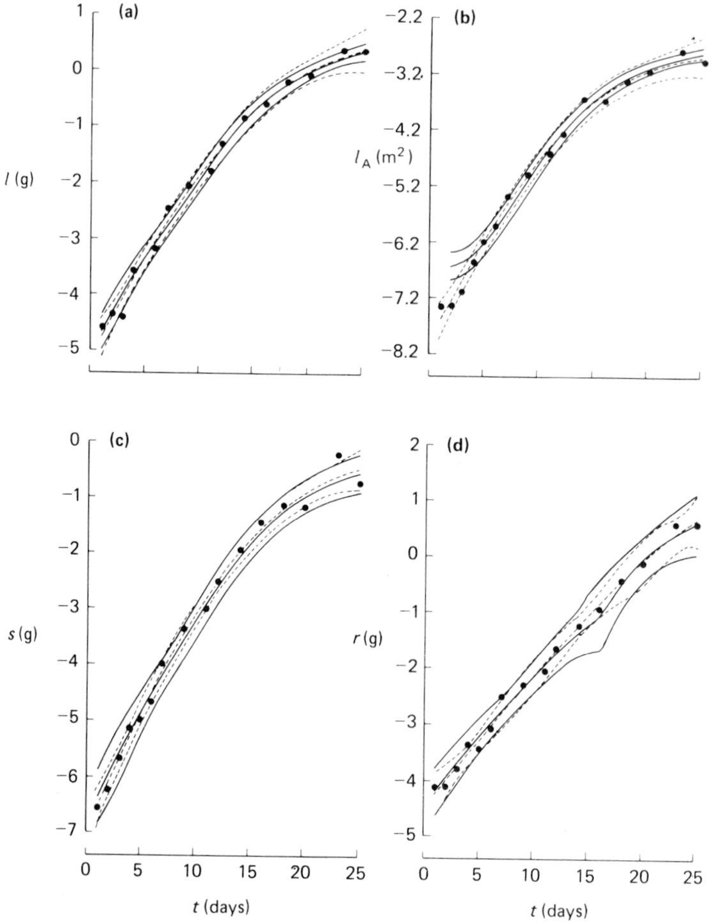

Fig. 7.10 Growth curves with 95% confidence bands for maize, of (**a**) foliage dry weight, (**b**) foliage area, (**c**) stem dry weight and (**d**) root dry weight. Model curves ——, Richards function curves ---- and primary data mean values ●.

relatively wide confidence intervals associated with both the model curves and the fitted Richards functions to the primary data (Fig. 7.10). Except for foliage area, there is good agreement between the primary data and both the model and the fitted Richards functions; the very serious divergence between the model and the data for foliage area (Fig. 7.10b) may be attributed to very poor fits provided by Richards functions to leaf areas 1 and 2, owing to lack of adequate data at the lower ends of these growth curves. The poorness of fit was particularly bad for leaf 1 area, which shows as a singularity on the leaf foliage ratio curves for area (Fig. 7.11b); leaf 1

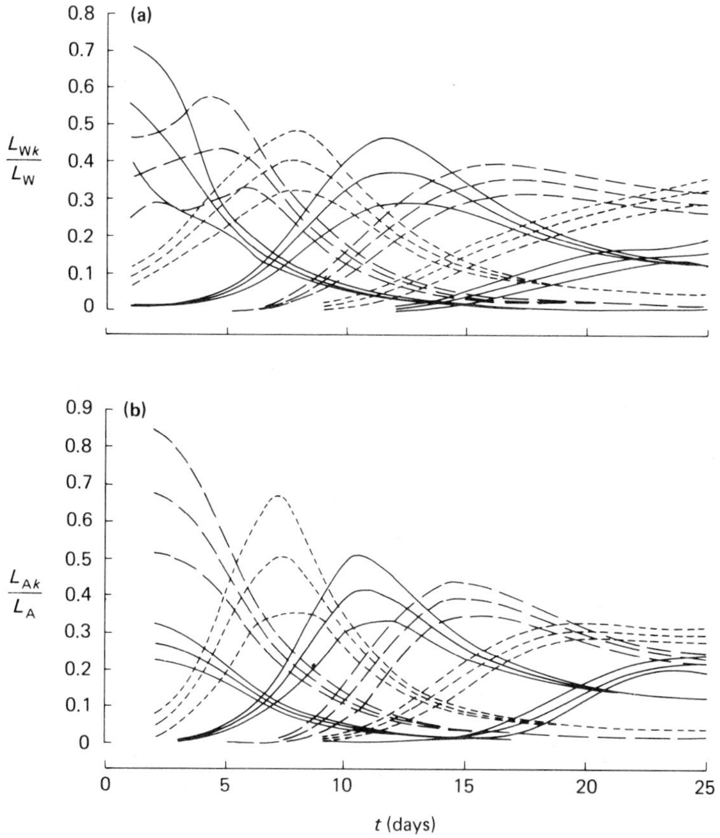

Fig. 7.11 Leaf/foliage ratio curves with 95% confidence bands for (**a**) dry weight and (**b**) leaf area, of leaves 1 ——, 2 — —, 3 ----, 4 ——, 5 — —, 6 ---- and 7 —— of maize.

always has a lower leaf foliage ratio than leaf 2 in this experiment, whereas in the early stages the reverse should be true.

Again, with whole plant dry weight (Fig. 7.12a), both model curve and fitted Richards function provide a good description of the original data, but the Richards function is slightly better at the beginning of the experiment, and has a narrower confidence interval. Because of the variability of the data, the confidence band associated with the relative growth rate curve is wide (Fig. 7.12b), but both the model and the derived Richards function produce a good description of relative growth rate trend.

Of the component entity ratios, only the stem weight ratio (Fig. 7.13b) model and derived Richards function are in good agreement with each other and with the directly calculated data. On the whole, the derived

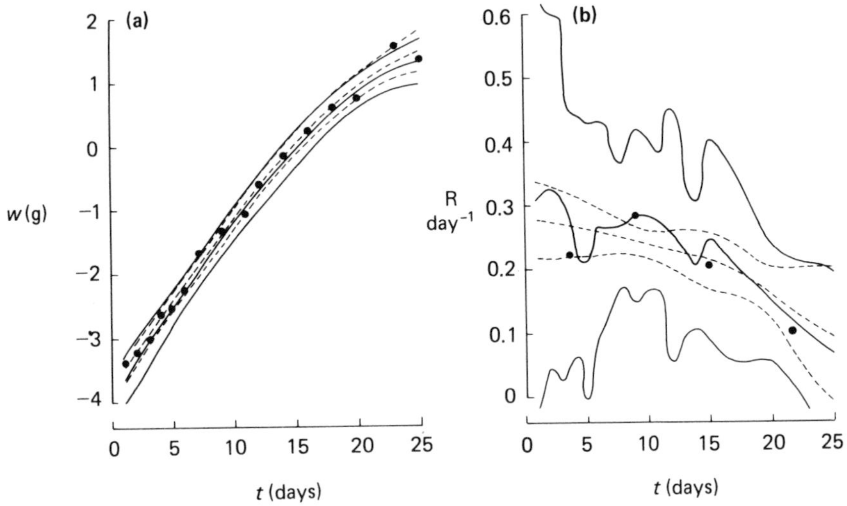

Fig. 7.12 Whole plant growth curves with 95% confidence bands for (**a**) dry weight and (**b**) relative growth rate, of maize. Model curves ——, Richards function curves ---- and primary data mean values and classical growth analysis values ●.

Richards function gives a somewhat better description of the leaf weight ratios than does the model (Fig. 7.13a), and the derived function is decidedly superior in the case of root weight ratio (Fig. 7.13c). The model area ratios are adversely affected at the beginning of the experiment by the poor fits to the areas of leaves 1 and 2, already mentioned, but otherwise the model provides a better description of the data, in that most of the directly calculated points lie within the confidence band of the model, but scarcely any do in the case of specific leaf area (Fig. 7.13d); the derived Richards function, however, provides an acceptable description of the leaf area ratio data (Fig. 7.13e). Finally, in the case of unit leaf rate, the model obviously gives a very good data simulation (Fig. 7.13f).

Birch

In birch, second degree polynomial functions were originally fitted to the Level 1a components and entity because there was insufficient curvature in the growth data for a Richards function to be used. The model and the polynomial curves are very similar for foliage dry weight, but the former shows an artificially early decline in growth rate at the end of the experiment (Fig. 7.14a); the foliage area and stem dry weight curves (Fig. 7.14b,c) also show the same divergence at their upper end. The model curve certainly follows the sudden changes in root growth in a way that a second degree polynomial could not (Fig. 7.14d), and this superior trend in the model

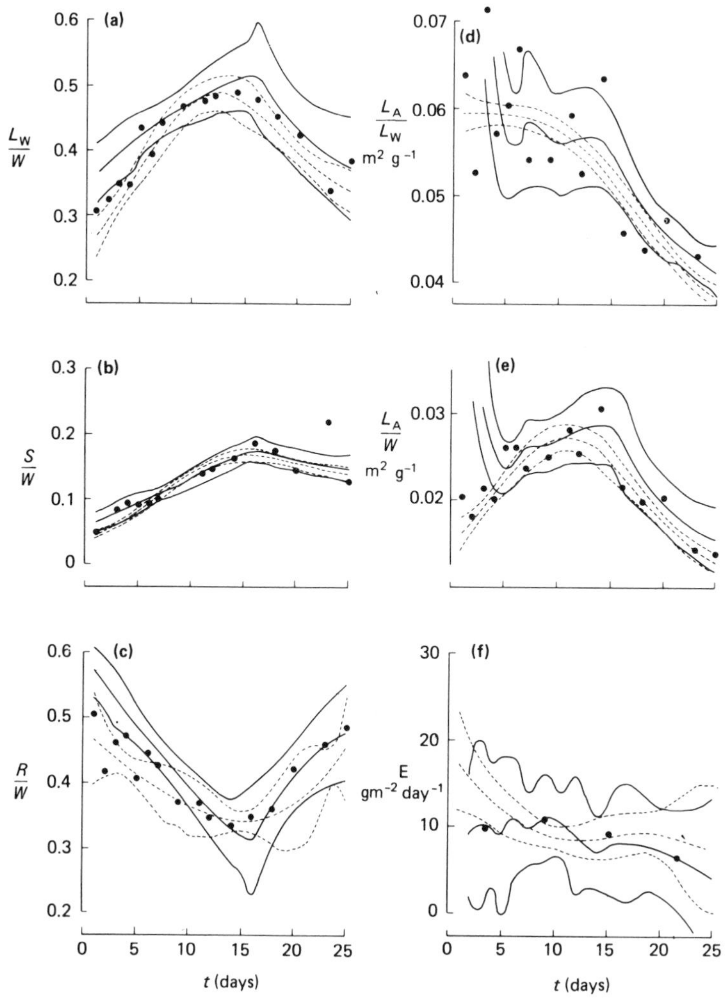

Fig. 7.13 Derived curves with 95% confidence bands for maize of (**a**) leaf weight ratio, (**b**) stem weight ratio, (**c**) root weight ratio, (**d**) specific leaf area, (**e**) leaf area ratio and (**f**) unit leaf rate. Model curves ——, Richards function curves ----, and classical growth analysis values ●.

results is the effect of the segmented allometric relationships between foliage and root (Fig. 6.5c,d, page 217).

Only alternate leaves, up to number 15, could be shown on the leaf foliage ratio graphs (Fig. 7.15), but the trend is clear. It will be realized that in species producing a large number of leaves, the contribution of each individual leaf to the foliage as a whole is not only small, but also changes

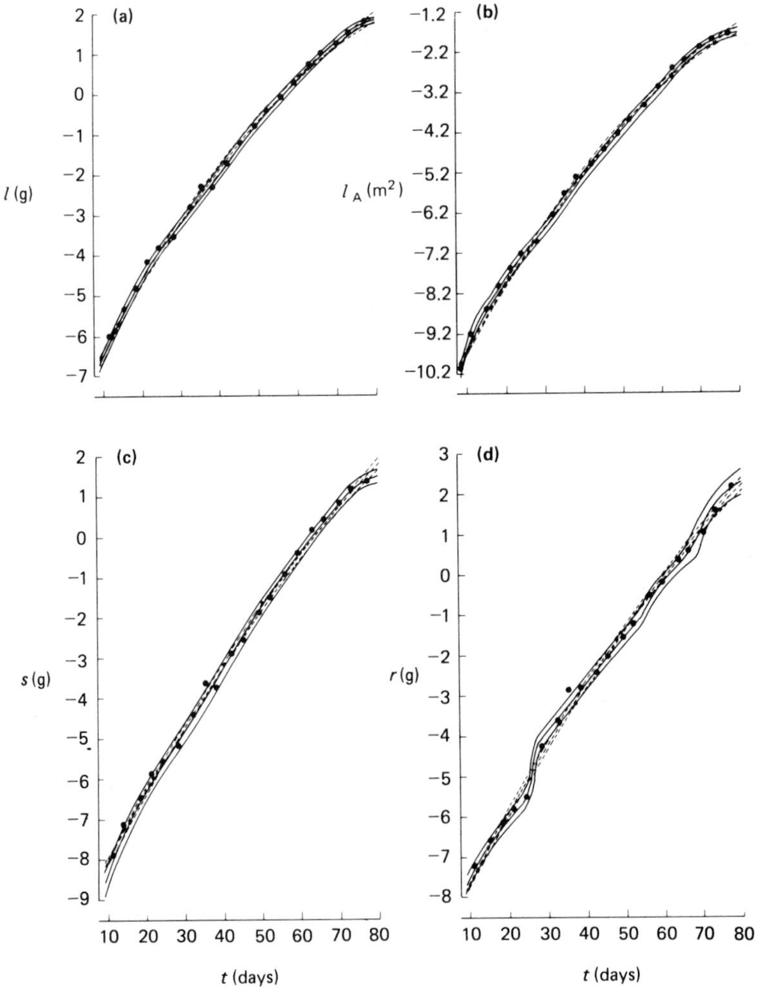

Fig. 7.14 Growth curves with 95 % confidence bands for birch, of (**a**) foliage dry weight, (**b**) foliage area, (**c**) stem dry weight and (**d**) root dry weight. Model curves ——, polynomial function curves ----- and primary data mean values ●.

little during the leaf's life span. This is not the case for the early leaves, and graphs such as those in Fig. 7.15 do serve to reinforce the importance of the early leaves on a plant, both from the viewpoint of their size and, indirectly, their activity.

Similar remarks apply in respect of whole plant dry weight (Fig. 7.16a) as for the component weights (Fig. 7.14). The model is undoubtedly superior in describing plant growth over most of the duration, but again the model

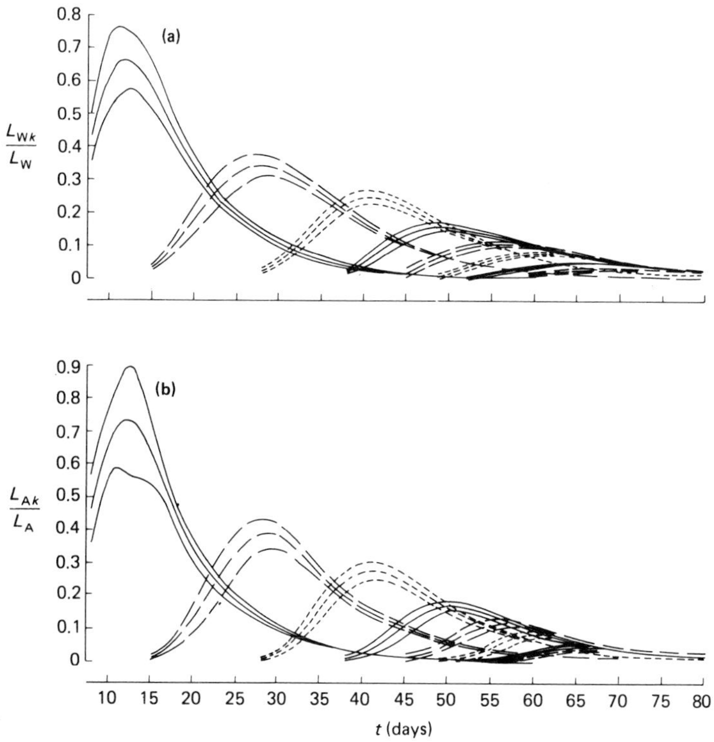

Fig. 7.15 Leaf/foliage ratio curves with 95% confidence bands for (**a**) dry weight and (**b**) leaf area, of leaves 1 ——, 3 — —, 5 ----, 7 ——, 9 — —, 11 ----, 13 —— and 15 — — of birch.

shows a tendency to a premature slowing of growth rate at the end of the experiment. On the other hand, the sudden increases of growth by the roots, which are reflected by similar sudden increases in the whole plant, are followed reasonably well by the model. Apart from one directly calculated observation, the model curve and confidence band describe the relative growth rate trend well (Fig. 7.16b), whereas the polynomial (which is now linear) appears very artificial, particularly in respect of the extremely narrow confidence band in the middle of the experimental period.

It is in the various ratios, where the sudden changes in root growth are revealed so markedly in the directly calculated data points, that the virtues of the model are highlighted. This is particularly so in the case of stem weight ratio (Fig. 7.17c) where, although the confidence intervals are narrow, most of the directly calculated data points lie within the band. The directly calculated points for the leaf weight and root weight ratios are rather more scattered (Fig. 7.17a, b), but nevertheless the model trend is

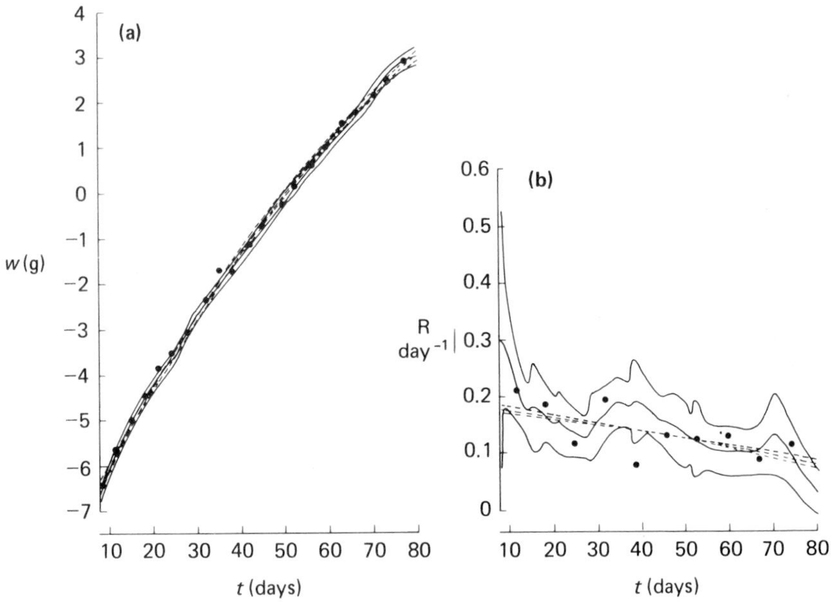

Fig. 7.16 Whole plant growth curves with 95% confidence bands for (**a**) dry weight and (**b**) relative growth rate, of birch. Model curves ——, polynomial function curves ---- and primary data mean values and classical growth analysis values ●.

obviously very satisfactory; similar remarks apply to the area ratios (Fig. 7.17d, e) where the confidence intervals of the polynomial curve are so small they cannot be shown on the graph. The model trend for unit leaf rate (Fig. 7.17f) is also, in the main, satisfactory except for the downward trend at the end of the experiment, which must be an artifact. The model description of the growth trends (Fig. 7.17) supplies the most convincing evidence that the model not only works in terms of telling us something about how component growth contributes to entity growth, but it also follows the classical analysis results better than the simple (empirical) model. With a large number of leaves, a relatively even aged population structure of leaves is soon approached, which implies that unit leaf rate should either remain constant over a considerable period of time, or fall only slowly. The data points, calculated by the classical method, indicate a constancy of unit leaf rate, and so neither the final rise shown by the polynomial derived curve nor the fall shown by the model curve at the end of the experiment is likely to be the true state of affairs. These convolutions in the confidence band are doubtless a consequence of the complexity of the computations involved, since until leaf rate is the last result to be obtained

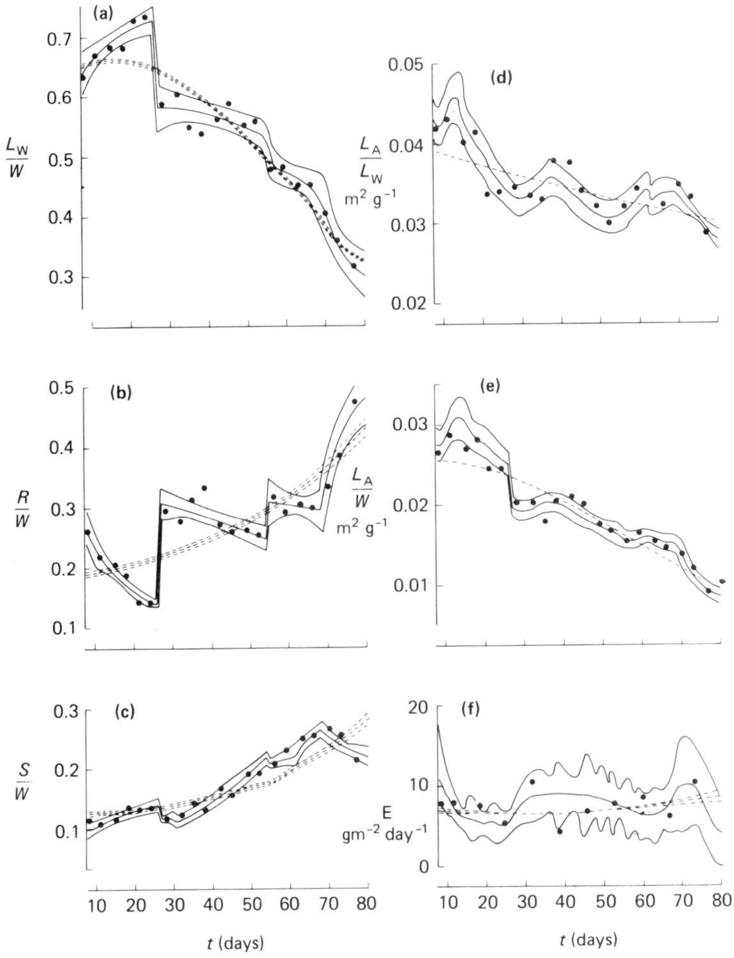

Fig. 7.17 Derived curves with 95% confidence bands for birch of (**a**) leaf weight ratio, (**b**) root weight ratio, (**c**) stem weight ratio, (**d**) specific leaf area, (**e**) leaf area ratio and (**f**) unit leaf rate. Model curves——, polynomial function curves——, and classical growth analysis values ●.

in the entire sequence of calculations employed in building the model.

The model: discussion

Two striking features of the model results are: (i) the ability of the curves to follow at least some of the oscillations in the relative growth and unit leaf rates that are also shown by the classical methods; and (ii) the overall difference in size of confidence intervals obtained on the one hand by the model, and on the other by derivations from the Richards, or polynomial, functions fitted to the primary data of the Level 1a components and entity.

It seems surprising that in a model of this kind, based on smooth mathematical curves, that oscillations in rates can occur in the results at all. Apart from the effects of sudden changes in allometric relationships, any fluctuations in whole plant relative growth rate, or indeed stem and root relative growth rates also, are almost certainly reflections of similar oscillations in foliage relative growth rate in this model, since stem and root growth (and hence plant growth) are derived from foliage growth through the mediation of smooth allometric relationships. Hence, we must look to the components of foliage growth, i.e. the growth of the individual leaves, to examine the possible causes of relative growth rate fluctuations in the Level 1a components or entity. In what follows, the term 'relative growth rate' will apply to any of the components or the entity of Level 1a, unless otherwise stated.

One way in which these relative growth rate oscillations can occur is by an irregularity in the production of leaves; this obviously occurs in plants growing in a natural or semi-natural environment if only because the rate of leaf production is very sensitive to temperature (Terry, 1968; Robson, 1972; Peacock, 1976). The plants in our experiments were grown in a glass house, and the concomitant temperature records show considerable fluctuations in the extremes, although the mean temperature from day to day was more stable. Another possible circumstance giving rise to relative growth rate fluctuations are the changing patterns of growth rate between the individual leaves themselves. If the growth of each individual leaf could be described by a Richards function having the same values of A, k, and n, and all leaves were produced at short intervals of time, there would be little fluctuation in foliage relative growth rate. But even with regular leaf production, we know that at least parameter A changes from one leaf to another, and this can cause some irregularity in the overall foliage relative growth rate trend.

The above speculations can, of course, be examined by manipulating the values of the parameters in the model. Changes may be made to the parameters of the Richards functions describing the growth of individual leaves, either alone or in suitable combinations, and the effects of these changes on the ultimate relative growth rate trends observed. The model may also be employed in the elucidation of the effects of hypothetical changes in the pattern of assimilate partitioning through the adjustment of the parameters of allometry.

Another advantage of employing the scheme described in this chapter to analyse growth data is particularly important when the plants are not growing in a controlled environment. Both environmental fluctuations and sampling variability may profoundly affect the results of a classical growth analysis made on experimental data, and this is particularly so in the cases of relative growth and unit leaf rates. It is obviously very desirable if the effects of environmental fluctuations and sampling variability could be separated, and the present model may go some way to achieving this goal. We shall return to this topic later.

The confidence intervals, shown as a band around each fitted and derived curve, must now be discussed in relation to their relative sizes as produced by the three different methods: the classical, the functional, and that of the present model. All confidence intervals in this book are 95% and this implies that the population expected value of the variate has a probability of 0.95 of lying within the confidence interval. Comparison of the superimposed bands in most of the graphs presented in this chapter shows that the confidence intervals can be very different according to the method of analysis employed, and the same can be said if one compares the confidence intervals produced by the classical and functional whole plant growth analysis methods (Chapter 2, Figs 2.5 to 2.7 with 2.9 to 2.11). In general, the intervals provided by the classical methods and the present model are comparable in size, but the intervals produced by the functional approach are decidedly smaller, often very much so in the case of the rates. We have already made some reference to these differences in confidence interval sizes between the two methods of whole plant growth analysis in Chapter 2 (page 61). For some of the model-derived curves, the methods of developing confidence intervals work quite well, for example Fig. 7.17; but in other situations the intervals are rather too big, see Fig. 7.8b and other rate curves. In contrast, the Richards function confidence intervals usually seem to be too small, with many data points lying outside. A simulation study would provide a clearer comparison of confidence intervals, given by the different methods, with the expected width. From the biological viewpoint we are sceptical about the small confidence intervals given by the Richards function, because this implies that plant growth is following the fitted function almost exactly throughout the whole experimental period. Obviously, this can never be exactly true for a very simple and usually empirical model embodied in a directly fitted function to the components and entity of Level 1a, and this is the reason why great care is necessary in the selection of functions for use in whole plant growth analysis.

The present model is, of course, an example of the functional approach to growth analysis, but it is very much more complicated, involving the simultaneous use of very many mathematical functions. On the other hand, all the functions are biologically relevent, and we employ statistical methods that are as rigorous as presently possible in their estimation (Chapters 4 and 6). Although a very large number of function parameters are involved in, say, whole plant growth (each Richards function has 4, and each linear allometric function has 2 parameters), the confidence intervals are not so large as to dismiss the method without further investigation. A direct polynomial fitting to whole plant dry weight data with the same number of parameters, even if there were sufficient harvests to allow the use of such a high order polynomial, would have prodigiously large confidence intervals, and such a fitted function would be utterly useless. In some instances the confidence intervals given by the model have been shown to be rather large, especially for the rate curves, but it is to be hoped that, with further information about the covariance terms, these intervals may be reduced to an acceptable size.

Our present model would thus seem to give us the best of both worlds; the advantages of the functional approach, but without the excessive smoothing and artificially reduced confidence intervals given by the empirical direct fitting of very simple models. Although we can offer no proof, it is hoped that the fluctuations in the trends of the model results are a reflection of real growth rate oscillations, that the biologist knows to occur, rather than sampling variability. If this is so, then the present model provides an advantage over the classical methods using what is essentially a functional approach and it should be possible to gain a much deeper insight into the kinetics of the plants grown in an experiment than by any other currently available method. Thus, we have a surer foundation on which to base hypotheses and formulate future experiments.

Postscript

In this book, then, we have come round in a full circle — commencing and ending with the whole plant. In the beginning, the quantitative aspects of whole plant growth were examined by existing methods; then a more detailed examination of the quantitative patterns of growth of some of the plant's parts was made, together with a consideration of the statistical methods required to investigate these patterns in experimental plant meterial. Finally, this intermediate material has been synthesized into the model described in this chapter. The hope is that, both in the biological and mathematical sense, our contribution has enhanced the understanding of the quantitative nature of plant growth, and that it will provide a basis for biologists, mathematicians, and statisticians to communicate with each other on this subject. Only by recognizing the quantitative basis of growth, and acting on this recognition, can the study of growth processes proceed in the most expeditious manner.

Appendix

Approximation formulae for the expected value and variance of a function of several variates, and the covariance of two functions of several variates.

These formulae are given in Kendall & Stuart (1977, pages 246–7) but, because of their extensive use throughout this book, they are reproduced here.

$$\mathscr{E}\{f(x_1, \ldots, x_k)\} \simeq f(\theta_1, \ldots, \theta_k) + \tfrac{1}{2} \cdot \sum_{i=1}^{k} \{f''_{ii}(\theta)\} \cdot \mathscr{V}(x_i)$$
$$+ \tfrac{1}{2} \cdot \sum_{i=1}^{k} \sum_{j=1}^{k} \{f''_{ij}(\theta)\}\ \mathscr{C}(x_i, x_j) \qquad i \neq j \qquad \text{(A.1)}$$

$$\mathscr{V}\{f(x_1, \ldots, x_k)\} \simeq \sum_{i=1}^{k} \{f''_i(\theta)\}^2 \cdot \mathscr{V}(x_i)$$
$$+ \sum_{i=1}^{k} \sum_{j=1}^{k} \{f'_i(\theta)\}\{f'_j(\theta)\} \cdot \mathscr{C}(x_i, x_j) \qquad i \neq j \qquad \text{(A.2)}$$

$$\mathscr{C}\{f(x_1, \ldots, x_k), g(x_1, \ldots, x_k)\} = \sum_{i=1}^{k} \{f'(\theta)\}\{g'_i(\theta)\} \cdot \mathscr{V}(x_i)$$
$$+ \sum_{i=1}^{k} \sum_{j=1}^{k} \{f'_i(\theta)\}\{g'_j(\theta)\} \cdot \mathscr{C}(x_i, x_j)$$
$$i \neq j \qquad \text{(A.3)}$$

where k is the number of variates, each represented as x_i; $f'_i(\theta)$ is the first order partial derivative of the function with respect to the ith variate, $f''_{ij}(\theta)$ is the second order partial derivative with respect to the ith and jth variates; and θ_i is the expected value of the ith variate.

Table A.1 Parameter estimates and their variances for Richards function curves fitted to sunflower entity growth data.

Plant part	Parameter	Estimate	Variance of estimate
Leaf weight	*a*	0.5977	0.0252
	b	6.6967	3.5606
	k	0.2496	0.0052
	n	1.4749	0.2173
Stem weight	*a*	1.2182	0.0438
	b	10.8685	10.2360
	k	0.3169	0.0132
	n	1.7013	0.2974
Root weight	*a*	0.7762	0.2212
	b	6.0926	8.8572
	k	0.2137	0.0122
	n	1.0415	0.3477
Leaf area	*a*	6.0871	0.0033
	b	5.7561	1.5128
	k	0.2887	0.0030
	n	1.2211	0.0785
Plant weight	*a*	2.4451	0.1871
	b	6.3531	4.8688
	k	0.2060	0.0064
	n	1.0719	0.1986

(N.B. area fittings are to cm^2 not m^2)

Table A.2 Parameter estimates and their variances for Richards function curves fitted to wheat entity growth data.

Plant part	Parameter	Estimate	Variance of estimate
Leaf weight	a	−0.1423	0.0290
	b	3.7266	1.5325
	k	0.2083	0.0029
	n	0.7120	0.0699
Stem weight	a	−1.4597	0.4089
	b	11.6482	75.3708
	k	0.4504	0.1293
	n	1.4488	1.4206
Root weight	a	−0.6653	0.0618
	b	14.8156	134.6073
	k	0.5894	0.2302
	n	3.3188	7.6299
Leaf area	a	5.5165	0.1226
	b	3.5489	10.2556
	k	0.2090	0.0191
	n	0.7516	0.5635
Plant weight	a	0.5681	0.0454
	b	6.0397	3.9692
	k	0.2696	0.0078
	n	1.2059	0.2079

(N.B. area fittings are to cm^2 not m^2)

Table A.3 Parameter estimates and their variances for Richards functions fitted to maize entity growth data.

Plant part	Parameter	Estimate	Variance of estimate
Leaf weight	*a*	0.8910	0.2599
	b	2.9022	6.0535
	k	0.1943	0.0138
	n	0.4998	0.1976
Stem weight	*a*	−2.0380	0.1531
	b	4.8916	8.1799
	k	0.2885	0.0234
	n	0.2672	0.2438
Root weight	*a*	1.5455	21.6567
	b	5.0545	230.7394
	k	0.1989	0.4653
	n	0.8418	10.1328
Leaf area	*a*	6.4837	0.0859
	b	3.1601	5.5186
	k	0.2362	0.0151
	n	0.6217	0.2343
Plant weight	*a*	1.8238	0.4874
	b	4.8595	15.6548
	k	0.2340	0.0354
	n	0.8762	0.6762

(N.B. area fittings are to cm^2 not m^2)

Table A.4 Parameter estimates and their variances for polynomial curves fitted to birch entity growth data.

Plant part	Parameter	Estimate	Variance of estimate
Leaf weight	a	−8.2481	3.73×10^{-3}
	b_1	0.1951	1.21×10^{-5}
	b_2	−0.0086	1.70×10^{-9}
Stem weight	a	−9.7849	5.80×10^{-3}
	b_1	0.1893	1.88×10^{-5}
	b_2	−0.0006	2.70×10^{-9}
Root weight	a	−9.4982	5.82×10^{-2}
	b_1	0.1943	1.89×10^{-5}
	b_2	−0.0006	2.70×10^{-9}
Leaf area	a	−2.2475	3.91×10^{-3}
	b_1	0.1910	1.27×10^{-5}
	b_2	−0.0009	1.80×10^{-9}
Plant weight	a	−7.7908	3.46×10^{-3}
	b_1	0.1895	1.12×10^{-5}
	b_2	−0.0007	1.60×10^{-9}

(N.B. area fittings are to cm^2 not m^2)

Table A.5 Richards function parameter estimates and their variances for sunflower, leaf pairs weight and area.

	Leaf number	a	b	k	n	$k/(n+1)$	$Ak/\{2(n+2)\}$	$2(n+2)/k$	$\mathscr{V}(a)$
WEIGHT	1	−2.5965	1.4665	0.3555	0.2930	0.2744	0.00058	13.01	0.00132
	2	−1.5592	6.6985	0.4937	0.6499	0.2982	0.0151	10.84	0.00109
	3	−1.0719	5.4816	0.3480	0.3581	0.2542	0.02499	13.87	0.00248
	4	−0.8589	5.0829	0.2612	0.2975	0.1978	0.02304	18.58	0.00248
	5	−1.1667	0.2263	0.0827	0.0363	0.0556	−0.1936	131.0	9.7973
AREA	1	3.6270	2.3889	0.4229	0.4026	0.3001	3.28	11.52	0.00330
	2	4.6131	12.9376	0.8391	1.3720	0.3495	12.32	8.27	0.00302
	3	4.9720	9.1513	0.5133	0.6633	0.3056	13.74	10.66	0.00176
	4	5.2429	6.6396	0.3016	0.4261	0.2096	11.53	16.60	0.00656
	5	5.1987	20.3815	0.6474	1.6672	0.2221	13.17	13.17	0.07270

(†$\mathscr{V}(k)$ are in fact estimated as $\mathscr{V}(K)$, as fitting method is for K)
(N.B. area fittings are to cm^2 not m^2)

$\mathscr{V}(b)$	$\mathscr{V}(k)$†	$\mathscr{V}(n)$	$\mathscr{V}(k/(n+1))$	$\mathscr{V}(Ak/\{2(n+2)\})$	$\mathscr{V}(2(n+2)/k\}$	Variance of curve	F-ratio
0.46475	0.00017	0.01152	0.000167	0.11×10^{-6}	0.896	0.07337	1.6 N.S [12, 128]
0.94799	0.00343	0.01099	0.000307	0.19×10^{-5}	0.748	0.08314	1.4 N.S [11, 120]
2.08394	0.00394	0.01490	0.000580	0.71×10^{-5}	3.160	0.11084	< 1
4.09040	0.00506	0.02448	0.000988	0.47×10^{-6}	13.150	0.18607	< 1
113.29300	0.01269	0.0855	0.007449	0.157	3605.0	0.34753	1.4 N.S [6, 80]
0.50646	0.00365	0.01315	0.000403	0.043	1.221	0.07551	3.4** [7, 88]
10.06190	0.03786	0.12731	0.000884	1.596	1.046	0.10007	2.9* [7, 88]
4.39650	0.01062	0.03021	0.000934	2.756	2.015	0.09358	1.2 N.S [7, 88]
3.20420	0.00402	0.02104	0.000566	0.993	5.990	0.06512	< 1
152.67000	0.15762	1.13782	0.002728	3.262	13.410	0.12428	1.05 N.S. [6, 80]

Table A.6 Richards function parameter estimates and their variances for wheat, single leaf weight and area.

	Leaf number	a	b	k	n	$k/(n+1)$	$Ak/\{2(n+2)\}$	$2(n+2)/k$
WEIGHT	1	−4.3491	0.3986	0.7056	0.2740	0.6180	0.00201	6.482
	2	−3.8749	1.8952	0.5123	0.2963	0.3950	0.00231	9.007
	3	−3.3595	21.1504	2.0779	2.2249	0.5663	0.00713	5.210
	4	−3.1113	10.4025	0.8659	0.5952	0.5402	0.00739	6.007
	5	−2.9149	26.2495	1.5250	1.6993	0.5537	0.01087	5.104
	6	−2.7425	14.0881	0.6826	0.8367	0.3684	0.0760	8.541
AREA	1	1.5192	−0.1810	0.7279	−0.2955	2.1350	1.055	4.494
	2	2.0391	3.8986	0.8417	0.5801	0.5282	1.238	6.289
	3	2.4646	32.9663	3.3631	3.4076	0.7340	3.401	3.535
	4	2.7033	12.9503	1.0567	0.8265	0.5774	2.770	5.453
	5	2.8067	28.4960	1.6967	1.9607	0.5486	3.327	5.167
	6	2.8896	22.6855	1.0427	1.5667	0.3995	2.540	7.181

(† $\mathscr{V}(k)$ are in fact estimated as $\mathscr{V}(K)$, as fitting method is for K)
(N.B. area fittings are to cm^2 not m^2)

Table A.7 Richards function parameter estimates and their variances for maize, single leaf weight and area.

	Leaf number	a	b	k	n	$k/(n+1)$	$Ak/\{2(n+2)\}$	$2(n+2)/k$
WEIGHT	1	−4.3818	6.1515	1.8996	10.1058	0.6540	0.00133	15.98
	2	−3.8245	11.5689	2.0234	5.2797	0.2997	0.00222	9.76
	3	−2.8669	3.3211	0.5013	0.6789	0.2881	0.00505	11.81
	4	−1.7438	4.5989	0.4590	0.5169	0.2976	0.01516	11.51
	5	−0.8027	0.9449	0.2441	0.0628	0.2268	0.02573	17.61
	6	−0.3359	−1.2185	0.1663	−0.0105	0.1566	0.02332	29.37
	7	−1.2214	2.1188	0.2433	0.0677	0.1220	−0.00577	43.22
AREA	1	} Insufficient data for fitting to be possible						
	2							
	3	3.1650	8.4058	1.2656	1.5748	0.4282	3.288	7.78
	4	4.2388	7.1027	0.7141	0.0815	0.3873	8.622	8.23
	5	4.8962	7.0115	0.5959	0.0496	0.3918	15.670	8.76
	6	5.1912	4.2400	0.3439	0.2427	0.2752	13.580	13.38
	7	4.8908	12.7327	0.6375	0.9660	0.2952	12.100	11.44

(† $\mathscr{V}(k)$ are in fact estimates as $\mathscr{V}(K)$, as fitting method is for K)
(N.B. area fittings are to cm^2 not m^2)

$\mathscr{V}(a)$	$\mathscr{V}(b)$	$\mathscr{V}(k)$†	$\mathscr{V}(n)$	$\mathscr{V}(k/(n+1))$	$\mathscr{V}(Ak/\{2(n+2)\})$	$\mathscr{V}(2(n+2)/k)$	Variance of curve	F-ratio
0.000371	9.27305	0.03780	0.37049	0.01663	0.24×10^{-7}	0.2895	0.2670	2.5* [13, 85]
0.000408	0.41514	0.00217	0.00942	0.00023	0.11×10^{-7}	0.2451	0.02822	4.1 [14, 90]
0.000150	332.569	2.90666	4.08222	0.01596	0.83×10^{-5}	1.9480	0.01104	1.9 N.S [11, 75]
0.000180	3.38381	0.01611	0.01969	0.00123	0.48×10^{-6}	0.3238	0.01395	2.6*** [8, 60]
0.000671	81.4128	0.05712	0.44043	0.00330	0.30×10^{-5}	0.6003	0.03929	6.8*** [4, 40]
0.004693	13.8456	0.02437	0.09951	0.00077	0.33×10^{-6}	1.0570	0.00691	< 1
0.000322	3.92396	0.06968	0.73545	0.83460	0.03953	0.8987	0.01954	1.2 N.S [12, 80]
0.000413	2.22516	0.03269	0.06187	0.00131	0.02073	0.5571	0.03118	1.7 N.S [12, 80]
0.000268	405.323	3.81918	4.94825	0.00408	0.38670	0.3077	0.02185	2.1* [11, 75]
0.000194	6.51829	0.03481	0.03738	0.00187	0.09085	0.3458	0.01447	2.1* [8, 60]
0.000810	222.684	0.72506	1.32567	0.00495	0.55810	1.0100	0.03171	< 1
0.003810	71.6468	0.12853	0.51887	0.00107	0.09254	1.0260	0.01896	< 1

$\mathscr{V}(a)$	$\mathscr{V}(b)$	$\mathscr{V}(k)$†	$\mathscr{V}(n)$	$\mathscr{V}(k/(n+1))$	$\mathscr{V}(Ak/\{2(n+2)\})$	$\mathscr{V}(2(n+2)/k)$	Variance of curve	F-ratio
0.00164	990.501	64.2658	3145.17	0.04566	0.50×10^{-6}	266.5	0.09529	< 1
0.00097	309.572	7.9012	69.1844	0.00369	0.80×10^{-6}	4.55	0.08185	1.0
0.00243	5.821	0.0464	0.2599	0.00197	0.13×10^{-5}	6.99	0.02393	< 1
0.00435	2.653	0.0145	0.0378	0.00181	0.63×10^{-5}	4.25	0.20658	< 1
0.01587	5.930	0.0032	0.0113	0.00099	0.53×10^{-5}	9.47	0.11691	< 1
0.17094	484.187	0.0081	0.0603	0.00265	0.81×10^{-5}	101.90	0.12169	2.6* {5, 45}
0.75869	347.377	0.1255	0.6965	0.02559	0.18×10^{-4}	325.80	0.2194	
0.00271	82.916	1.3918	3.6771	0.01369	2.676	5.49	0.10295	< 1
0.00197	5.127	0.0355	0.0810	0.00217	1.747	1.75	0.11628	< 1
0.00163	6.076	0.0236	0.0459	0.00229	6.208	2.13	0.07453	1.8 N.S [5, 45]
0.00230	3.794	0.0049	0.0347	0.00059	1.708	2.94	0.04649	9.6*** [5, 45]
0.03152	74.793	0.1573	0.5729	0.00625	9.227	11.84	0.08896	1.0

Table A.8 Richards function parameter estimates and their variances for birch, single leaf weight.

	Leaf number	a	b	k	n	$k/(n+1)$	$Ak/\{2(n+2)\}$	$2(n+2)/k$	$\mathscr{V}(a)$
WEIGHT	1	−5.4752	0.7447	0.2012	−0.2233	0.2847	0.000234	18.11	0.000651
	2	−4.4964	2.3610	0.1942	0.2049	0.1604	0.000487	23.06	0.001251
	3	−3.4889	0.6070	0.1123	−0.0760	0.1213	0.000884	34.77	0.001312
	4	−2.8721	3.3459	0.1457	0.1678	0.1244	0.001888	30.20	0.000914
	5	−2.3981	3.6856	0.1505	0.0920	0.1378	0.003248	28.16	0.000796
	6	−2.1279	8.5235	0.2059	0.4476	0.1420	0.004272	24.13	0.000393
	7	−1.9364	3.8013	0.1560	0.0269	0.1527	0.005471	26.62	0.000960
	8	−1.8397	7.4720	0.1803	0.1860	0.1509	0.006465	24.91	0.001241
	9	−1.6913	4.4985	0.1545	0.0231	0.1544	0.006895	26.99	0.002535
	10	−1.5619	4.3772	1.2222	−0.0845	0.1381	0.006751	31.30	0.001918
	11	−1.5502	5.3283	0.1261	−0.1428	0.1509	0.007087	30.22	0.002115
	12	−1.5657	5.5047	0.1217	−0.1719	0.1538	0.006779	31.15	0.003527
	13	−1.6825	11.1591	0.1938	0.3414	0.1445	0.007579	24.79	0.001508
	14	−1.7460	7.4739	0.1805	0.0189	0.1856	0.007654	23.06	0.001765
	15	−1.8479	19.3154	0.2845	1.0391	0.1488	0.007253	21.99	0.001444
	16	−1.9846	27.1726	0.3892	1.4223	0.1658	0.007561	18.48	0.000957
	17	−2.0383	2.0924	0.2049	$\simeq 0$	0.2049	0.006653	19.83	0.001275
	18	−2.2817	−13.0579	0.3025	0.0022	0.3018	0.007724	13.24	0.000537
	19	−2.2766	3.5641	0.2421	$\simeq 0$	0.2421	0.006193	16.84	0.001300
	20	−2.3453	1.7969	0.2091	$\simeq 0$	0.2091	0.004988	19.45	0.002104
	21	−2.4717	2.8116	0.2299	$\simeq 0$	0.2299	0.004815	18.07	0.003320
	22	−2.7691	−16.9560	0.1379	0.0163			29.22	0.001938

(†$\mathscr{V}(k)$ are in fact estimated as $\mathscr{V}(K)$, as fitting method is for K)

$\mathscr{V}(b)$	$\mathscr{V}(k)$†	$\mathscr{V}(n)$	$\mathscr{V}(k/(n+1))$	$\mathscr{V}(Ak/\{2(n+2)\})$	$\mathscr{V}(2(n+1)/k)$	Variance of curve	F-ratio
0.70661	0.000339	0.14002	0.003331	0.54×10^{-9}	3.658	0.09498	1.5 N.S [17, 189]
1.37531	0.000872	0.01381	0.000113	0.17×10^{-8}	5.415	0.12252	1.6 N.S [17, 189]
1.13027	0.000300	0.01455	0.000045	0.47×10^{-8}	11.240	0.10046	< 1
2.74652	0.000514	0.01951	0.000054	0.24×10^{-7}	8.407	0.07969	1.2 N.S [14,162]
6.85466	0.000522	0.02329	0.000062	0.57×10^{-7}	6.067	0.06196	1.7 N.S [12, 144]
4.69832	0.001231	0.04697	0.000070	0.17×10^{-6}	4.805	0.03685	1.2 N.S [11, 135]
118.338	0.001228	00.05974	0.000095	0.26×10^{-6}	8.656	0.04241	1.9 N.S [8, 117]
8.56605	0.001423	0.03207	0.000138	0.51×10^{-6}	10.350	0.06046	1.1 N.S [9, 117]
270.077	0.001702	0.10724	0.000301	0.37×10^{-6}	11.180	0.04160	2.1 N.S [7, 99]
0.66993	0.000590	0.03173	0.000552	0.52×10^{-4}	5.460	0.04709	1.5 N.S [10, 126]
0.33938	0.000913	0.06887	0.000316	0.37×10^{-6}	12.460	0.05732	1.1 N.S [9, 117]
0.50395	0.001286	0.10771	0.000489	0.37×10^{-6}	16.990	0.05669	1.7 N.S [8, 108]
14.3532	0.001993	0.08547	0.000127	0.54×10^{-6}	8.462	0.06038	< 1
600.410	0.002756	0.15976	0.000802	0.60×10^{-6}	8.170	0.06367	1.2 N.S [7, 99]
84.6363	0.013060	1.09358	0.000604	0.61×10^{-6}	7.234	0.04117	< 1
186.336	0.031399	1.47535	0.000448	0.11×10^{-5}	6.446	0.05663	1.4 N.S. [6.90]
9814.95	0.000653	$\simeq 0$	0.000653	0.49×10^{-6}	5.927	0.05155	1.4 N.S [5, 81]
$4. \times 10^{12}$	$8. \times 10^{8}$	4×10^{4}	0.000969	0.19×10^{-6}	5.718	0.10247	20.3*** [5, 81]
198.745	0.001143	$\simeq 0$	0.001143	0.57×10^{-6}	5.323	0.05850	2.2 N.S [4, 72]
37.2034	0.000737	$\simeq 0$	0.000737	0.27×10^{-6}	6.168	0.06344	1.2 N.S [4, 72]
−5249.3	0.002043	$\simeq 0$	0.002043	0.61×10^{-6}	11.700	0.09091	.3 N.S [3, 63]
3×10^{10}	6×10^{6}	43194.4				0.03680	33.6***

Table A.9 Richards function parameter estimates and their variances for birch, single leaf area.

	Leaf number	a	b	k	n	$k/(n+1)$	$Ak/\{2(n+2)\}$	$2(n+2)/k$	$\mathscr{V}(a)$
AREA	1	0.3034	6.5176	0.5836	0.9632	0.2887	0.128	10.87	0.000536
	2	1.2164	4.9975	0.3078	0.4559	0.2087	0.208	16.53	0.000876
	3	2.1951	1.8652	0.1647	0.0745	0.1528	0.353	25.65	0.000785
	4	2.7964	9.6650	0.2969	0.6248	0.1813	0.918	18.08	0.000521
	5	3.2650	7.0764	0.2249	0.1969	0.1873	1.330	19.89	0.000496
	6	3.5446	10.8652	0.2673	0.4413	0.1845	1.878	18.66	0.000362
	7	3.6984	13.0821	0.2881	0.4945	0.1930	2.305	17.71	0.000428
	8	3.8173	13.7833	0.2884	0.4478	0.1974	2.646	17.47	0.000692
	9	3.9424	17.7134	0.3351	0.6902	0.1997	3.160	16.55	0.000644
	10	3.9493	20.0528	0.3669	0.6715	0.2176	3.499	15.15	0.000439
	11	3.9556	27.5208	0.4682	1.1342	0.2155	3.685	14.76	0.000441
	12	3.9186	21.3114	0.3605	0.6586	0.2118	3.253	16.03	0.000549
	13	3.8373	23.3631	0.3807	0.7676	0.2131	3.125	15.21	0.000534
	14	3.7870	2.1989	0.2584	$\simeq 0$	0.2584	2.846	15.72	0.000637
	15	3.6958	3.2964	0.2481	$\simeq 0$	0.2481	2.493	16.39	0.000726
	16	3.5850	2.5446	0.2583	$\simeq 0$	0.2583	2.325	15.07	0.000562
	17	3.4919	2.9046	0.2818	$\simeq 0$	0.2818	2.310	14.46	0.000726
	18	3.3864	3.9222	0.2644	$\simeq 0$	0.2644	1.950	15.29	0.000763
	19	3.2779	4.5373	0.3104	$\simeq 0$	0.3104	2.054	13.15	0.000962
	20	3.1814	2.9043	0.2565	$\simeq 0$	0.2565	1.541	15.79	0.001236
	21	3.0483	4.6117	0.2835	$\simeq 0$	0.2835	1.487	14.63	0.002171
	22	2.9937	2.2637	0.2062	$\simeq 0$	0.2062	1.018	20.12	0.004590

(† $\mathscr{V}(k)$ are in fact estimated as $\mathscr{V}(k)$, as fitting method is for K)
(N.B. area fittings are to cm^2 not m^2)

$\mathcal{V}(b)$	$\mathcal{V}(k)$†	$\mathcal{V}(n)$	$\mathcal{V}(k/(n+1))$	$\mathcal{V}(\mathrm{A}k/\{2(n+2)\})$	$\mathcal{V}(2(n+2)/k)$	Variance of curve	F-ratio
9.3272	0.04678	0.27903	0.000659	0.00058	3.621	0.09121	1.2 N.S [17, 189]
3.4382	0.000529	0.04422	0.000439	0.00087	5.980	0.11275	1.1 N.S [17, 189]
8.2126	0.00082	0.02282	0.000094	0.00178	7.430	0.09519	< 1
4.8925	0.00311	0.0355	0.000233	0.00113	4.617	0.07824	1.2 N.S [15, 171]
5.5615	0.00152	0.02596	0.000179	0.01840	4.547	0.06196	1.7 N.S. [12, 144]
7.9252	0.00266	0.04489	0.000198	0.04081	4.483	0.03533	1.2 N.S [11, 135]
15.7679	0.00420	0.10687	0.000205	0.05784	3.633	0.04117	1.2 N.S [9, 117]
14.0499	0.00395	0.04515	0.000302	0.11510	5.387	0.05983	< 1
34.4642	0.00848	0.22358	0.000468	0.14710	4.453	0.04471	1.3 N.S [7, 99]
40.0892	0.01019	0.13380	0.000483	0.26990	4.958	0.04731	< 1
219.706	0.05381	0.87835	0.000851	0.69500	8.688	0.04749	< 1
94.7654	0.02058	0.34731	0.000786	0.42250	9.142	0.04736	1.1 N.S [8, 108]
59.4859	0.01273	0.17141	0.000566	0.25080	5.679	0.05288	1.1 N.S [8, 108]
−8893.7	0.00105	$\simeq 0$	0.001049	0.10850	3.765	0.05234	< 1
20815.1	0.00103	$\simeq 0$	0.001026	0.08210	4.333	0.04077	< 1
−1542.3	0.00091	$\simeq 0$	0.000909	0.06183	3.267	0.04499	1.8 N.S [6, 90]
37.2300	0.00149	$\simeq 0$	0.001499	0.08491	3.803	0.04499	1.2 N.S [5, 81]
−605.37	0.00076	$\simeq 0$	0.000759	0.02897	2.485	0.02897	1.3 N.S [5, 81]
−81.66	0.00194	$\simeq 0$	0.001944	0.07152	3.351	0.05369	< 1
26.2340	0.00081	$\simeq 0$	0.000812	0.02204	3.001	0.05921	< 1
−345.50	0.00928	$\simeq 0$	0.002982	0.06543	7.386	0.09104	< 1
33.7668	0.00159	$\simeq 0$	0.001590	0.02297	14.070	0.10708	< 1

Table A.10 Polynomial curve parameter estimates and their variances for single leaves.

Species	Leaf number		a	b_1	b_2	(a)	(b_i)	(b_2)	Variance of curve
Sunflower	cotyledons	weight	−3.5471	−0.0017	–	1.04×10^{-3}	2.29×10^{-6}	–	0.03867
		area	1.8772	0.0147	–	7.42×10^{-4}	8.03×10^{-6}	–	0.4667
	6	weight	−14.0998	0.3765	–	0.1265	1.89×10^{-4}	–	0.67296
		area	−3.2189	−0.0602	0.0087	4.1442	0.0294	1.14×10^{-5}	0.31625
	7	weight	−15.4247	0.3751	–	0.5590	6.71×10^{-4}	–	0.53407
		area	9.4404	−1.0464	0.0255	43.6962	0.2426	8.03×10^{-5}	0.58000
	8	weight	−15.2677	0.3269	–	0.6063	7.40×10^{-4}	–	0.66944
		area	5.4825	−0.7098	0.0173	303.2116	1.3678	3.82×10^{-4}	0.38961
	9	weight	−15.8885	0.3251	–	1.9481	0.0021	–	0.59194
		area	−9.9689	0.3019	–	1.4389	0.0015	–	0.33167
Wheat	7	weight	−11.4306	0.7416	−0.0139	0.4468	0.0049	3.19×10^{-6}	0.09817
		area	−5.4149	0.7513	−0.0144	0.4099	0.0044	2.68×10^{-6}	0.09774
Maize	1	area	1.2373	0.0097	–	0.0303	3.91×10^{-4}	–	0.15378
	2	area	2.1513	0.0088	–	0.0161	7.87×10^{-4}	–	0.07458
	8	weight	−23.2498	1.6684	−0.0344	11.1052	0.1217	7.94×10^{-5}	0.39167
		area	−12.9012	1.1551	−0.0199	60.3339	0.6415	3.94×10^{-4}	1.02100
	9	weight	−13.6934	0.3895	–	1.9242	39.8249	–	0.39900
		area	−10.4197	0.5184	–	1.8801	39.8071	–	0.42542
Birch	cotyledons	weight	−7.8349	0.0176	–	0.0011	1.82×10^{-6}	–	0.14036
		area	−1.7659	0.0098	–	0.0016	2.28×10^{-6}	–	0.11693
	23	weight	−13.3083	0.1315	–	1.9092	2.55×10^{-4}	–	0.38483
		area	−10.5484	0.1624	–	4.4123	5.79×10^{-4}	–	0.56967

(N·B· area fittings are to cm^2 not m^2)

Table A.11 Parameter estimate and their variances for Richard function curves fitted to wheat and maize leaf and stem growth data (data modified so that leaf sheaths are included with stem) – for model.

WHEAT

Plant part	Parameter	Estimate	Variance of estimate
Leaf weight	*a*	−0.8082	0.0092
	b	5.2323	2.1615
	k	0.2957	0.0052
	n	1.1422	0.1296
Stem weight	*a*	1.3517	4.2729
	b	0.2125	16.1634
	k	0.0720	0.0048
	n	0.0839	0.0654

MAIZE

Plant part	Parameter	Estimate	Variance of estimate
Leaf weight	*a*	0.6882	0.3303
	b	2.6971	6.4834
	k	0.1824	0.1396
	n	0.4743	0.2143
Stem weight	*a*	−0.1716	0.2949
	b	2.6289	4.9336
	k	0.1838	0.0105
	n	0.3946	0.1159

(N.B. area fittings are to cm^2 not m^2)

Glossary of symbols

Many composite symbols in the book are constructed according to definite rules, which are summarized below; and so not every symbol encountered in the text will be found in the Glossary. Where a composite symbol has a special use, however, it is defined here. Conventional mathematical notation is not given.

The variates of plant or plant part sizes at Levels 1 and la (see Table 1.1) are given single italicized capital (upper case) letters, and the natural logarithms of these quantities are given corresponding small (lower case) letters. Any supplementary information on the nature of the size, for example dry weight or area, is given as a suffix in the form of a non-italicized capital letter. All other additional information concerning component number (k), harvest number (i), and replicate number (j) are given in lower case italicized letters.

Derived quantities (rates and ratios of the plant and plant part sizes) are denoted by non-italicized capital letters, having an italicized subscript denoting the plant part, or no plant part subscript if the symbol relates to the whole plant. Other suffices, detailed in the previous paragraph, may be added as necessary.

A 'hat' (^) placed over a quantity means that it is a statistical (especially maximum likelihood) estimate of that quantity. Similarly, a 'bar' (¯) placed over a variate symbol implies that it is the arithmetic mean of several replicate values of that variate. The bar notation is also used to imply the mean value of a rate of change of a quantity over an interval of time or over a range of sizes, but these quantities will be explicitly defined in the Glossary.

Where the component suffix is missing, the whole entity is implied. For example, l_k denotes the dry weight of leaf k, whereas l is the dry weight of all leaves (foliage). If the replicate suffix is missing, this implies that replicate number is not relevant in that particular use, or that all replicates are taken into account; in particular, we have

$$\bar{l}_i = N^{-1} . \sum_{j=1}^{N} l_{ij}$$

If the time or harvest suffix is missing, this again implies that harvest number is not important in that particular use, or that all harvests are taken into account; of special note is

$$\bar{l} = (Nh)^{-1} . \sum_{i=1}^{h} \sum_{j=1}^{N} l_{ij}$$

which should be compared with the previous equation.

The symbol (S) denotes a special use of an otherwise standard item of notation.

Throughout the book (except where noted) S.I. units are used, but the day is used as the unit of time rather than the second.

Roman

$\bar{\mathrm{A}}$	Mean absolute growth rate in size from 0 to A as given by the Richards function. $\bar{\mathrm{A}} = A\kappa/\{2(\nu+2)\}$.
a	1. The least squares estimate of parameter α in a straight line equation, and so, generally, the intercept in a straight line equation. 2. The least squares estimate of parameter α in the Richards function.
$\mathbf{a}_i$	Column vector of parameter estimates of the Richards function obtained from the ith Newton-Raphson iteration. Starting values are in vector $\mathbf{a}_0$.
b	1. The least squares estimate of parameter β in a straight line equation, and so, generally, the gradient in a straight line equation. 2. The least squares estimate of parameter β in the Richards function. 3(S). The width of a leaf (page 121).
b_i	The value of b in the ith main iteration of Hadley's method of acquiring starting values for Richards function estimation; b_0 is the initial value.
b_k	The least squares estimate of β_k.
C	Shoot dry weight – all above-ground parts.
c	Natural logarithm of shoot dry weight.
$\mathscr{C}(,)$	The covariance of two variates.
D	The duration of growth of a single leaf as given by the Richards function. $\mathrm{D} = 2(\nu+2)/\kappa$.

D — 1. The denominator of the fraction defining the increment of K, ΔK, between the K_{ij}th and the $K_{i(j+1)}$th sub-iteration of Hadley's method of acquiring starting values for Richards function estimation.
2. The largest absolute value of d_j in the Kolmogorov-Smirnov test.

d_j — $d_j = P_j - E_j$ in the Kolmogorov-Smirnov test.

E — Instantaneous unit leaf rate (leaf area basis).

$\overline{\mathrm{E}}$ — Mean unit leaf rate over an interval of time.

E_A — Instantaneous unit leaf rate (leaf area basis).

E_W — Instantaneous unit leaf rate (leaf dry weight basis).

$\overline{\mathrm{E}}_j$ — (S). Mean unit leaf rate over an interval of time of the jth replicate data set in the simulation study on unit leaf rate.

$\hat{\overline{\mathrm{E}}}_m$ — Mean value of the N estimated values of $\hat{\overline{E}}_j$ in the simulation study on unit leaf rate.

$\overline{\mathrm{E}}_u$ — Underlying mean unit leaf rate over an interval of time in the simulation study.

E_j — The jth expected cumultive proportion of the N replicate observations in the Kolmogorov-Smirnov test.

$\mathscr{E}(\ )$ — The expected value of a variate. Occasionally used as the estimate of the expected value for typographical convenience.

F — The variance ratio.

f — 1. The length of a leaf (page 121).
2. Index number for component, when k is already in use.

h — The number of harvest times in an experiment; hence, the number of distinct x-values in a regression, and the number of distinct ξ-values in a functional relationship.

$\mathbf{I}$ — 1. The information matrix of second order partial derivatives of the logarithm of the likelihood of a set of observations with respect to all the population parameters. Thus $-\mathbf{I}^{-1}$ is the maximum likelihood estimate of the variance-covariance matrix of the parameter estimates.
2. The unit matrix.

i — 1. Index number for time or harvest.
2. Index number for degree of polynomial.
3(S). Index number for component in the theorems of Chapters 6 and 7.

j	1. Index number for replicate. 2(S). Index number for sub-component in Theorem 6.3.
K	The least squares estimate of parameter K (Greek kappa) in the Richards function.
K_{ij}	The value of K in the ith main iteration and jth sub-iteration of Hadley's method of acquiring starting values for Richards function estimation; K_{00} is the initial value.
k	1. Index number for component. 2. Index number for parameter in a function. 3. $k = -K$ in the Richards function.
L	1. Foliage size (dry weight, unless otherwise stated). 2. The likelihood of all h or Nh observations.
L_A	Foliage area.
L_W	Foliage dry weight.
L_I	Leaf size at the point of inflexion of a sigmoid curve.
l	1. Natural logarithm of foliage or leaf size (dry weight, unless otherwise stated). 2. The dimension of length.
l_A	Natural logarithm of foliage area.
l_W	Natural logarithm of foliage dry weight.
l_m	$l_m = l_{ij} - (l_{ij} - \bar{l}_i)\sqrt{V_s}/s_i$ in the point moving technique.
$Lf\{\mathrm{R}:t\}$	A logistic function of the form $Lf\{\mathrm{R}:t\} = \mathrm{R}/(1 + ce^{\mathrm{R}t})$, where c is exp(constant of integration).
M	Symmetric matrix containing the second order partial derivatives associated with the least squares estimation of the $(i-1)$th iteration.
M_B	Between groups mean square.
M_D	Deviations from regression mean square.
M_R	Regression mean square.
M_W	Within groups mean square.
m	1. The dimension of mass. 2. A parameter in the Von Bertalanffy function, $\frac{2}{3} \leqslant m \leqslant 1$.

N	1. Number of replicate observations in a sample.
	2(S). Number of replicate data sets used in the simulation study on unit leaf rate.
N_g	Number of $\mathscr{V}(\hat{\bar{E}}_j)$ greater than $\mathscr{V}(\hat{\bar{E}}_m)$ in the simulation study on unit leaf rate.
n	1. The least squares estimate of parameter v in the Richards function.
	2. The degree of a polynomial function.
	3(S). The power of L_A in Whitehead & Myerscough's equation 2.24.
P	Non-foliar plant parts dry weight. $P = W - L_W$.
P_j	(S). The jth observed cumulative proportion of the N replicate observations in the Kolmogorov-Smirnov test.
p	The number of parameters in a function.
Q	The number of components in an entity.
q	The number of sub-components in an entity; specifically, the number of leaves comprising the foliage.
R	Instantaneous whole plant relative growth rate.
$\overline{R}$	Mean relative growth rate over an interval of time.
R_I	Instantaneous relative growth rate at the point of inflexion of the $R(t)$ curve derived from a sigmoid function.
R_L	Instantaneous foliage weight relative growth rate.
R_{LA}	Instantaneous foliage area relative growth rate.
R_R	Instantaneous root relative growth rate.
R_S	Instantaneous stem relative growth rate.
R_{max}	The maximum instantaneous value of R over a specified interval of time.
R	Root dry weight.
r	1. Natural logarithm of root dry weight.
	2. Sample correlation coefficient.
S	Stem dry weight.
s	1. Natural logarithm of stem dry weight.
	2. Sample standard deviation, hence s^2 is sample variance.
s_{lA}	Standard deviation of a sample of l_{Aj}.

s_w Standard deviation of a sample of w_j.

$s_{\delta\delta}^{(i)}, s_{\varepsilon\varepsilon}^{(i)}, s_{\delta\varepsilon}^{(i)}$ Sample estimate of the variance of x, y or covariance of x and y for the ith population in a functional relationship.

$s_{\delta\delta}, s_{\varepsilon\varepsilon}, s_{\delta\varepsilon}$ Mean of the variance (or covariance) estimates for all populations involved in a functional relationship.

$\mathscr{S}^2$ The sum of the squares of deviations of all h or Nh data points from a regression line.

$\mathscr{S}_i^2$ The deviations sum of squares given by the ith main iteration of Hadley's method of acquiring starting values for Richards function estimation.

t 1. Time; the dimension of time.
2. Student's-t.

t_I Time at the point of inflexion of a sigmoid function.

t_m $t_m = (l_m - \bar{l}_i)/b + t_i$ in the point moving technique.

V_s 'Structural' variance of the replicates of a single leaf.

$\mathbf{v}$ Column vector of first order partial derivatives for the least squares estimation of the Richards function, evaluated with the parameter values of the $(i-1)$th iteration.

$\mathscr{V}(\ )$ The variance of a variate.

W 1. Whole plant dry weight.
2. Dry weight of an entity.

W_p Whole plant dry weight gain over an interval of time by photosynthesis less photorespiration.

W_r Whole plant dry weight loss over an interval of time by dark respiration.

$W_{i,j}$ The dry weight of the jth sub-component of the ith component of an entity.

w Natural logarithm of whole plant dry weight.

x Dark respiration rate per unit whole plant dry weight at time t, divided by leaf area ratio. $\mathrm{x} = \rho/(L_A/W)$.

$\bar{\mathrm{x}}$ Mean of x over an interval of time.

x_i The ith component entity ratio.

x The regressor (or independent) variable, known without

error, in regression; also, one variate in a functional relationship.

x_i — 1(S). The ith iterative value of x (ordinary variable in a univariable Newton-Raphson technique; starting value is $i = 0$.
2(S). The ith component entity ratio. W_i/W, in Theorems in chapters 6 and 7.

y — The response (or dependent) variate in regression; also, one variate in a functional relationship.

z — Standard normal deviate.

Greek

A — The population value of the upper asymptote of a sigmoid function.

α — 1. The population value of the intercept of a straight line equation.
2. The population value of the natural logarithm of the upper asymptote of a sigmoid function.
3. The population coefficient of x^β in the allometric function.

B — The part of the information matrix concerned with the function parameters (not the incidental parameters) in a functional relationship.

β — 1. The population value of the gradient of a straight line equation.
2. The population parameter associated with the position in time of a sigmoid function.
3. The allometric constant (linear allometry).

β_k — The kth parameter in a mathematical function.

δ_i — A deviation of the x-variate of an observation from the population mean of x at the ith harvest. $\delta_i = x_i - \xi_i$.

ε_i — The deviation of the y-value from a regression value at the ith value of x; also, a deviation of the y-variate of an observation from the population mean of y at the ith harvest. $\varepsilon_i = y_i - \eta_i$.

η — (S). Rate of anabolism in the Von Bertalanffy function.

η_i — The regression line value of variate y for the ith value of x; also, the population value of variate y at the ith harvest in a functional relationship.

θ — Any population parameter.

K	The negative of the population parameter associated with the rate of increase of a sigmoid function. $K = -\kappa$.
κ	1. The population parameter associated with rate of increase of a sigmoid function. 2(S). The rate of catabolism in the Von Bertalanffy function.
λ	1. The ratio of the variances of the two variates in a functional relationship. $\lambda = \sigma_{\varepsilon\varepsilon}/\sigma_{\delta\delta}$. 2. An 'adjustment' factor in Marquardt's iterative method.
μ	The population mean (univariate).
ν	The population parameter governing the position of the point of inflexion in the Richards function.
$\mathbf{\Xi}$	The part of the information matrix concerned with the incidental parameters in a functional relationship.
ξ_i	The population value of variate x at the ith harvest in a functional relationship.
π	Instantaneous gross photosynthetic rate (less photorespiration) per unit leaf area. $\pi = (1/L_A).(dW_p/dt)$.
$\bar{\pi}$	Mean of π over an interval of time.
ρ	Instantaneous dark respiration rate per unit whole plant dry weight.
$\bar{\rho}$	Mean of ρ over an interval of time.
ρ	A population correlation coefficient.
$\mathbf{\Sigma}$	Population variance-covariance matrix of the variates in a functional relationship.
σ	Population standard deviation, hence σ^2 is the population variance.
$\sigma_{\delta\delta}$, $\sigma_{\varepsilon\varepsilon}$, $\sigma_{\delta\varepsilon}$	Population variance of x, y, and covariance of x and y in a functional relationship.
χ^2	The distribution of Σz^2.
$\mathbf{\Omega}$	The part of the information matrix concerned with both the incidental and other parameters in a functional relationship.

Bibliography

Acock, B., Charles-Edwards, D. A. & Hearn, A. R. (1977). Growth response of Chrysanthemum crops to the environment. I. Experimental techniques. *Annals of Botany*, **41**, 41–48.

Acton, F. S. (1959). *Analysis of straight-line data*. Dover Publications Inc. New York.

Amer, F. A. & Williams, W. T. (1957). Leaf growth in *Pelargonium zonale*. *Annals of Botany* N. S., **21**, 339–342.

Ashby, E. (1948). Studies in the morphogenesis of leaves. II. The area, cell size and cell number of leaves of *Ipomoea* in relation to their position on the shoot. *New Phytologist*, **47**, 177–195.

Ashby, E. & Wangermann, E. (1950a). Studies in the morphogenesis of leaves. IV. Further observations on area, cell size and cell number of leaves of *Ipomoea* in relation to position on shoot. *New Phytologist*, **49**, 23–35.

Ashby, E. & Wangermann, E. (1950b). Studies in the morphogenesis of leaves. V. A note on the origin of differences in cell size among leaves at different levels of insertion on the stem. *New Phytologist*, **49**, 189–192.

Audus, L. J. (1939). Mechanical stimulation and respiration in the green leaf. II. Investigations on a number of Angiospermic species. *New Phytologist*, **38**, 284–288.

Auld, B. A., Dennett, M. D. & Elston, J. (1978). The effect of temperature changes on the expansion of individual leaves of *Vicia faba* L. *Annals of Botany*, **42**, 877–888.

Austin, R. B. (1964). A study of the growth and yield of red-beet from a long-term manurial experiment. *Annals of Botany* N. S., **28**, 637–646.

Austin, R. B., Nelder, J. A. & Berry, G. (1964). The use of a mathematical model for the analysis of manurial and weather effects on the growth of carrots. *Annals of Botany* N. S., **28**, 153–162.

Avery, G. S. (1933). Structure and development of the tobacco leaf. *American Journal of Botany*, **20**, 595–592.

Ball, N. G. & Dyke, I. J. (1954). An endogenous 24-hour rhythm in the growth of the *Avena* coleoptile. *Journal of Experimental Botany*, **5**, 421–433.

Ballard, L. A. T. & Petrie, A. H. K. (1936). Physiological ontogeny in plants and its relation to nutrition. I. The effect of nitrogen supply on the growth of the plant and its parts. *Australian Journal of Experimental Biology and Medical Science*, **14**, 135–163.

Barnes, A. (1979). Vegetable plant part relationships. II. A quantitative hypothesis for shoot/storage root development. *Annals of Botany*, **43**, 487–499.

Barnett, V. D. (1970). Fitting straight lines – the linear functional relationship with replicated observations. *Applied Statistics*, **19**, 135–144.

Bartlett, M. S. (1937). Properties of sufficiency and statistical tests. *Proceedings of the Royal Society*, A., **160**, 268–282.

Bazzaz, F. A. & Harper, J. L. (1977). Demographic analysis of the growth of *Linum usitatissimum*. *New Phytologist*, **78**, 193–208.

Bertalanffy, L. von (1941). Stoffwechseltypen und Wachstumstypen. *Biologisches Zentralblatt*, **61**, 510–532.

Bertalanffy, L. von (1957). Quantitative laws for metabolism and growth. *Quarterly Review of Biology*, **32**, 217–231.

Bidwell, R. G. S. (1974). *Plant Physiology*. Macmillan.

Blackman, G. E. & Black, J. N. (1959). Physiological and ecological studies in the analysis of plant environment. XI. A further assessment of the influence of shading on the growth of different species in the vegetative phase. *Annals of Botany* N. S., **23**, 51–63.

Blackman, G. E. & Rutter, A. J. (1948). Physiological and ecological studies in the analysis of plant environment. III. The interaction between light intensity and mineral nutrient supply in leaf development and net assimilation rate of the bluebell (*Scilla non-scripta*). *Annals of Botany* N. S., **12**, 1–26.

Blackman, G. E. & Wilson, G. L. (1951a). Physiological and ecological studies in the analysis of plant environment. VI. The constancy for different species of a logarithmic relationship between net assimilation rate and light intensity and its ecological significance. *Annals of Botany* N. S., **15**, 63–94.

Blackman, G. E. & Wilson, G. L. (1951b). Physiological and ecological studies in the analysis of plant environment. VII. An analysis of the different effects of light intensity on the net assimilation rate, leaf-area ratio, and relative growth rate of different species. *Annals of Botany*, N. S., **15**, 373–408.

Blackman, G. E. & Wilson, G. L. (1954). Physiological and ecological studies in the analysis of plant environment. IX. Adaptive changes in the vegetative growth and development of *Helianthus annuus* induced by alteration in light level. *Annals of Botany* N. S., **18**, 71–94.

Blackman, V. H. (1919). The compound interests law and plant growth. *Annals of Botany*, **33**, 353–360.

Blackman, V. H. (1920). The significance of the efficiency index of plant growth. *New Phytologist*, **19**, 97–100.

Bohning, R. H. & Burnside, C. A. (1956). The effect of light intensity on rate of apparent photosynthesis in leaves of sun and shade plants. *American Journal of Botany*, **43**, 557–561.

Borrill, M. (1959). Inflorescence initiation and leaf size in some Gramineae. *Annals of Botany* N.S. **23**, 217–227.

Bowen, M. R. & Wareing, P. F. (1971). Further investigations into

hormone-directed transport in stems. *Planta*, **99**, 120–132.
Box, M., Davies, D. & Swann, W. (1969). *Non-linear optimisation techniques*. ICI Monograph No. 5. Oliver & Boyd, Edinburgh.
Bradshaw, A. D., Lodge, R. W., Jowett, D. & Chadwick, M. J. (1958). Experimental investigations into mineral nutrition of several grass species. I. Calcium level. *Journal of Ecology*, **46**, 749–757.
Briggs, C. E. (1928). A consideration of some attempts to analyse growth curves. *Proceedings of the Royal Society Series B*, **102**, 280–285.
Briggs, G. E., Kidd, F. & West, C. (1920a). A quantitative analysis of plant growth. I. Relative growth curve. *Annals of Applied Biology*, **7**, 103–123.
Briggs, G. E., Kidd, F. & West, C. (1920b). A quantitative analysis of plant growth. II. Unit leaf rate. *Annals of Applied Biology*,**7**, 202–223.
Brouwer, R., Jenneskens, P. J. & Borggreve, G. J. (1961). Growth responses of shoots and roots to interruptions of the nitrogen supply. *Jaarboek van het Instituut voor biologisch en scheikundig onderzoek van landbouwgewassen*, 29–36.
Bulmer, M. G., (1965). *Principles of Statistics*. Oliver & Boyd.
Causton, D. R. (1967). *Some mathematical properties of growth curves and applications to plant growth analysis*. PhD thesis. University of Wales.
Causton, D. R. (1969). A computer program for fitting the Richards function. *Biometrics*, **25**, 401–409.
Causton, D. R. (1970). Growth functions, growth analysis and plant physiology. *Forestry Commission Research Division, Statistics Section Paper No. 151*.
Causton, D. R. (1977). *A Biologists Mathematics*. Contemporary Biology Series. Edward Arnold, London. 326 pp.
Causton, D. R., Elias, C. O. & Hadley, P. (1978). Biometrical studies of plant growth. I. The Richards Function and its application in analysing the effects of temperature on leaf growth. *Plant, Cell and Environment*, **1**, 163–184.
Chambers, J. M. (1973). Fitting non-linear models: numerical techniques. *Biometrika*, **60**, 1–13.
Chanter, D. O. (1977). Fitting a linear relationship between specific respiration and growth rates using time-course data. *Journal of Applied Ecology*, **14**, 269–278.
Charles-Edwards, D. A. (1976). Shoot and root activities during steady state plant growth. *Annals of Botany*, **40**, 767–773.
Conway, E. R., Glass, N. R. & Wilcox, J. C. (1970). Fitting nonlinear models to biological data by Marquardt's Alogarithm. *Ecology*, **51**, 503–507.
Coombe, D. E. (1960). An analysis of the growth of *Trema guineensis*. *Journal of Ecology*, **48**, 219–231.
Cooper, A. J. (1959). Observations on the growth of leaves of glasshouse tomato plants between March and August. *Journal of Horticultural Science*, **34**, 104–110.
Cooper, A. J. & Thornley, J. H. M. (1976). Response of dry matter

partitioning growth and C & N levels in the tomato plant to changes in root temperature. *Annals of Botany*, **40**, 1139–1152.

Cox, M. G. (1972). The numerical evaluation of β-splines. *Journal of the Institute of Mathematics and its Applications*, **10**, 134–149.

Crowther, F. (1934). Studies in growth analysis of the cotton plant under irrigation in the Sudan. I. The effects of different combinations of nitrogen applications and water-supply. *Annals of Botany*, **48**, 877–913.

Crowther, F. (1937). Experiments in Egypt on the interaction of factors in crop growth. 7. The influence of manuring on the development of the cotton crop. *Royal Agricultural Society, Egypt, Bulletin 31.*

Currah, I. E. & Barnes, A. (1979). Vegetable plant part relationships. I. Effects of time and population density on the shoot and storage root weights of carrot (*Daucus carota* L.) *Annals of Botany*, **43**, 475–486.

Curry, R. B., Baker, C. H. & Streeter, J. G. (1975). SOYMOD 1: a dynamic simulation of soybean growth and development. *Transactions of the American Society of Agricultural Engineers*, **18**, 963–968.

Cutter, E. G. (1971). *Plant anatomy; experiment and interpretation. Part 2. Organs*. Contemporary Biology Series. Edward Arnold, London.

Davies, O. L. & Ku, J. Y. (1977). Re-examination of the fitting of the Richards growth function. *Biometrics*, **33**, 546–547.

Dennett, M. D., Auld, B. A. & Elston, J. (1978). A description of leaf growth in *Vicia faba* L. *Annals of Botany*, **42**, 223–232.

Dennett, M. D., Elston, J. & Milford, J. R. (1979). The effect of temperature on the growth of individual leaves on *Vicia faba* L. in the field. *Annals of Botany*, **43**, 197–208.

De Wit, C. T., Brouwer, R. & Penning de Vries, F. W. T. (1970). The simulation of photosynthetic systems. *Proceedings of the IBP/PP Technical Meeting, Prediction and Measurement of Photosynthetic Productivity 1969, Trebon*. 47–70.

Dixon, L. C. W. (1972). *Non Linear Optimisation*. E. U. P., London.

Doodson, J. K., Manners, J. G. & Myers, A. (1964). Distribution pattern of ^{14}C assimilation by 3rd leaf of wheat. *Journal of Experimental Botany*, **15**, 96–103.

Dormer, K. J. (1972). *Shoot organisation in vascular plants*. Chapman & Hall, London.

Draper, N. R. & Smith, H. (1966). *Applied Regression Analysis*. Wiley, New York.

Eagles, C. F. (1969). Time changes of relative growth rate in two natural populations of *Dactylis glomerata*. *Annals of Botany*, **33**, 937–946.

Eagles, C. F. (1971). Changes in net assimilation rate and leaf-area ratio with time in *Dactylis glomerata* L. *Annals of Botany*, **35**, 63–74.

Elias, C. O. & Causton, D. R. (1975). Temperature and the growth of *Impatiens parviflora* D. C. *New Phytologist*, **75**, 495–505.

Elias, C. O. & Causton, D. R. (1976). Studies on data variability and the use of polynomial to describe plant growth. *New Phytologist*, **77**, 421–430.

Erickson, R. O. & Michelini, F. J. (1957). The plastochron index. *American Journal of Botany*, **44**, 297–305.

Erickson, R. O. (1966). Modelling of plant growth. *Annual Review of Plant Physiology*, **27**, 407–435.

Evans, G. C. (1972). *The Quantitative Analysis of Plant Growth.* Blackwell Scientific Publications, Oxford.

Evans, G. C. & Hughes, A. P. (1961). Plant growth and the aerial environment. I. Effect of artificial shading on *Impatiens parviflora. New Phytologist*, **60**, 150–180.

Evans, G. C. & Hughes, A. P. (1962). Plant Growth and the aerial environment. III. On the computation of unit leaf rate. *New Phytologist*, **61**, 322–327.

Evans, L. T. & Rawson, H. M. (1970). Photosynthesis and respiration by the flag leaf and components of the ear during grain development in wheat. *Australian Journal of Biological Science*, **23**, 245–254.

Fisher, R. A. (1921). Some remarks on the methods formulated in a recent article on the quantitative analysis of plant growth. *Annals of Applied Biology*, **7**, 367–372.

Friend, D. J. C., Helson, V. A. & Fisher, J. E. (1962). The rate of dry weight accumulation in Marquis wheat as affected by temperature and light intensity. *Canadian Journal of Botany*, **40**, 939–955.

Gifford, R. M. (1977). Growth pattern, carbon dioxide exchange and dry weight distribution in wheat growing under differing photosynthetic environments. *Australian Journal of Plant Physiology*, **4**, 99–111.

Gillis, P. R. & Ratkowsky, D. A. (1978). The behaviour of estimators of the parameters of various yield-density relationships. *Biometrics*, **34**, 191–198.

Godwin, H. (1935). The effect of handling on the respiration of cherry laurel leaves. *New Phytologist*, **34**, 403–406.

Goldsworthy, P. R. (1970). The growth and yield of tall and short season sorghums in Nigeria. *Journal of Agricultural Science, Cambridge*, **75**, 109–122.

Gompertz, B. (1825). On the nature of the function expressive of the law of human mortality. *Philosophical Transactions of the Royal Society, London*, **36**, 513–585.

Good, R. E. & Good, N. F. (1976). Growth analysis of pitch pine seedlings under three temperature regimes. *Forest Science*, **22**, 445–448.

Gregory, F. G. (1918). Physiological conditions in cucumber houses. *Experimental Research Station, Turner's Hill, Cheshunt, Herts, Annual Report*, **3**, 19–28.

Gregory, F. G. (1921). Studies in the energy relation of plants. I. The increase in area of leaves and leaf surface of *Cucumis sativus. Annals of Botany*, **35**, 93–123.

Gregory, F. G. (1926). The effect of climatic conditions on the growth of barley. *Annals of Botany*, **40**, 1–26.

Gregory, F. G. (1928). The analysis of growth curves – a reply to a criticism.

Annals of Botany, **42**, 531–539.

Grime, J. P. (1965). Shade tolerance in flowering plants. *Nature*, **208**, 161–163.

Grime, J. P. (1979). *Plant Strategies and Vegetation Processes*. John Wiley & Sons.

Grime, J. P. & Hunt, R. (1975). Relative growth rate: its range and adaptive significance in a local flora. *Journal of Ecology*, **63**, 393–422.

Haberlandt, G. (1884). *Physiologische pflanzenanatomie*. Wilhelm Engelmann, Leipzig.

Hackett, C. & Rawson, H. M. (1974). An exploration of carbon economy of tobacco plants. II. Patterns of leaf growth and dry matter partitioning. *Australian Journal of Plant Physiology*, **1**, 271–281.

Hadley, P. (1978). *Growth and photosynthetic activity of individual leaves of plants during growth*. PhD thesis, University of Wales.

Hall, A. J. (1977). Assimilate source-sink relationships in *Capsicum annuum L*. I. The dynamics of growth in fruiting and deflorated plants. *Australian Journal of Plant Physiology*, **4**, 623–636.

Hardwick, K., Wood, M. & Woolhouse, H. W. (1968). Photosynthesis and respiration in relation to leaf age in *Perilla frutescens*. *New Phytologist*, **67**, 79–86.

Hartley, H. O. (1948). The estimation of non-linear parameters by 'internal least squares'. *Biometrika*, **35**, 32–45.

Heading, J. (1963). *Mathematical methods in science and engineering*. Edward Arnold, London.

Heath, O. V. S. (1937a). The effect of age on net assimilation and relative growth rates in the cotton plant. *Annals of Botany* N. S., **1**, 565–566.

Heath, O. V. S. (1937b). A study in soil cultivation. The effects of varying soil consolidation on the growth and development of rain-grown cotton. *Journal of Agricultural Science*, **27**, 511.

Heath, O. V. S. (1938). Drift of net assimilation rate in plants. *Nature*, **141**, 288.

Heath, O. V. S. & Gregory, F. G. (1938). The constancy of mean net assimilation rate and its ecological importance. *Annals of Botany* N. S., **2**, 811–818.

Higgs, D. E. B. & James, D. B. (1969). Comparative studies on the biology of upland grasses. I. Rate of dry matter production and its control in four grass species. *Journal of Ecology*, **57**, 553–563.

Hinkley, D. V. (1969). Inference about the intersection in two-phase regression. *Biometrika*, **56**, 495–504.

Ho, L. C. (1976). Variation in the carbon/dry material ratio in plant material. *Annals of Botany*, **40**, 163–165.

Hover, J. M. & Gufstafson, F. G. (1926). Rate of respiration related to age. *Journal of General Physiology*, **10**, 33–39.

Hughes, A. P. (1965a). Plant growth and the aerial environment. VI. The apparent efficiency of conversion of light energy of different spectral compositions by *Impatiens parviflora*. *New Phytologist*, **64**, 48–54.

Hughes, A. P. (1965b). Plant growth and the aerial environment. VII. The growth of *Impatiens parviflora* in very low light intensities. *New Phytologist*, **64**, 55–64.

Hughes, A. P. (1965c). Plant growth and the aerial environment. VIII. The effects of (a) blue light and (b) low temperature on *Impatiens parviflora. New Phytologist*, **64**, 323–329.

Hughes, A. P. (1965d). Plant growth and the aerial environment. IX. A synopsis of the autecology of *Impatiens parviflora. New Phytologist*, **64**, 399–413.

Hughes, A. P. & Cockshull, K. E. (1972). Further effects of light intensity, carbon dioxide concentration, and day temperature on the growth of *Chrysanthemum morifolium* c.v. Bright Golden Anne, in controlled environments. *Annals of Botany*, **36**, 535–550.

Hughes, A. P. & Evans, G. C. (1962). Plant growth and the aerial environment. II. Effects of light intensity on *Impatiens parviflora. New Phytologist*, **61**, 154–174.

Hughes, A. P. & Evans, G. C. (1963). Plant growth and the aerial environment. IV. Effects of daylength on *Impatiens parviflora. New Phytologist*, **62**, 367–388.

Hughes, A. P. & Evans, G. C. (1964). Plant growth and the aerial environment. V. The effects of (a) rooting condition and (b) red light on *Impatiens parviflora. New Phytologist*, **63**, 194–202.

Hughes, A. P. & Freeman, P. R. (1967). Growth analysis using frequent small harvests. *Journal of Applied Ecology*, **4**, 553–560.

Hunt, R. (1975). Further observations on root-shoot equilibria in perennial ryegrass. *Annals of Botany*, **39**, 745–755.

Hunt, R. (1978a). Demography versus plant growth analysis. *New Phytologist*, **80**, 269–272.

Hunt, R. (1978b). *Plant Growth Analysis.* Studies in Biology No. 96. Edward Arnold, London. 64 pp.

Hunt, R. (1979). Plant Growth Analysis. The rationale behind the use of the fitted mathematical function. *Annals of Botany*, **43**, 245–249.

Hunt, R. (1982). *Plant Growth Curves: An Introduction to the Functional Approach to Plant Growth Analysis.* Edward Arnold, London.

Hunt, R. & Bazzaz, F. A. (1980). The biology of *Ambrosia trifida* L. V. Response to fertilizer, with growth analysis at the organismal and sub-organismal levels. *New Phytologist*, **84**. (In press.)

Hunt, R. & Burnett, J. A. (1973). The effects of light intensity and external potassium level on root/shoot ratio and rates of potassium uptake in perennial ryegrass (*Lolium perenne* L.). *Annals of Botany*, **37**, 519–537.

Hunt, R. & Parsons, I. T. (1974). A computer program for deriving growth functions in plant growth analysis. *Journal of Applied Ecology*, **11**, 297–307.

Hunt, R. & Parsons, I. T. (1977). Plant growth analysis: further applications of a recent curve-fitting program. *Journal of Applied Ecology*, **14**, 965–968.

Hurd, R. G. (1977). Vegetative plant growth analysis in controlled environments. *Annals of Botany*, **41**, 779–787.

Hurd, R. G. & Thornley, J. H. M. (1974). An analysis of the growth of young tomato plants in water culture at different light integrals and CO_2 concentrations. I. Physiological aspects. *Annals of Botany*, **38**, 375–388.

Hutchinson, T. C. (1976). Comparative studies on the ability of species to withstand prolonged periods of darkness. *Journal of Ecology*, **55**, 291–299.

Hutchinson, T. C. (1968). A physiological study of *Teucrium scorodonia* ecotypes which differ in their susceptibility to lime-induced chlorosis and iron-deficiency chlorosis. *Plant and Soil*, **28**, 81–90.

Huxley, J. S. (1924). Constant differential growth rates. *Nature*, **114**, 895–896.

Huxley, J. S. (1932). *Problems of Relative Growth*. Methuen & Co. Ltd.

Jarvis, P. G. & Jarvis, M. S. (1964). Growth rates of woody plants. *Physiologia Plantarum*, **17**, 654–666.

Jefferies, R. L. & Willis, A. J. (1964). Studies on the calcicole-calcifuge habit. II. The influence of calcium on the growth and establishment of four species in soil and sand cultures. *Journal of Ecology*, **52**, 691–707.

Jolicoeur, P. & Heusner, A. A. (1971). The allometry equation in the analysis of the standard oxygen consumption and body weight of the white rat. *Biometrics*, **27**, 841–855.

Jones, H., Martin, R. V. & Porter, H. K. (1959). Translocation of ^{14}C in tobacco following assimilation by one leaf. *Annals of Botany* N. S., **23**, 493–508.

Kemp, C. D. (1960). Methods of estimating leaf area of grasses from linear measurements. *Annals of Botany* N.S., **24**, 491–499.

Kendall, M. G. & Stuart, A. (1977). *The Advanced Theory of Statistics, Vol I*, 4th Edition. Griffin & Co, London.

King, R. W., Wardlaw, I. F. & Evans, L. T. (1967). Effect of assimilate utilisation on photosynthetic rate in wheat. *Planta*, **77**, 261–276.

Krause, G. F., Siegel, P. B. & Hurst, D. C. (1967). A probability structure for growth curves. *Biometrics*, **23**, 217–225.

Kreidemann, P. G., Kliewer, W. M. & Harris, J. M. (1970). Leaf age and photosynthesis in *Vitis vinifera* L. *Vitis*, **9**, 97–104.

Kreusler, U., Prehn, A. & Hornberger, R. (1879). Beobachtungen uber das wachsthum der maispflanze. (Bericht uber die versuche vom jahre 1878.) *Landwirtschaftliche Jahrbucher*, **8**, 617–622.

Laird, A. K., Tyler, S. A. & Barton, A. D. (1965). Dynamics of normal growth. *Growth*, **29**, 233–248.

Laird, A. K. (1965). Dynamics of relative growth. *Growth*, **29**, 249–263.

Lamoreaux, R. J., Chaney, W. R. & Brown, K. M. (1978). The plastochrom index: a review after two decades of use. *American Journal of Botany*, **65**, 586–593.

Larcher, W. (1969). The effect of environmental and physiological variables on the carbon dioxide gas exchange of trees. *Photosynthetica*, **3**, 167–198.

Ledig, F. T. & Perry, T. O. (1969). Net assimilation rate and growth in loblolly pine seedlings. *Forest Science*, **15**, 431–438.

Leech, F. B. & Healy, M. J. R. (1959). Analysis and interpretation of experiments on growth rates. *Biometrics*, **15**, 152–153.

Lindley, D. V. (1947). Regression lines and the linear functional relationship. *Journal of the Royal Statistical Society*, B, **9**, 218–244.

Loach, K. (1967). Shade tolerance in tree seedlings. I. Leaf photosynthesis and respiration in plants raised under artificial shade. *New Phytologist*, **66**, 607–621.

Lovell, P. H. (1971). Translocation of photosynthates in tall and dwarf varieties of pea, *Pisum sativum. Physiologia Plantarum*, **25**, 382–385.

Lovell, P. H. & Moore, K. G. (1970). A comparative study of cotyledons as assimilatory organs. *Journal of Experimental Botany*, **21**, 1017–1030.

Lovell, P. H., Oo, H. T. & Sagar, G. R. (1972). An investigation into the rate and control of assimilate movement from leaves in *Pisum sativum. Journal of Experimental Botany*, **23**, 255–266.

Madansky, A. (1959). The fitting of straight lines when both variables are subject to error. *Journal of the American Statistical Association*, **54**, 173–205.

Maggs, D. H. (1964). Growth-rates in relation to assimilate supply and demand. *Journal of Experimental Botany*, **15**, 574–583.

Maksymowych, R. (1973). *Analysis of Leaf Development*. Cambridge University Press.

Marquardt, D. W. (1963). An algorithm for the estimation of non-linear parameters. *Journal of the Society for Industrial and Applied Mathematics*, **11**, 431–441.

Mather, K. (1967). *The Elements of Biometry*. Metheun, London.

McIntyre, G. A. & Williams, R. F. (1949). Improving the accuracy of growth indices by the use of ratings. *Australian Journal of Scientific Research*, B, **2**, 319.

McKinion, J. M., Jones, J. W. & Hesketh, J. D. (1975). A system of growth equations for the continuous simulation of plant growth. *Transactions of the American Society of Agricultural Engineers*, **18**, 975–979.

Mead, R. & Pike, D. J. (1975). A review of response surface methodology from a biometric viewpoint. *Biometrics*, **31**, 803–851.

Medawar, P. B. (1940). Growth, growth energy and ageing of the chicken's heart. *Proceedings of the Royal Society*, B, **129**, 332–355.

Meddis, R. (1975). *Statistical Handbook for Non-Statisticians*. McGraw-Hill.

Milthorpe, F. L. & Newton, P. (1963). Studies on the expansion of the leaf surface. III. The influence of radiation on cell division and leaf expansion. *Journal of Experimental Botany*, **14**, 483–495.

Monteith, R. L. (1965). Light distribution and photosynthesis in field crops. *Annals of Botany* N. S., **29**, 113.

Moorby, J. (1970. The production, storage, and translocation of carbohydrates in developing potato plants. *Annals of Botany*, **34**, 297–308.

Moran, P. A. P. (1971). Estimating structural and functional relationships. *Journal of Multivariate Analysis*, **1**, 232–255.

Myerscough, P. J. & Whitehead, F. H. (1967). Comparative biology of *Tussilago farfara* L., *Chamaeneiron angustifolium* (L.) Scop., *Epilobium montanum* L., and *Epilobium adenocaulon* Hausskn. II. Growth and ecology. *New Phytologist*, **66**, 785–823.

Namkoong, G. & Matzinger, D. F. (1975). Selection for annual growth curves in *Nicotiana tabacum*. *Genetics*, **81**, 377–386.

Neales, T. F. & Incoll, L. D. (1968). Control of leaf photosynthesis rate by levels of assimilate in the leaf: a review of the hypothesis. *Botanical Review*, **34**, 107–125.

Neales, T. F. & Nicholls, A. O. (1978). Growth responses of young wheat plants to a range of ambient CO_2 levels. *Australian Journal of Plant Physiology*, **5**, 45–59.

Nelder, J. A. (1961). The fitting of a generalisation of the logistic curve. *Biometrics*, **17**, 89–110.

Nelder, J. A. (1962). An alternative form of a generalised logistic equation. *Biometrics*, **18**, 614–616.

Nelder, J. A. (1963). Quantitative genetics and growth analysis. In: *Statistical Genetics and Plant Breeding*. (Eds W. D. Hanson & H. F. Robinson) National Academy of Science (NRC 982) Coûncil, Washington D.C., 445–454.

Nelder, J. A., Austin, R. B., Bleasdale, J. K. A. & Salter, P. J. (1960). An approach to the study of yearly and other variation in crop yields. *Journal of Horticultural Science*, **35**, 73–82.

Nelder, J. A. & Mead, R. (1965). A simplex method for function minimisation. *Computer Journal*, **7**, 308–313.

Nelkon, R. & Parker, P. (1977). *Advanced Level Physics*. Heinemann.

Neyman, J. & Scott, E. L. (1951). On certain methods of estimating the linear structural relationship. *Annals of Mathematical Statistics*, **22**, 352–361.

Nicholls, A. O. & Calder, D. M. (1973). Comments on the use of regression analysis for the study of plant growth. *New Phytologist*, **72**, 571–581.

O'Neill, M., Sinclair, I. G. & Smith, F. J. (1969). Polynomial curve fitting when abscissas and ordinates are both subject to error. *Computer Journal*, **12**, 52–56.

Parsons, I. T. & Hunt, R. (1980). Plant growth analysis: A program for the fitting of lengthy series of data by the method of β-splines. *New Phytologist*. (In press.)

Patefield, W. M. & Austin, R. B. (1971). A model for the simulation of growth in *Beta vulgaris*. *Annals of Botany*, **35**, 1227–1250.

Peacock, J. M. (1976). Temperature and leaf growth in four grass species. *Journal of Applied Ecology*, **13**, 225–232.

Pearl, R. & Reed, L. J. (1923). Skew growth curves. *Proceedings of the National Academy of Sciences, Washington*, **11**, 16–22.

Pearsall, W. H. (1927). Growth studies. VI. On the relative sizes of growing plant organs. *Annals of Botany*, **41**, 549–556.

Petrie, A. H. K. & Arthur, J. I. (compiled by Wood, J. G.) (1943). Physiological ontogeny in the tobacco plant. The effect of varying water supply on the drifts in dry weight and leaf area and on various components of the leaves. *Australian Journal of Experimental Biology and Medical Science*, **21**, 191–200.

Petrie, A. H. K., Watson, R. & Ward, E. D. (1939). Physical ontogeny in the tobacco plant. 1. The drifts in dry weight and leaf area in relation to phosphorus supply and topping. *Australian Journal of Experimental Biology and Medical Science*, **17**, 93–122.

Pielou, E. C. (1965). *Mathematical Ecology*. John Wiley & Sons, Inc.

Pieters, G. A. (1974). The growth of sun and shade leaves of *Populus euramericana* 'Robusta' in relation to age, light intensity and temperature. *Mededelingen van de Landbouwhoogeschool de Wageningen*, 00, 74–111.

Pollard, D. F. W. & Wareing, P. F. (1968). Rates of dry matter production in forest tree seedlings. *Annals of Botany*, **32**, 573–591.

Portsmouth, G. B. (1937). Variation in the leaves of cotton plants grown under irrigation in the Sudan Gezira. *Annals of Botany* N. S., **1**, 277–291.

Powell, M. J. D (1964). An efficient method for finding the minimum of a function of several variables without calculating derivatives. *Computer Journal*, **7**, 155–162.

Quebedeaux, B. & Chollet, R. (1977). Comparative growth analyses of *Panicum* species with differing rates of photorespiration. *Plant Physiology*, **59**, 42–44.

Radford, P. J. (1967). Growth analysis formulae: their use and abuse. *Crop Science*, **7**, 171–175.

Rajan, A. K., Betteridge, B. & Blackman, G. E. (1971). Interrelationships between nature of light source, ambient air temperature and vegetative growth of different species within growth cabinet. *Annals of Botany*, **35**, 323–343.

Rajan, A. K. & Blackman, G. E. (1975). Interacting effects of light, and day and night temperatures on the growth of four species in the vegetative phase. *Annals of Botany*, **39**, 733–743.

Rao, C. R. (1965). The theory of least squares when the parameters are stochastic, and its application to the analysis of growth curves. *Biometrika*, **52**, 447–504.

Reed, H. S. (1920a). Slow and rapid growth. *American Journal of Botany*, **7**, 327–333.

Reed, H. S. (1920b). The dynamics of a fluctuating growth rate. *Proceedings of the National Academy of Sciences, Washington*, **6**, 397–410.

Reed, H. S. (1920c). The nature of the growth rate. *Journal of General Physiology*, **2**, 545–561.

Reed, H. S. & Holland, R. H. (1919). The growth rate of an annual plant *Helianthus*. *Proceedings of the National Academy of Sciences, Washington*, **5**, 135–144.

Rees, A. R. & Chapas, L. C. (1963). An analysis of growth of oil palms under nursery conditions. I. Establishment and growth in the wet season. *Annals of Botany* N. S., **27**, 607–614.

Richards, F. J. (1934). On the use of similar observations on successive leaves for the study of physiological change in relation to leaf age. *Annals of Botany*, **48**, 497–504.

Richards, F. J. (1959). A flexible growth function for empirical use. *Journal of Experimental Botany*, **10**, 290–300.

Richards, F. J. (1969). The quantitative analysis of growth. *Plant Physiology*, Vol. VA (Ed. Steward, F. C.), pp. 3–76.

Robertson, T. B. (1923). *The Chemical Basis of Growth and Senescence.* Lippincott, Philadelphia & London.

Robson, M. J. (1972). The effect of temperature on the growth of S170 tall fescue (*Festuca arundinacea*) I. Constant temperature. *Journal of Applied Ecology*, **9**, 643–653.

Rogan, P. G. & Smith, D. L. (1975). Rates of leaf initiation and leaf growth in *Agropyron repens* L. Beauv. *Journal of Experimental Botany*, **26**, 70–78.

Rorison, I. H. (1968). The response to phosphorus of some ecologically distinct plant species. I. Growth rates and phosphorus absorption. *New Phytologist*,**67**, 913–923.

Ryle, G. J. A. & Powell, C. E. (1972). The export and distribution of ^{14}C–labelled assimilates from each leaf on the shoot of *Lolium temulentum* during reproductive and vegetative growth. *Annals of Botany*, **36**, 363–375.

Sandland, R. L. & McGilchrist, C. A. (1979). Stochastic growth curve analysis. *Biometrics*, **35**, 255–273.

Sauer, W. & Possingham, J. V. (1970). Studies on the growth of spinach leaves (*Spinaceae oleracea*). *Journal of Experimental Botany*, **21**, 151–158.

Schoch, P. G. (1974). Reprise d' activité des premieres feuilles de *Vignia sinensis* L. à la suite de defoliations. *Physiologie Végétale*, **12**, 289–298.

Shibles, R., Anderson, I. C. & Gibson, A. H. (1975). Soybean. In: *Crop Physiology* (Ed. Evans, L. T.). Cambridge University Press, London.

Sinnott, E. W. (1936). A developmental analysis of inherited shape differences in Cucurbit fruits. *American Naturalist*, **70**, 245–254.

Sivakumar, M. V. K. & Shaw, R. H. (1978). Methods of growth analysis in field-grown soya beans (*Glycine max* (L.) Merrill). *Annals of Botany*, **42**, 213–222.

Smith, R. I. L. & Walton, D. W. H. (1975). A growth analysis technique for assessing habitat severity in tundra regions. *Annals of Botany*, **39**, 831–845.

Sokal, R. R. & Rohlf, F. J. (1969). *Biometry, the Principles and Practice of Statistics in Biological Research.* W. H. Freeman, San Francisco.

Solari, M. E. (1969). The 'maximum likelihood solution' of the problem of estimating a linear functional relationship. *Journal of the Royal*

Statistical Society, B., **31**, 372–375.
Sprent, P. (1961). Some hypotheses concerning two phase regression lines. *Biometrics*, **17**, 634–645.
Sprent, P. (1966). A generalised least-squares approach to linear functional relationships. *Journal of the Royal Statistical Society*, B, **28**, 278–297.
Sprent, P. (1967). Estimation of mean growth curves for groups of organisms. *Journal of Theoretical Biology*, **17**, 159–173.
Sprent, P. (1968). Linear relationships in growth and size studies. *Biometrics*, **24**, 639–656.
Sprent, P. (1969). *Models in Regression*. Methuen & Co. Ltd., London.
Stanhill, G. (1977a). Allometric growth studies of the carrot crop. I. Effects of plant development and cultivar. *Annals of Botany*, **41**, 533–540.
Stanhill, G. (1977b). Allometric growth studies of the carrot crop. II. Effects of cultural practices and climatic environment. *Annals of Botany*, **41**, 541–552.
Steer, B. T. (1971). Dynamics of leaf growth and photosynthetic capacity in *Capsicum frutescens*. *Annals of Botany*, **35**, 1003–1015.
Steer, B. T. (1972). Leaf growth parameters and photosynthetic capacity in expanding leaves of *Capsicum frutescens*. *Annals of Botany*, **36**, 377–384.
Sunderland, N. (1960). Cell division and expansion in the growth of the leaf. *Journal of Experimental Botany*, **11**, 68–80.
Sweet, G. B. & Wareing, P. F. (1966a). Role of plant growth in regulating photosynthesis. *Nature*, **210**, 77.
Sweet, G. B. & Wareing, P. F. (1966b). The relative growth rates of large and small seedlings in forest tree species. *Forestry Supplement*, 110–117.
Sweet, G. B. & Wareing, P. F. (1968a). A comparison of the seasonal rates of dry matter production of three coniferous species with contrasting patterns of growth. *Annals of Botany*, **32**, 721–734.
Sweet, G. B. & Wareing, P. F. (1968b). A comparison of the rates of growth and photosynthesis in first-year seedlings of four provenances of *Pinus contorta*. *Annals of Botany*, **32**, 735–751.
Tanner, J. W. & Ahmed, S. (1974). Growth analysis of soybeans treated with TIBA. *Crop Science*, **14**, 371–374.
Terry, N. (1968). Developmental physiology of sugar-beet. I. The influence of light and temperature on growth. *Journal of Experimental Botany*, **19**, 795–811.
Terry, N. & Mortimer, D. C. (1972). Estimation of the rates of mass carbon transfer by leaves of sugar beet. *Canadian Journal of Botany*, **50**, 1049–1054.
Thiagarajah, M. R. & Hunt, L. A. (1974). Effects of temperature on leaf growth in corn. *Canadian Journal of Plant Science*, **54**, 449.
Thorne, G. N. (1960). Variations with age in net assimilation rate and other growth attributes of sugar-beet, potato, and barley in a controlled environment. *Annals of Botany* N. S., **24**, 356–371.

Thorne, G. N. (1961). Effects of age and environment on net assimilation rate of barley. *Annals of Botany*, N. S., **25**, 29–38.

Thornley, J. H. M. (1972a). Model to describe partitioning of photosynthate during vegetative plant growth. *Annals of Botany*, **36**, 419–430.

Thornley, J. H. M. (1972b). Balanced model for root:shoot ratios in vegetative plants. *Annals of Botany*, **36**, 431–441.

Thornley, J. H. M. (1976). *Mathematical Models in Plant Physiology* (Experimental Botany: An International Series of Monographs No. 8). Academic Press.

Thornley, J. H. M. & Hesketh, J. D. (1972). Growth and respiration in cotton bolls. *Journal of Applied Ecology*, **9**, 315–317.

Thrower, S. L. (1977). Translocation into mature leaves – the effect of growth pattern. *New Phytologist*, **78**, 361–365.

Tiver, N. S. (1942). Studies of the flax plant. I. Physiology of growth, stem anatomy and fibre development in fibre flax. *Australian Journal of Experimental Biology and Medical Science*, **20**, 149–160.

Tiver, N. S. & Williams, R. F. (1943). Studies of the flax plant. 2. The effect of artificial drought on growth and oil production in a linseed variety. *Australian Journal of Experimental Biology and Medical Science*, **21**, 201–209.

Troughton, A. (1955). The application of the allometric formula to the study of the relationship between the roots and shoots of young grass plants. *Agricultural Progress*, **30**, 59–65.

Troughton, A. (1956). Studies on the growth of young grass plants with special reference to the relationship between the root and shoot systems. *Journal of the British Grassland Society*, **11**, 56–65.

Troughton, A. (1960). Further studies on the relationship between shoot and root systems of grasses. *Journal of the British Grassland Society*, **11**, 41–47.

Troughton, A. (1967). The effect of mineral nutrition on the distribution of growth in *Lolium perenne*. *Annals of Botany* N.S., **31**, 447–454.

Turgeon, R. & Webb, J. A. (1975). Leaf development and phloem *transport in Cucurbitta pepo*: carbon economy. *Planta*, **123**, 53–62.

Vernon, A. J. & Allison, J. C. S. (1963). A method of calculating net assimilation rate. *Nature*, **200**, 814.

Venus, J. C. (1978). *The Quantitative Analysis of Plant Growth with Special Reference to the Richards Function*. PhD thesis, University of Wales.

Venus, J. C. & Causton, D. R. (1979a). Plant growth analysis: The use of the Richards function as an alternative to polynomial exponentials. *Annals of Botany*, **43**, 623–632.

Venus, J. C. & Causton, D. R. (1979b). Plant growth analysis: A re-examination of the methods of calculation of relative growth and net assimilation rates without using fitted functions. *Annals of Botany*, **43**, 633–638.

Venus, J. C. Causton, D. R. (1979c). Confidence intervals for the Richards function. *Journal of Applied Ecology*, **17**, 939–947.

Voldeng, H. D. & Blackman, G. E. (1973). The interrelated effects of stage of development and seasonal changes in light and temperature on the components of growth in Zea mays. *Annals of Botany*, **37**, 895–904.

Vyvyan, M. C. (1957). An analysis of growth and of form in young apple trees. I. Relative growth and net assimilation rates in 1- and 2-year old trees of the apple rootstock variety M.XIII. *Annals of Botany* N.S., **21**, 479–497.

Wald, A. (1940). The fitting of straight lines of both variables are subject to error. *Annals of Mathematical Statistics*, **11**, 284–300.

Warren Wilson, J. (1960). Observations on net assimilation rates in arctic environments. *Annals of Botany* N.S., **24**, 372–381.

Warren Wilson, J. (1966a). An analysis of plant growth and its control in arctic environments. *Annals of Botany* N.S., **30**, 383–402.

Warren Wilson, J. (1966b). High net assimilation rates of sunflower plants in an arid climate. *Annals of Botany* N.S., **30**, 745–751.

Warren Wilson, J. (1966c). Effect of temperature on net assimilation rate. *Annals of Botany* N.S., **30**, 753–761.

Warren Wilson, J. (1967). Effects on seasonal variation in radiation and temperature on net assimilation and growth rates in an arid climate. *Annals of Botany* N.S., **31**, 41–57.

Warren Wilson, J. (1972). Control of crop processes. In: *Crop Processes in Controlled Environments* (Rees, A. R., Cockshull, K. E., Hand, D. W. and Hurd, R. G., Eds), pp. 7–30. Academic Press.

Wassink, E. C., Richardson, S. D. & Pieters, G. A. (1956). Photosynthetic adaptation to light intensity in leaves of *Acer pseudoplatanus*. *Acta Botanica Neerlandica*, **5**, 247–256.

Watson, D. J. (1947a). Comparative physiological studies on the growth of field crops. I. Variation in net assimilation rate and leaf area between species and varieties and within and between years. *Annals of Botany* N.S., **11**, 41–76.

Watson, D. J. (1947b). Comparative physiological studies on the growth of field crops. II. The effect of varying nutrient supply on net assimilation rate and leaf area. *Annals of Botany* N. S., **11**, 375–407.

Watson, D. J. (1952). The physiological basis of variation in yield. *Advances in Agronomy*, **4**, 101–145.

Watson, D. J. & Baptiste, E. C. D. (1938). A comparative physiological study of sugar beet and mangold with respect to growth and sugar accumulation. I. Growth analysis of the crop in the field. *Annals of Botany* N. S., **2**, 437–480.

Watson, D. J. & Ken-Ichi, Hayashi (1965). Photosynthetic and respiratory components of the net assimilation rates of sugar beet and barley. *New Phytologist*, **64**, 38–47.

Watson, D. J. & Witts, K. J. (1959). The net assimilation rates of wild and cultivated beets. *Annals of Botany* N. S., **23**, 431–439.

Watson, R. & Petrie, A. H. K. (1940). Physiological ontogeny in the tobacco plant. 4. The drift in nitrogen content of the parts in relation to phosphorus supply and topping with an analysis of the determination of ontogenetic changes. *Australian Journal of Experimental Biology and Medical Science*, **18**, 313–340.

Weber, C. A. (1879). *Uber specifische assimilationsenergie.* Inaugural-Dissertation der philosophischen Facultat der Kgl. Maximilians-Universitat zu Wurzburg. Stabel'schen Buchdruckerei, Wurzburg.

Weber, C. A. (1882). Uber specifische assimiliationsenergie. *Arbeiten aus dem Botanischen Institut in Wurzburg*, **2**, 346–352.

West, C., Briggs, G. E. & Kidd, F. (1920). Methods and significant relations in the quantitative analysis of plant growth. *New Phytologist*, **19**, 200–207.

Whaley, W. G. & Whaley, C. Y. (1942). A developmental analysis of inherited leaf patterns in *Tropaeolum. American Journal of Botany*, **29**, 195–200.

Whitehead, F. H. & Mycersough, P. J. (1962). Growth analysis of plants. Ratio of mean relative growth rate to mean growth rate of increase of leaf area. *New Phytologist*, **61**, 314–321.

Williams, R. F. (1936). Physical ontogeny in plants and its relation to nutrition. II. The effect of phosphorus supply on the growth of the plant and its parts. *Australian Journal of Experimental Biology and Medical Science*, **14**, 165–185.

Williams, R. F. (1937). Drift of net assimilation rate in plants. *Nature*, **140**, 1099.

Williams, R. F. (1939). Physiological ontogeny in plants and its relation to nutrition. 6. Analysis of the unit leaf rate. *Australian Journal of Experimental Biology and Medical Science*, **17**, 123–132.

Williams, R. F. (1946). The physiology of plant growth with special reference to the concept of net assimilation rate. *Annals of Botany*, **10**, 41–72.

Williams, R. F. (1975). *The Shoot Apex and Leaf Growth.* Cambridge University Press.

Wilson, D. & Cooper, J. P. (1969). Apparent photosynthesis and leaf characters in relation to leaf position and age, among contrasting *Lolium* genotypes. *New Phytologist*, **68**, 645–655.

Wilson, D. & Cooper, J. P. (1970). Effects of selection for mesophyll cell size on growth and assimilation in *Lolium perenne. New Phytologist*, **69**, 233–245.

Wishart, J. (1938). Growth rate determinations in nutrition studies with the bacon pig, and their analysis. *Biometrika*, **30**, 11–15.

Wold, S. (1974). Spline functions in data analysis. *Technometrics*, **16**, 1–11.

Wolf, F. A. (1947). Growth curves of oriental tobacco and their significance. *Bulletin of the Torrey Botanical Club*, **74**, 199–214.

Woodward, R. G. (1976). Photosynthesis and expansion of leaves of soybean grown in two environments. *Photosynthetica*, **10**, 274–279.

Yeoman, M. M. (1976). *Cell division in higher plants.* Academic Press, London.

Plant Index

Bold numbers indicate a main section, which may extend over more than one page. *Italicized* numbers refer to pages containing relevant text figures or tables.

Author Index

Italicized numbers refer to pages containing relevant text figures or tables. Page numbers refer to occurrences in the text only.

Subject Index

Bold numbers indicate a main section, which may extend over more than one page. *Italicized* numbers refer to pages containing relevant text figures or tables.